Natural Gas Installations and Networks in Buildings

Natural Gas Installations and Networks in Buildings

Authored by

Alexander V. Dimitrov

CRC Press is an imprint of the
Taylor & Francis Group, an **informa** business

First edition published 2021
by CRC Press
6000 Broken Sound Parkway NW, Suite 300, Boca Raton, FL 33487-2742
and by CRC Press
2 Park Square, Milton Park, Abingdon, Oxon, OX14 4RN

First edition published by CRC Press 2021

CRC Press is an imprint of Taylor & Francis Group, LLC

ISBN: 9780367536725 (hbk)
ISBN: 9781003082835 (ebk)

Typeset in Times LT Std
by KnowledgeWorks Global Ltd.

Contents

Foreword ix
About the Author xi

Chapter 1 Introduction 1

1.1 Gas Pipe Networks and Systems as Elements of the Building Logistics 1
1.2 Fuel Gas, Current State and Perspectives 3

Chapter 2 Theoretical Foundations of Gas Pipe Networks and Installations 9

2.1 Physiomechanical Properties of Fuel Gas as a Primary Energy Source 9
2.2 Transmission Parameters of Fuel Gas 14
2.2.1 Operational Pressure 15
2.2.2 Pressure Drop 15
2.3 Basic Terms and Laws 17
2.3.1 Gas Flow Rate and Thermal Power. The Link between Gas Flow Rate and Decrease of Gas-Dynamic Losses – Kirchhoff's Law for Gas Pipe Networks 17
2.3.2 Law of Continuity 19
2.3.3 Law of Balance between Intake Flow Rate and Exhaust Flow Rate at a Knot 19
2.3.4 Bernoulli's Principle (for Real Fluids) 21
2.3.5 Law the of Gas-Dynamic Head in a Closed Loop 22
2.3.6 Darcy's Law (Concerning Gas-Dynamic Losses of Pressure in Linear Pipeline Sections) 24
2.3.7 Calculation of the Parameters of Natural Gas – Examples 26
2.4 Fuel Gas Thermodynamics 30
2.4.1 Law of Gas State 30
2.4.2 Charles's Law 33
2.4.3 Boyle's Law 34
2.4.4 Graham's Law for Gas Diffusion 35
2.5 Fuel Gas Combustion 35
2.5.1 Complete Combustion of Methane 36
2.5.2 Incomplete Combustion of Methane 37
2.5.3 Controlled Oxidation of Methane 38
2.5.4 Fuel Cell Classification and Characteristics. Advantages and Disadvantages 40

2.5.5 Fuel Gas Cells in Household Micro-Cogeneration Systems (mHPS) ... 43
2.5.6 Technical Characteristics of mHPS ... 47
2.6 Gas Flame and Combustion Devices ... 49
2.7 A Flameless Combustion of Fuel Gas and Combustion Devices ... 55

Chapter 3 Gas Supply of Urbanized Regions ... 57

3.1 Territorial (State) Pipe Network of Gas Transport ... 57
3.2 Urban and Regional Gas Networks. Classification. Structure ... 60
3.2.1 Classification of Urban Gas-Distributing Networks ... 60
3.3 Decentralization of Building Power Supply. Energy Centers ... 64
3.3.1 Gas Supply of Buildings. Types of Gas-Supplied Energy Centers (GEC) of Buildings. Classification ... 64
3.3.2 Structure of the Gas Regulating Station in an EC ... 69
3.3.3 Gas Boiler/mHPS Room ... 72
3.3.3.1 Structure of EC with Gas Water Heaters/mHPS Station (Boiler Room) ... 72
3.3.3.2 General Requirements. Operative and Indoor Environment of the Boiler Rooms. Calculation of Warehouse Gas Amount ... 74
3.4 Systems Exhausting Gas Waste ... 79
3.4.1 Open System Exhausting Gas Combustion Products via Natural Convection ... 84
3.4.2 Open Systems with Forced Convection ... 89
3.4.3 Centralized Systems Exhausting Gas Combustion Products by Means of Forced Convection ... 90
3.4.3.1 Modular Systems ... 91
3.4.3.2 Apartment Systems ... 93
3.4.4 Releasing Devices (Terminals) Installed on the Building Envelope ... 93
3.4.4.1 Terminals Installed on Roof Structures ... 94
3.4.4.2 Terminals Installed on Facades ... 95
3.5 Design of a Pipe Network ... 101
3.5.1 Structure of Gas Pipe Networks in Residential and Public Buildings ... 101
3.5.2 Calculation of Pipe Networks ... 106
3.5.2.1 Pipe Network Preliminary Calculation ... 108

3.5.2.2 Final Pipe Network Calculation by the Estimation of the Gas-Dynamic Pressure Losses in Pipes Supplying the Remotest End Users and Knowing Pipe Network Geometry and Topology 122

Chapter 4 Main Building Systems Operating with Fuel Gas 129

4.1 System For Thermal Comfort Control 129
4.1.1 Need for Thermal Comfort and Accessible Levels of Thermal Loads of Occupied Areas 130
4.1.2 Classification of Systems Providing Thermal Comfort 134
4.1.2.1 "Water-Air" Fuel Gas Systems 136
4.1.2.2 Convective Systems "Gas-Air"................ 139
4.1.2.3 Radiant Heating Systems "Gas-Air" 150
4.1.3 Advantages and Disadvantages of Gas Radiant Heating Installations 162
4.2 Gas Equipment and Control of Indoor Air Quality.............. 163
4.2.1 Indoor Air Quality and Methods of Its Attainment.. 163
4.2.2 Building Ventilation and IAQ 170
4.2.3 Gas Equipment and Building Ventilation 174
4.3 System For Visual Comfort and Gas Appliances (Gas Lamps)... 181
4.3.1 Visual Comfort in Buildings.................................. 181
4.3.2 Visual Comfort-Gas Lighting Fixtures................... 184
4.3.3 Design Steps in the Calculation of a Visual Control System 188
4.4 Gas Equipment in a System for Food Preparation and Conservation 193
4.4.1 A System for Food Preparation and Conservation... 193
4.4.2 Household Kitchens and Gas Appliances 200
4.4.3 Commercial Kitchen Arrangement..........................200
4.4.3.1 Ventilation of Commercial Kitchens 203
4.4.3.2 Design of Local Canopy Ventilation 204
4.5 Hygiene Maintenance Systems(Domestic Hot Water System.) 206
4.5.1 Structure of a Hygiene Maintenance System. 206
4.5.2 Gas Water Heaters of a Hygiene Maintenance System 209
4.5.2.1 Rating of Heat Generators in a Hygiene Maintaining System 212
4.5.3 Combined Schemes of Water Heating in Systems for Hygiene Maintenance 213

4.5.4 Gas Dryers in Systems for Hygiene Maintenance 215
4.5.5 Calculation of Hygiene Maintenance Systems 216
4.5.5.1 Water Heater 221
4.5.5.2 Recirculation Pump 221
4.6 Gas Heat Pumps In Engineering Installations of Buildings 221
4.6.1 Heat Pumps Principle of Action 221
4.6.2 Gas Vapor-Compression Heat Pumps 225
4.6.3 Operation of a Gas Heat Pump 228
4.6.4 Heat Pumps Driven by Natural Gas and Available on the Market – an Investment Point of View 230
4.7 Necessity of Increasing the Share of Gas Devices in a Building 232
4.7.1 Gas Micro-Cogenerating Systems in Domestic Autonomous Energy Centers 233
4.7.2 Importance of Gas Micro-Cogenerating Systems for the National Energy Strategy 240

Chapter 5 Conclusions and Acknowledgment 243

Appendix 245

References 261

Foreword

The book "**Natural Gas Installations and Networks in Buildings**" by D.Sc., Ph.D., Eng. Alexander V. Dimitrov, Professor at the European Polytechnical University, is a current study on modern power systems in Bulgaria. It outlines the means and measures of providing buildings and interior space with living comfort, addressing specialists in research, development, design and applications in the field. Moreover, it can be a reference book to a vast majority of graduate and postgraduate students, architects and structural designers.

Prior to the present study, Prof. Dimitrov authored 12 books on heating and power technology, 5 of them were translated into English and published in the USA and Great Britain. He tackles basic laws of nature involving fluids, energy, thermo- and illumination technologies and aerodynamics.

Prof. Dimitrov started teaching at the High Military Civil Engineering School (HMCES), Sofia, Bulgaria, in 1981, having defended a Ph.D. thesis on aerodynamics before a jury on power engineering and technologies and enrolled at HMCES as an assistant professor. He joined the High Technical School "T. Kableshkov", Sofia, Bulgaria, in 1989 as an associate professor on building engineering installations. Dr. Dimitrov defended a D.Sc. thesis in 2012 in professional trend 5.7 "Architecture, Civil Engineering and Geodesy", specialization "Heating Technology". He later enrolled at the European Polytechnic University, Department of Civil Engineering and Architecture, Pernik, Bulgaria, as a professor on building engineering installations.

Prof. Dimitrov delivered lectures on: Building engineering installations, Ecology of buildings, Energy efficient architecture, Mechanical components of HVAC systems, Theoretical, Applied and Fluid mechanics. He started his research in 1980 as a graduate student at "Lykov" Institute of Heat and Mass Transfer, Acad. Sci. of Belarus, Minsk. He was a scientific consultant for the Energy Research Center, Mechanical Department at College of Engineering, UNLV. He was later a visiting researcher at Lawrence Berkeley National Laboratory, Environmental Energy Technology Division, Indoor Environment Department. Prof. Dimitrov is an author of more than 100 research papers. He participated in 45 research projects and has 15 years of experience in the design of building engineering systems, which make him one of the prominent specialists in the field.

With its simple but technically precise style, the book puts forward the use of so-called "blue fuel" in buildings. It presents in a concise form the theoretical foundations of systems – principles of operation, general structure, scheme types, basic components and solutions. It also outlines the main principles of system design, giving various practical calculation examples. The book conforms to Bulgarian normative documents and standards.

An in-building fuel gas pipework is a component of the logistic structure of a building energy system. It is designed and assembled to convey and distribute fuel gas between end users, which build the basic or subsidiary energy subsystems. Fuel gas can feed either systems for thermal and air comfort, food preparation and preservation and hygiene maintenance or even a system for visual comfort.

True merit of the book is the enclosed numerical solutions of 40 practical problems, concerning the calculation of the fuel gas state parameters and dimensioning of fuel gas and smog conveying systems. The examples are skillfully selected involving two phases of design. The author encloses a number of tables and standard monograms, as well as original auxiliary tables, facilitating the calculations.

The book may also be of interest to all specialists involved in real estate investment and control, climatologists and ecologists. Note in particular the visionary character of Paragraph 4 "Main Building Systems Operating with Fuel Gas", advancing the idea that fuel gas consumption should amount to 30% of the total energy balance, thus decreasing the electric power share. This would improve the air quality; thanks to the decrease of toxic emissions from the combustion of liquid and solid fuel – nitric and sulfur oxides and blue particles.

Professor Dr. Arch. Jordan Radev
Sofia, June 2017

About the Author

Alexander V. Dimitrov is a professional lecturer with more than 35 years of experience at different universities in Bulgaria. Dr. Dimitrov has lectured and studied at leading scientific laboratories and institutes in different countries, including the Institute of Mass and Heat Exchange "Likijov", BAS, Minsk; Lawrence Berkeley National Laboratory, Environmental Energy Technology Division, Indoor Environment Department; UNLV, College of Engineering, Center for Energy Research, Stanford University, California, Mechanical Department.

Professor D.Sc., Ph.D., Eng. Dimitrov has conducted systematical research in aerodynamics and gas dynamics of the pile lines systems, energy efficiency, computer simulations of energy consumption in the buildings, leaks in ducts of HVAC systems and application of Gas fuel cells in mHPS and Gas Heat pumps in the buildings energy centers.

Professor Dimitrov has defended two scientific degrees: Doctor of Philosophy (Ph.D. in 1980) in Aerodynamics and Doctor of Science (D.Sc. in 2012) in Thermodynamics. With significant audit experience in the energy systems of buildings and their subsystems, he has developed an original method for evaluating the performance of the building envelope and energy labeling of buildings. He also has experience in the assessment of energy transfer through the building envelope and gas dynamics pressure losses in the ducts of HVAC systems. He is a scientific consultant of the biggest Bulgarian gas distributor company "Overgas Ink".

Professor Dimitrov's methodology has been applied in several tens projects with great success. He has several patents for useful models in HVAC systems. Professor Dimitrov has developed a mathematical model for assessment of the environmental sustainability of buildings, named BG_LEED. He earned the professor degree in "Engineering installations in Buildings" with dissertation "The building's energy systems in the conditions of environment sustainability" at the European Polytechnical University in 2012. Subsequently, he has authored more than 100 scientific articles and 9 books, including 5 in English.

EPU,
Department of
"Building Constructions and Architecture"
Address: 2600 Pernik, bul. "Cyril and Methodi" 23
Cell: 001 702 426 2300
Internet address: dimitrov_epu@Yahoo.com
Web site: in Likedin
Place, Date of birth: Pernik, 07.30.1948
Income in EPU: 2012 г.

1 Introduction

The growing interest of the engineering community to fuel gas applications necessitated a second revised edition of the present book with additions. More specifically, the interest nowadays focuses on fuel gas usage in in-building systems for air conditioning and micro heat and power generators, the latter being intensively applied in countries with increasingly advancing power technologies such as Japan, Korea and Germany – see the additions in Section 2.7, 4.1, 4.2 and 4.5.

1.1 GAS PIPE NETWORKS AND SYSTEMS AS ELEMENTS OF THE BUILDING LOGISTICS

Architecture-construction science reflects the public trend toward the exploitation of cheap, accessible and user-friendly sources of fossil and renewable energy carriers (fuel gas in particular) such as natural gas (NG), gas fractions of oil cracking, methane, shale gas and synthetic gas.

Change in the character of building design started back in the 1970s. It involved not only design philosophy and aesthetics but also the "methodology" of the development of architectural projects. As a result, a building is treated now not only as an occupied volume and depository but also as a complex system. Besides physiology (i.e. the need for air, water, food), an appropriate atmosphere of occupants' safety, devotion and communication should also be guaranteed, and occupants should be able to satisfy their abstract needs for truth and beauty [10].

An integral building project, using this new systematical approach [9], treats the following principle components:

- Technological – fulfillment of the building principal mission;
- Envelope – with aesthetic functions; protects against environmental impacts;
- Structural – guaranteeing building strength and stability;
- Energy – transforming and distributing energy to the end users;
- Building comfort – thermal, air and visual comfort;
- Food preparation and preservation;
- Protection – against intrusion, fire, gas or bio attacks etc.;
- Building monitoring and control (BMC) – process and data regulation and control;
- Logistic – for transport of water, materials and waste and power supply.

Following this systematical description, it is assumed that a logistic system consists of the following subsystems:

- Power supply;
- Gas supply;

- Water supply;
- Sanitary drain – removal of liquid waste;
- Removal of waste;
- Elevators, lifts and pneumatic transport.

These components form a so-called in-building logistic infrastructure, known as "engineering installations". At the same time, the first two components assist the energy system and serve as its logistic basis of the other building subsystems.

The implementation of the described view to building structure is an innovative designer's methodological tool for synthesis and constructing of their onsite engineering systems and installations. From that point of view, central gas pipelines and power supplying installations constitute one of the two types of internal power installations, being part of the building energy system. They are designed, assembled, regulated and set in operation in conformity with the general requirements for:

- Gas-mechanical compatibility – gas devices often operate in chains subjected to urban pipeline pressure; they can also be connected to systems releasing combustion products – the combustion systems should be of one and the same type;
- Randomization of gas appliances;
- Fire and explosion protection.

However, gas appliances belong to different systems and have different functions (see Chapter 4. for more details)[1]. Hence, they should meet general normative requirements and standards. For instance, gas heat generators (steam boilers, water heaters) are components of the building energy system, while devices such as stoves, furnaces, grills and barbecues, as well as gas absorption fridges, are components of a system for food preparation and storage. Other devices, such as convectors, heaters and radiators, belong to a system for thermal comfort (heating/air conditioning system), while capacitive and tankless water heaters and dryers belong to a system for hygiene maintenance and hot water supply.

The characteristic features of gas devices are high power efficiency and low thermal inertness, making them attractive and user-friendly. Due to liberate state policy, the infrastructure for central gas household supply is well developed in Europe. In Bulgaria, the Government supports of centralized electric power supply offering the application of electrical equipment and gas devices are less popular. Yet, regardless of, household end users of fuel gas regularly increase in number, approaching the impressive 1.6 million.

[1] Besides their function pursuant to BDS EN 437A1[38], gas devices are classified with respect to fuel gas type and pressure as follows:
- Class I – fuel gas belonging to a single group – see Table P28;
- Class II – having three subclasses (H, L and E); fuel gas belonging to two groups – see Table P28;
- Class III – having three subclasses (3B/P, P and B); fuel gas belonging to three groups – see Table P28.

On the other hand, the spontaneous increase in the number of fuel gas end users carries risks of failures, fire, explosions and destruction and casualties[2]. Hence, gas pipe networks should be strictly monitored and controlled.

The regulations and normative acts concerning gas installations in Bulgaria are [1÷7] and [8]. Their application to the design, installation and exploitation of gas devices and pipe networks is formally a prerogative of the Design Chamber, but it does not bear legal responsibility for eventual damages.

This book, which discusses the described systematical approach to buildings and their structure, can be used as a successful engineering handbook of building designers in their everyday effort.

1.2 FUEL GAS, CURRENT STATE AND PERSPECTIVES

History proves that NG was known, extracted and used far back in 9th century B.C., but nobody knows when and how methane deposits emerged in prehistoric times[3].

A Chinese parchment describes the use of "red-hot coke" gas[4] in 900 B.C. to extract salt from seawater in an "industrial" installation. Ancient engineers piped 1000 wells and transported gas to a salt extracting installation through a bamboo pipeline. The year 1859 is considered to be the starting year of fuel gas consumption in Europe (England), but later gas extraction was terminated due to logistic and technical problems arising during transportation. In modern times, the first gas installations were built in the early 1960s in Great Britain and the USA, when the installment of central heating in new buildings became mandatory. The first installations conveyed fuel gas to convectors, radiators, stoves, furnaces and water heaters in buildings with central heating.

In the era of the industrial revolution, fuel gas was extracted via coal incineration (at 1000°C) under limited air supply and water steam blowing. Industrial volumes of NG were extracted for the first time in 1825 in Fredonia, USA, but the product did not gain popularity due to its small power density and ineffective transport technologies. It was often treated as an unwanted "coproduct" (see Figure 1.1).

Following the so-called "**first oil embargo**" during the late 1950s, a **new era** of fuel gas consumptions on large scale started, and fuel gas became one of the three main power sources – together with coal and oil. Parity between the three classical energy carriers (oil, NG and coal) is typical for the modern economy[5], but a tendency to the increase in gas output is observed. One of the basic arguments is that the main oil fields would be exhausted in the next 20–30 years, and the oil price would rise.

Until recently, the available deposits of **natural gas** in Bulgaria (Southeast Europe) were considered to be moderate (amounting to 2×10^9 m^3). Deposits were

[2] Regarding similar accidents (for instance, the 1968 gas explosion in Canning Town, Eastern London, where a 22-storey tower block partly collapsed), draconic measures were taken and voluntary gas units were created such as the Confederation of Registered Gas Installers (CORGI) founded in 1970 in Great Britain.

[3] In fact, methane is a coproduct of organic matter decay, but there are no proofs whether the gigantic gas deposits available worldwide result from that process.

[4] Consists of hydrogen – 45% and carbon monoxide – 55%.

[5] Worldwide, the ratio between oil, natural gas and coal is **32:22:21** pursuant to data from 2001.

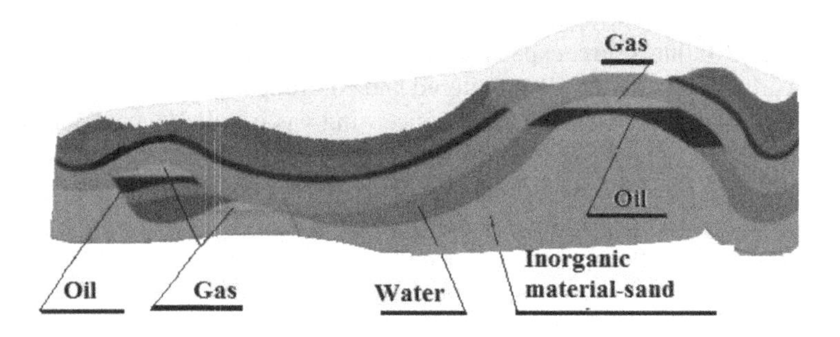

FIGURE 1.1 Structure of earth's crust with locations of natural gas deposits.

localized in the basin of river Kamchia, nearby Chiren, Devetaki, Buganovtsi and Selanovtsi. Local gas extraction amounted to 4×10^6 m^3 in 2012 (see Figure 1.2).

For now, fuel gas is extracted near Kavarna and Kaliakra by the Scottish company Melrose Resources, which extracted 0.360×10^9 m^3 in 2014 and sold the fuel to the state at a price of \$290/thousand m^3.

Results of prospecting performed by "Chevron" and "Direct Petroleum" in the Pleven region and in the so-called "Etropole argelith formation" were made public in 2011–2012. It was proved in 2014 that nearby resources including deposits near Etropole and Lovech amount to about 22×10^9 m^3. They would cover local fuel gas needs in the following 15 years. Herein, fuel gas extraction was planned to start till the middle of 2015 while production was to reach 1.5×10^9 m^3 in 2016. Yet, there is no evidence that this plan has been realized.

The American company Colorado Direct Petroleum Exploration, Denver, Colorado, found gas deposits nearby Cherven Bryag 6 years ago, amounting to about 10×10^9 m^3. The amount of NG deposits as proved in 2014, including deposits nearby Etroplole and Lovech, is about 22×10^9 m^3. In the same year, consortium "Total,

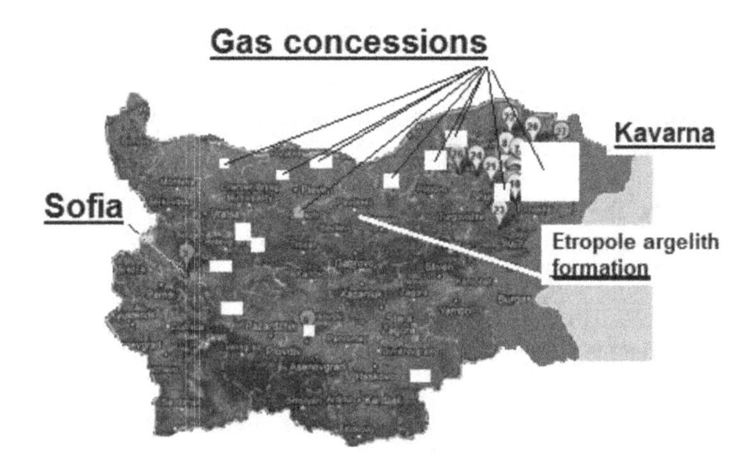

FIGURE 1.2 Map of concession offers of shale gas blocs in Bulgaria.

OMV and Repsol" proved the existence of significant gas deposits in the Black Sea territorial waters of Bulgaria amounting to 100×10^9 m^3. More gas deposits were found in the Black Sea continental shelf, in Galata, Izgrev, Kaliakra South, Kaliakra, Obzor, Ropotamo and Chaika. Their development and exploitation is forthcoming.

According to a report by the US Energy Information Agency (EIA) on shale gas deposits in 41 states outside the USA, Bulgaria has at its disposal deposits amounting to 0.49×10^{12} m^3, as well as 200×10^6 barrels of shale oil. They are sufficient to satisfy local needs in the next 100 years. Figure 1.2 shows BG areas of gas extraction concessions.

Shale gas is not only a basic oil coproduct initially extracted in the USA and Canada after 2000 and later in Poland, China and South Africa, but it is also extracted from shale via "flexible probing".

Shale gas extraction was known in the USA since the 1970s, but it intensified only 5–6 years ago. At present there are 400 000 functioning wells[6] in the USA, and the price of liquid fuel dropped 5–6 times.

Worldwide, China[7] is considered to be the country with the largest shale gas deposits – about $31{,}000\times10^9$ m^3, and its government planned to extract 6.5×10^9 m^3 in 2016 and between 60 and 100×10^9 m^3 annually by 2020. The European Union (EU) intensified shale gas extraction to compensate for the consumption of comparatively expensive fossil fuel imported from Russia. Most EU members declare a serious interest in shale gas, having overcome ecology issues.

Worldwide, the present consumption of shale gas is estimated to be $3–4\times10^{15}$ m^3 annually. Although the volumes of 80×10^{15} m^3 extracted till now, reserves of "blue fuel" will be sufficient for electric power generation for the next 250 years[8].

Fuel gas is extracted via two industrial technologies – in-depth oil cracking and Fischer-Tropsch technology. For instance, ethane, propylene, propane[9] and butane are products of the first method, while others like coalbed methane are extracted via plant and wood processing applying the second technology. So-called "hydrocarbonate gases" (methanol, methane, ammonia, hydrogen sulfide gas, "landfill" gas consisting of methane and hydrogen sulfide gas, mercaptan etc.) can be added to fuel gas. Yet, only methanol is used as an energy carrier in some fuel cells for now.

Bulgaria considers the import of liquefied gas in containers to be a significantly more expensive method of supply as compared to pipeline gas transport. World

---

[6] The USA already extracted 40% of natural gas and 30% of oil via "flexible probing".

[7] According to EIA report from June 2013, countries with the largest shale gas deposits are China – 31 Tm^3, the Argentina – 22.7 Tm^3, Algeria – 20 Tm^3, USA – 18.8 Tm^3, Canada – 16.2 Tm^3, Mexico – 15.4 Tm^3, Australia – 12.4 Tm^3, South Africa – 11 Tm^3, Russia – 8.1 Tm^3 and Brazil – 6.9 Tm^3.

[8] Regarding different estimations, Russia has the largest reserves of natural gas (47.6×10^{15} m^3), while Gazprom is the largest company of gas extraction and trade. Next are Iran with $33.1–33.8\times10^{15}$ m^3 in 2013, Turkmenistan with 17.5×10^{15} m^3 in 2013 and the USA with 8.5×10^{15} m^3 in 2013. A deposit of about 900×10^{15} m^3 of "nonconventional gas" such as shale gas exists, from which 180×10^{15} m^3 can be extracted. The largest gas deposit is in the offshore zone South Pars/North Dome between Iran and Qatar. It amounts to 51.0×10^{15} m^3 of natural gas and 50 billion barrels of gas condensate; Studies by MTI, Back & Veatch and DOE estimate that the larger amount of thermal and electric power in the following decades will come mainly from natural gas.

[9] Pure butane and propane are only used in experiments. Butane and propane offered on the market are in fact gas mixes with ingredients, which improve their operational characteristics. They are supplied only in a liquid form, and are known as liquefied petroleum gas (LPG).

experience and that of neighboring Balkan states prove that the end cost of gas supply, including liquefaction, container transport, regasification, insurance etc., is about \$50 per 1000 m^3, slightly depending on the remoteness of the end user[10]. Yet, there are several advantages:

- Flexibility and diversification of gas supply[11];
- Market principle of pricing[12];
- National independence and security[13].

Note that the passage of tankers through the Bosporus strait is prohibited. Yet, the construction of liquefied natural gas (LNG) gas terminals along the Black Sea coast is not a productive strategy as compensation. However, if Georgia builds a corresponding infrastructure, tanker transport of Caspian gas across the Black Sea will become possible. Greece especially reconstructed its liquefied gas terminal in Revythousa (45 km west of Athens) in 2011 with the aim to facilitate gas supply to Bulgaria and Turkey. Its capacity is 5.2×10^9 m^3 per annum. The reconstruction included a third reservoir, increasing gas capacity by 40% to 225 000 m^3 LNG. Construction of a new terminal in Northern Greece is also envisaged, where large gas amounts can be directed to Bulgaria. The new system interconnector between Bulgaria and Greece with NG capacity 5 bcm-y and length 182 km now is under construction. Note that an alternative import can proceed via the functioning terminals in Turkey and Greece.

Interconnectors in Southeast Europe between Bulgaria and Romania and Greece were built in 2016 (see Figure 3.1). Greek pipes are still empty, but Bulgaria and Greece signed a contract for liquefied gas supply. The productivity of Romanian pipes is still low due to low pressure[14]. A Turkish interconnector is also to be built, but its construction drags on despite the significance of the project.

Following Directive (EO) 2009/73, the national independent gas transfer operator Bulgartransgas[15] prepared a 10-year plan (2014–2023) to satisfy the needs for NG. It foresees an annual consumption of NG of 3.15×10^9 m^3 in 2016, which is to increase to 4.65×10^9 m^3 by 2023. Moreover, Russia is not to be the main supplier, and 30% of the gas supply should be from local sources (1.55×10^9 m^3). The plan also envisages alternative supply from Greece, Turkey and Serbia.

[10] If pipe networks are very long and pass through rough terrain, under water or through a number of countries, the price may be exceeded. The price of LNG is quite high owing to the strong demand, especially in Japan after the Fukushima accident, but there is a tendency of price drop. The price of liquefied gas was about \$400 per 1000 m^3 by the end of 2014, being lower than that by the end of 2013. It is expected to reach \$250 per 1000 m^3 in 1–2 years.

[11] For instance, Turkey imported gas from Algeria, Nigeria, Egypt and Australia last year via an LNG terminal, since supply through direct gas pipe networks is difficult.

[12] The access to the terminal enables each country to negotiate a lower price of gas transport through its pipe networks.

[13] Russia sold smaller amounts of gas (by 25%) to the EU and at a lower price (by 25%). The main reason was that the EU intensified the construction of new LNG terminals and increased the power of the existing ones. There were over 20 LNG terminals in Europe, including Turkey, by 2014, with capacity of about 210×10^9 m^3 annually.

[14] Romania plans the construction of a compressor station.

[15] Sole trading company founded on 15.01.2007 year., owned by BEH. The company covers 5 exploitation regions.

The gas pipe network to Romania[16] – IBR, only 24 km long (15 km on Bulgarian territory) and connecting Ruse and Giurgiu is with a capacity of 1.5×10^9 m^3. Thus, Bulgaria can import gas from its northern neighbor. Romania extracted 10.9×10^9 m^3 gas and consumed 13.5×10^9 m^3 in 2012, but its proven gas reserves are 80×10^9 m^3. Moreover, its reserve of shale gas is estimated to be 1.400×10^9 m^3, thus standing third in the EU after France and Poland.

The first gas connection between Kulata (Bulgaria) and Valovishta (Greece) is a reversive one, with a daily capacity of 3×10^6 m^3, and it has been in operation since January 2017 year. Yet, negotiations stumbled on the controversial barter "**Bulgarian electricity for Greek natural gas**" disagreeing with the annual export of NPS electric power by local dealers. The problem is that Greece has installed a significant number of photovoltaic power stations (PPS) standing second in the EU after Italy. Thus, it needs electric power from Bulgaria to keep its power balance, only if photovoltaics do not operate or power shortage occurs[17].

The second connection to Greece – IGB (Komotini-Haskovo-Stara Zagora) is at a standstill due to financial issues – it is 10 times more expensive than the first one. Pipeline capacity is $3–5\times10^9$ m^3 annually and the pipeline is of reversive type, 182 km long (150 km on Bulgarian territory). It is expected to start operation after 2020 and it will connect all suppliers of liquefied gas and deposits in Cyprus and Israel.

Greece planned to start the construction of the TAP gas pipe network in 2017 with an initial capacity of 10×10^9 m^3 annually, eventually increasing it by further 100%. It is entirely reversive, allowing gas transport from Italy to Greece and Bulgaria and supplying Caspian gas to Bulgaria via the local connection to IGB at Komotini. TAP capacity is $10–20\times10^9$ m^3 annually; it is to be completed in 2020 and will transport Caspian gas to Europe.

The gas pipeline from Turkey – ITB may supply gas even from Azerbaijan and provide an indirect connection to the terminal for liquefied gas. Its capacity is planned to be $1–3\times10^9$ m^3 with an increase to 5×10^9 m^3. The infrastructure is needed to establish a free-gas market in the Southeast of EU in compliance with Directive (EO) 2009/73 and the third EU energy package. Each connection can provide the necessary 1.5×10^9 m^3. Local gas extraction will satisfy 40% of the consumption and it will enhance our national energy security. In addition, a free-competitive gas market can be established on the Balkans similar to those in Europe and the USA, and problems of gas supply will be avoided, thus focusing on the current stock market quotes only.

The International Energy Agency (IAE) predicts that the increase in NG consumption worldwide will proceed. Gasification is stimulated by the enormous amounts of available NG – 150×10^{12} m^3 proved, and by the improvement of distant gas transfer. Keeping the present tendencies, consumption of NG by 2030 is expected to rise to 3300–4100 bsm^3 (billion standard m^3).

[16] Set in operational readiness in 2016; full scale operation from 2019.

[17] NEC is in a similar condition. The share of power from PPS and WEPP in 2012–2013 amounted to 40% of the total power capacity of the system. Thus, NPS stands for a buffer with a share of only 6% of the power mix. The residual power is a warm reserve for the Balkan states.

Gas pipe networks were first regulated in the 1970s with respect to safety and fire resistance. The first health and safety-ergonomic requirements (H&SER) to all involved in the gas industry were introduced in 1991. CORGI register of gas engineers operated till 2009. Since then, Gas Safe Register has been the official list of gas engineers who are qualified to work legally on gas appliances.

The interest in the design and construction of gas pipe networks grew just during the period 2001–2009. Yet, due to the high price of the NG supplied by Russia the interest diminished[18]. However, a serious expansion of gas pipe networks is expected in the next 5–8 years. This can happen if local gas extraction from Black Sea fields proves to be perspective and if a free-gas market emerges in the EU South-West[19].

[18] Prices in Bulgaria exceed by 26% those in Central Europe and by 30% those in Germany.

[19] The central regulation of the fuel gas price, applied by the Commission for Energy and Water Regulation, operates irresponsibly, discouraging the development of local gas industry as a result of the abrupt increase of the final gas price by 30% in the second quarter of 2017.

2 Theoretical Foundations of Gas Pipe Networks and Installations

2.1 PHYSIOMECHANICAL PROPERTIES OF FUEL GAS AS A PRIMARY ENERGY SOURCE

There are three classes of fuel gas according to a popular classification [35]:

Class	Group	Wobbe number – N_w, MJ/m³
I	A	22.4–24.8
II	H	45.7–54.7
	L	39.1–44.8
	E	40.9–54.7
III	B/P	62.9–81.3
	P	72.9–76.8
	B	81.8–87.3

Here N_W is the Wobbe number or Wobbe index accounting for two basic factors – calorific value (higher heating value) and specific gravity[1] which are essential characteristics of fuel gas.

Hence, natural gas belongs to Class II, group H, subgroup G2 with N_W=50.68 MJ/m³, and propane belongs to Class III with N_W=64.47 MJ/m³. (see Table 2.1).

Wobbe number serves also as a specific criterion of the design of new gas burners or choice of regular burners and combustion chambers of gas appliances. It can also predict fuel output under device's known heating capacity.

Weak adhesion forces act between the individual molecules, and this is the basic characteristic of fuel gases. Hence, molecules follow complex 3D trajectories moving as free particles. The different chemical composition of gases implies their different chemical properties.

Quite often, fuel mixes consist of different flammable and non-flammable components pursuant to the gas deposit or the technology of gas extraction. Note that a large

---

[1] Wobbe index is calculated via the empirical formula:

$$N_w = CV/\sqrt{SG}, \text{ MJ/m}^3 \qquad (2.1)$$

Where: CV, MJ/m³ – Calorific (Higher heating) value (see Table 2.2); SG – Specific gravity (see Table 2.2).

TABLE 2.1
Fuel gases, used in the building installations

Fuel Gas	Composition	Natural Gas (Second Group)	Propane (Third Group)
Methane	CH_4	90.0%	–
Ethane	C_2H_6	5.3	1.5
Propylene	C_3H_6	–	12.0
Propane	C_3H_8	1.0	85.9
Butane	C_4H_{10}	0.4	0.6
Nitrogen	N_2	2.7	–
Carbon dioxide	CO_2	0.7	–
		100.0%	100.0%
CV, MJ/m^3		38.6	86.0
SG, –		0.58	1.78
N_W, MJ/m^3		**50.68**	**64.47**

number of gases are offered in a liquid form and they are known as LG (liquid gases), despite their gaseous state under normal conditions.

Considering an engineering point of view, the most important characteristics of a fuel gas are higher calorific value, density, viscosity, condensity, specific temperature etc. Markets offer a variety of gases with different thermal-physical properties. Tables 2.1 and 2.2 give the main characteristics of the two most common fuel gases – natural gas and liquefied petroleum gas (LPG) (propane)[2]. Besides chemical composition (chemical formulas), they also contain characteristic temperatures (of boiling, flammability/ignition, combustion), stoichiometric relations (air-combustion number, oxygen-combustion number), calorific value, flame ranges and velocity, specific gravity, gaseous phases and operational pressure.

The most important property of fuel gases is their calorific value (CV)[3], which is the energy contained in a fuel, determined by measuring the heat produced by the complete combustion of a specified quantity (1 nm^3) of it. Table 2.2 shows its variation within wide ranges.

Considering different deposits or technological lines, data vary around the specified values. The higher CV of natural gas varies within ranges 39.0 MJ/m^3 (≈10.8 kWh/m^3) – 49 MJ/kg (≈13.5 kWh/kg) (density about 0.8 kgm^{-3}). The CV of propane is 86.0 MJ/m^3. The CV of fuel gases depends on their physicochemical properties, on the design of combustion devices and on conditions of operation (pressure, for

[2] Other industrial fuel gases such as: blast-furnace gas – 4–5 MJ/m^3; coke gas – 17–20 MJ/m^3; convertor gas – 9.0 MJ/m^3; ferroalloy gas – 9.7 MJ/m^3; generator gas – 5.5 MJ/m^3.

[3] The calorific value of fuel gas is higher and lower. These values differ with respect to the heat released during combustion of a gas unit volume in a calorimetric chamber, after liquefaction of the released water vapor. The higher calorific value includes heat released during water vapor condensation. It amounts to 2.26 MJ/kg.

TABLE 2.2
[19] Physico-chemical properties of gas fuels

	Qualitative Characteristics	Natural Gas-Methane	LG Propane	LG Butane	LG Ethane
1	Higher calorific value	38.5 MJ/m^3	86.0 MJ/m^3	113.0 MJ/m^3	52.5 MJ/m^3
2	Specific gravity	–	0.5	0.57	0.54
3	Specific gravity of gas vapors	0.58	1.78	2.0	–
4	$T_{boiling}$	−162°C	−40°C	− 2°C	− 88.5°C
5	Flame margins	5–15%	1.7–10.9%	1.4–9.3%	2.9–13%
6	Air-gas number	9.81:1	23.8:1	30.9:1	–
7	Oxygen-gas number	2:1	5:1	6.5:1	3.5:1
8	Flame velocity	0.36 m/s	0.48 m/s	0.38 m/s	–
9	$T_{flammability}$	704°C	530°C	500°C	472°C
10	Max $T_{flammability}$	1000°C	1980°C	1996 °C	–
11	Wall pressure	210–20 kPa	370–60 kPa	280–50 kPa	–
12	Vapor pressure in the vessel	–	0.6–0.7 MPa	0.15–0.2 MPa	3.84 MPa
13	Enthalpy of condensation	512.4 kJ/mol	428.4 kJ/mol	390.6 kJ/mol	1561–1590 kJ/mol
14	Density under 20°C and 1 Bar	0.73 kg/m^3	2.019 kg/m^3	2.703 kg/m^3	

instance). The CV is accounted for by an efficiency coefficient η, specified in the device technical passport.

Table 2.2 gives the temperature of phase transition or "liquid to gas" evaporation designated in Table 2.2 as boiling temperature - $T_{boiling}$. Fuel gases such as methane, propane and butane liquefy under high pressure and low temperature. Under atmospheric pressure, methane is gaseous above −162°C, propane – above −40°C, and butane – above −2°C.

Fuel gases have different specific weight (density, respectively) yielding different characteristics of transmission and different behavior under leak. It is assessed by a characteristic called "specific gravity" – SG[4]. Methane, whose share in the fuel gas reaches 92–98%, has SG=0.58, i.e. its density amounts to 58% of the air density, thus being lighter than air. If methane leaks in a confined space, it accumulates in high-lying areas. Any gas (like propane and butane, for instance) with SG > 1 accumulates in low-lying areas upon leak in a confined space. Appliances in the building basements and on the ground floor, consuming fuel gas with SG > 1.0 (heavier than air), should be mechanically ventilated for safety.

[4] "Specific gravity" is defined as the ratio between gas density and air density under standard conditions (293.16 K and 0.101325 MPa). SG of air is equal to 1.0.

The specific gravity SG is also involved in:

- Graham's effect where lighter gases mix with each other faster than other gases (Graham's law of diffusion);
- Classification of fuel gases into different groups (Wobbe index is in a non-linear relation with the specific gas gravity – see Eq. (2.1)).

The specific mass (density) of a fuel gas is the product of the specific gravity and the air density under standard conditions. For instance, methane density is:

$$\rho_{NG} = SG_* \rho_a = 0.55_* 1.2 = 0.66, \text{kg/m}^3. \quad (2.2)$$ [5]

Note that engineering calculations employ a mean value of the density of the natural gas, which contains methane, hard mechanical particles and moisture. Moisture content is 0.73 kg/m^3 [15, 18]. Comparative analyses or predictions deal with amounts of fuel gas (including natural gas) reduced via Eq. (2.36) or Eq. (2.37) and measured in "**normal m^3**".

In air, natural gas with a concentration of 5–15% ("flammability limits") can ignite spontaneously and explode if the temperature exceeds 704°C. Different fuel gases have different flammability limits and temperatures: propane – 1.7–10.1% and 530°C, butane – 1.4–9.3% and 500°C, ethane – 2.9–13.0% and 472°C. These data are important for the safe usage and storage of fuel gas, especially under leaks and spill. For instance, the **lower flammability limit** of a fuel gas presents the degree of an admissible fuel leak in a confined space and the probability of an explosion. If the current gas concentration in a confined space under a leak is lower than a particular value (<1.4% for butane, for example), the probability of gas explosion is small.

The increase of gas concentration in a confined space and channels under continued leak demands forced mechanical ventilation to decrease the concentration under the lower flammability limit.

A massive gas leak where the concentration exceeds the "**higher flammability limit**" also decreases the probability of spontaneous ignition, but it deteriorates the air comfort ousting breathing oxygen. Most fuel gases have no odor but providers add scent to them (diethyl sulfide and ethyl butyl mercaptan) to make them easily identifiable at leak. No heat-generating agent (electric or piezo-mechanical spark etc.) should be introduced in the space, since gas temperature may rise above the flammability limit $T_{flammability}$.

The fuel-air mixture is prepared, keeping strict stoichiometric proportions to guarantee its stable combustion in household or technological appliances. "Air-gas" or "oxygen-gas" mixes are prepared keeping the mix proportions – see **line 7** and **line 8** in Table 2.2. Data show that 1 volume of natural gas needs 9.81 air volumes (see Figure 2.1), 1 volume of propane – needs 23.8 air volumes and 1 volume of butane – needs 30.9 air volumes for stable and economic combustion. Note that two

[5] Air density under normal conditions is ρ_a=1.2 kg/m^3.

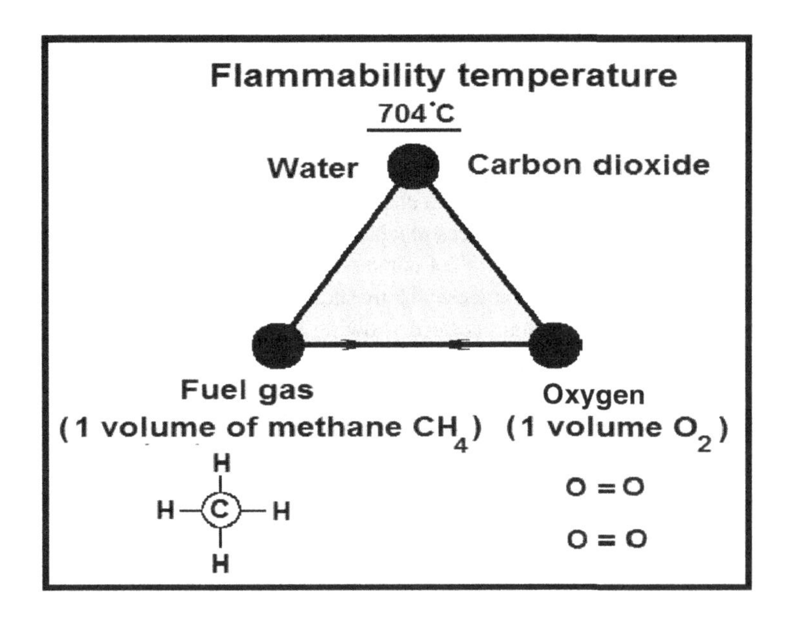

FIGURE 2.1 Combustion of fuel gas (methane)[6] [19].

unfavorable effects arise, violating the air-gas proportion and decreasing the amount of fresh air:

- Incomplete combustion – at natural gas-air proportion ≥ 1:10 where air (oxygen) is insufficient; release of carbon monoxide takes place, which is strongly poisonous.
- Decrease of the efficiency coefficient of the appliance – at proportion ≤ 1:10; excess of air (oxygen) occurs affecting fuel utilization.

The velocity of flame propagation is another important characteristic of fuel combustion affecting the type of the burner and hence – the combustion chamber and the appliance. Data given in Table 2.2 show that the flame velocity for natural gas and butane is of one and the same order (0.36 and 0.38 m/s, respectively), while it is by 26% higher for propane. There are fuel gases (hydrogen, for example) whose flame velocity is so high that it may reach the sound barrier. Besides the burner type, flame velocity affects also burner's parameters of operation such as pressure and flow rate of the gas-air mix.

When the hot front goes through high total pressure, flame "breaks off". Opposite, if pressure is low, the flame returns to the burner and burning terminates. One should control pressure at the burner nozzle to realize stable combustion with maximal efficiency, while keeping the combustion regime.

[6] It is assumed that the ratio nitrogen/oxygen in air is 1:5, i.e. 1 volume of natural gas needs 2 volumes of oxygen; 1 volume of propane needs 5 volumes of oxygen; 1 volume of butane needs 6 volumes of oxygen.

If the temperature of the fuel gas mix is lower than the temperature of fuel flammability – $T_{flammability}$, combustion cannot start. To increase the mix temperature over $T_{flammability}$, one needs to introduce a thermal agent near the burner's head to ignite easily the outflowing mix.

Fuel components oxidize as a result of ignition. The oxidation reaction is exothermic and heat released within the combustion chamber increases gas temperature. Burning temperature – $T_{burning}$ in gas appliances reaches values within the range 1000°C (T^0 of natural gas combustion) – 1996°C (T^0 of butane combustion: see **line 10** of Table 2.2).

The operation of gas appliances within such temperature ranges is a strong challenge to designers and manufacturers, owing to the need for thermal-resistant materials, whose cost would not rise drastically.

2.2 TRANSMISSION PARAMETERS OF FUEL GAS

Designers select transmission parameters of fuel gas including pressure, temperature and velocity, depending on the distance to the end user (see Table 2.3). For instance, during natural gas transmission to close destinations (urban transmission) pressure should be in the range of 3.0–7.0 kPa. If the distance is long (3000 km or more) pressure should be 1000 times higher, and it can reach 25 MPa.

Liquefied natural gas (LNG) is carried in tanks of cryogenic type at low temperatures and under high pressure. LNG installations operating in EU and the US prove that such transport is profitable. This is so, since LNG energy density increases twice during gas liquefaction as compared to natural gas compression, and LNG volume decreases 600 times.

Pressure maintained in tanks with liquefied gas depends on the fuel type. The typical pressure of storage and transport of propane is 0.6–0.7 MPa, while that of butane is 0.1–0.2 MPa. Tanks with liquefied fuel gas are secured against internal super pressure. Hence, their volume is filled by no more than 80–85%, and the gap is left to compensate for fuel's thermal expansion. The gap also gathers vapors released by the liquid phase, which are ready for use. Important components of the safety system of tanks are the balance valves enabling the release of part of the fuel (in overflow pits, in the atmosphere etc.).

Under atmospheric pressure, natural gas keeps its gaseous form at temperatures above –162°C, while at standard temperature (20°C) it is stable, regardless of pressure.

TABLE 2.3
Operating pressures of the natural gas networks

	High Pressure	Intermediate Pressure	Medium Pressure	Low Pressure
Maximal	>0.7 MPa	0.7 MPa	0.2 MPa	7.5 kPa
Minimal	0.7 MPa	0.2 MPa	7.5 kPa	3.0 kPa

Gas temperature in the pipeline should be within the interval 40°C ÷ 120°C (Par. 12, Sub-par. 1 [2]). Gas maximal velocity should not exceed 25 m/s (Par. 12, Sub-par. 2 [2]).

2.2.1 Operational Pressure

As known, pressure is due to forces of interaction between the fluid and the tank/ pipe walls. Pressure units are 1 Pa=1 N/1 m^2 and derivatives 1 MPa and 1 Bar (1 Bar=0.1 MPa=10^5 Pa).

Pressure maintained in pipe networks of residential, business and industrial buildings is called operational pressure. Fuel gas "operates" under different pressure. For instance, gas pressure in regional pipe networks is within the ranges 0.7–0.4 MPa (see Table 2.2), while that in internal pipe networks is within ranges 7.5–3.0 kPa. Table P24 gives pressure at some points of gas input of Bulgaria's national gas pipe network.

Pressure in propane tanks is maintained within ranges of 0.6 ± 0.7 MPa, while that in butane tanks – within ranges of 0.15 ± 0.2 MPa[7].

The typical input operational gas pressure is within ranges 2.1 ± 0.2 kPa. Minimal gas output pressure at the safety balance valve is 1.9 kPa under peak supply (see Figure 2.2). The operational pressure of gas appliances is within ranges 1.0–2.0 kPa, conforming to the requirements of the manufacturer. Operational pressure in propane pipe networks is reduced to 3.7 ± 0.5 kPa, while that of butane pipe networks – to 2.8 ± 0.5 kPa.

2.2.2 Pressure Drop

If a pipe network is full but not operating, pressure is constant and equal to the static operational pressure. Gas starts flowing after opening the valve and pressure drops along the pipe (see Figure 2.2). This is due to "gas-dynamic" losses due to viscous and turbulent forces of friction within the pipe network, which oppose gas flow. Engineering practice proves that this effect depends on:

- Pipeline length
  It is directly proportional to gas-dynamic losses – triple length increase yields triple loss increase. Cut of the pipe network length results in a proportional decrease of pressure gas-dynamic losses;
- Pipe diameter

[7] Pursuant to DVGW-G 464 for low-pressure gas pipe networks (up to 10^4–10^5 Pa), pipework dimensions are calculated such as to attain a stable regime of transmission. Admissible gas-dynamic losses of pressure are about 250.0 Pa, if there are no other requirements. Admissible gas-dynamic losses of pressure in gas appliances depend on the Wobbe number- N_w (see Section 2.1):

- For gases of group I: 800.0 Pa;
- For gases of group II: 2000.0 Pa;
- For gases of group III: 1500.0 Pa;

and on the admissible losses in the main distributor (respectively, on the output pressure of the domestic regulator and on the pressure at the regulator installed before the flowmeter).

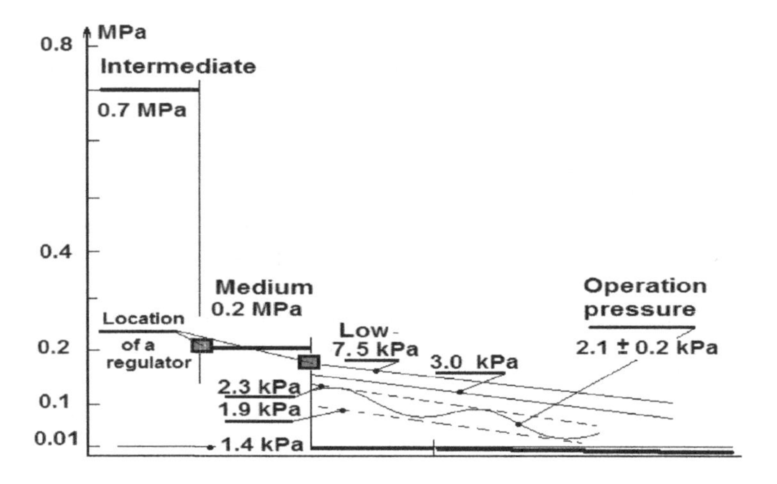

FIGURE 2.2 Pressure distribution within the pipework [19].

Gas-dynamic losses are inversely proportional to the inner pipe diameter via an exponent (with factor 5), i.e. increase of pipe diameter by 100% (twice) yields 32 times decrease of gas-dynamic losses;

- Pipe material[8]
  Rougher pipes produce greater gas-dynamic losses. New seamless pipes are with roughness of Δ=0.00002 m; copper, brass and lead pipes are with roughness Δ=0.0000015–0.00001 m; aluminum pipes are with roughness Δ=0.000015–0.00006 m, and polyethylene pipes PE-HD – with roughness Δ=0.000002 m. Steel smooth-wall pipes attain roughness Δ=0.00012 m after 1 year of exploitation, and Δ=0.00004–0.0002 m – in several years. Welded steel pipes of fuel gas pipe networks have significantly higher roughness – new ones are with Δ=0.0005 m and the older ones (after 20 years of exploitation) – with – Δ=0.0011 m. Asperity height Δ increases the value of the coefficient of linear resistance λ, resulting in an increase of gas-dynamic losses. The increase may reach 20% [11];
- Fuel gas output
  The dependence between gas-dynamic losses and fuel output is quadratic. Hence, 2 times increase in fuel output yields 4 times increase of losses;
- Specific gravity
  Practice proves that gas-dynamic losses are in a direct proportion with the "specific gravity – SG" and with fuel density ρ, respectively. Pressure losses in pipe networks with natural gas and in those with butane differ by 245% from one another, since butane specific gravity is by 245% larger than that of natural gas (see Table 2.2). Hence, higher pressure should act at the entry of an LG pipework as compared to a pipework conveying natural gas under identical other conditions.

[8] Rolled seamless steel, aluminum, internally tin-plated copper and brass pipes are used (avoid non-plated copper and brass pipes).

- Fuel gas viscosity
  The coefficient of linear resistance and pressure drop in pipe networks depend on the dynamic ($\mu = (Fl)/(sv)$) or kinematic ($\nu = \mu/\rho$) viscosity[9], respectively;
  Designers recommend the following values of gas-dynamic losses:
- In transmission pipelines: 0.3×10^2 Pa;
- In branches: 0.8×10^2 Pa;
- In a flowmeter: 1.0×10^2 Pa;
- In a pipework: 0.3×10^2 Pa;
- In branches to appliances: 0.5×10^2 Pa.

Maximal gas-dynamic losses in a fuel gas installation should be 0.1 kPa, while those in a liquefied gas installation (LG) – 0.25 kPa. Gas appliance manufacturers specify pressure in network branches. Yet, it quite often belongs to a particular range, for instance 1.4 ± 0.1 kPa. Figure 2.2 shows a pressure diagram regarding the sections of a linear pipe network.

2.3 BASIC TERMS AND LAWS

Fuel gas obeys all physical laws governing the operation of closed and semi-closed gas pipe systems. We present here a short survey and systematization of those laws employed in pipe network design. We also offer numerical examples tackling some practical issues that face pipeline designers, manufacturers and controllers.

2.3.1 Gas Flow Rate and Thermal Power. The Link between Gas Flow Rate and Decrease of Gas-Dynamic Losses – Kirchhoff's Law for Gas Pipe Networks

Gas flow rate and thermal power are the two basic characteristics accounted for in the design of gas installations. Gas flow rate[10] $\dot{V}$ is gas quantity having passed through the cross-section of a pipeline per unit time. It is measured in m^3/s. Thermal power is the quantity of thermal energy generated during the combustion of a specific volume of fuel gas per unit time. A unit of power is **watt** where 1 W=1 J/1 s (derivatives are 1 kW; 1 MW and 1 GW).

The thermal power of a gas appliance is the product of the flow rate $\dot{V}$ and the caloric value CV of the inflowing gas:

$$Q = \eta_* CV_* \dot{V}, W. \tag{2.3}$$

Here **η** is the efficiency coefficient of the burner.

---

[9] The kinematic viscosity of natural gas under standard conditions is 14.3×10^{-6} m^2/s. Yet, it depends on the temperature of operation, reaching 2.8×10^{-6} m^2/s at 120°C.

[10] Various flowmeters are used to record gas flow rate (see Chapter 4). One should fix the initial reading of the device and its final reading after an hour of operation. For instance, if the reading after 1 hour of operation is 30 m^3, the flow rate is 0.0083 m^3/s.

Equation (2.3) can be modified by substituting CV for $CV = N_w\sqrt{SG}$, – see Eq. (2.1). Hence, we obtain an expression containing number N_W, specific gravity SG and gas flow is convenient in calculations:

$$Q = \eta_* \dot{V}_* N_{W*} 10^6 \sqrt{SG}, W. \quad (2.4)$$

Example 2.1

Find the power of a **propane** burner (CV=86.0 MJ/m³) if the fuel gas flow rate is 0.00018 m³/s. The efficiency coefficient is η E 0.94. Find the power if the burner operates on natural gas.

SOLUTION

Using Expr. (2.3) we find that the power of a propane burner is:

$$Q = 0.94_* 86.0_* 0.00018 = 0.01455, MW = 14.55, kW = 14550, W.$$

Using a **natural gas** burner (38.5 MJ/m³), its power will be significantly lower – 6.93 kW, while the power of a **butane** burner (113.0 MJ/m³) will be by 31.4% greater – 19.12 kW. We could obtain the same power (14550,W) if the flow rate of natural gas is 0.0004 m³/s, or 0.00014 m³/s for butane flow rate.

Introduce an efficiency coefficient of 0.9 when using old gas devices in order to obtain a more reliable result. The real output power of a propane burner will be 0.9× 86.0×0.00018=13.93 kW.

Example 2.2

Find the gas flow rate of a propane burner with a power of 30 kW (burner efficiency coefficient is $\eta = 0.85$).

SOLUTION

Redoing Expr. (2.4) with respect to the flow rate of propane $\dot{V}$ and inserting propane data (N_W=64.47 MJ and SG=1.78 – Table 2.2), we find:

$$\dot{V} = \frac{Q}{\eta_* N_{W*} \sqrt{SG}} = \frac{30000}{0.85_* 64.47 * 10^6 \sqrt{1.78}} = 0.00041, m^3/s. \quad (2.5)$$

A number of numerical examples given below illustrate the wide application of formula (Eq. (2.5)) in practical problems solving.

Fuel gas is conveyed to appliances through semi-open[11] centralized pipe networks. Basic physical laws of mass and energy conservation operate therein, and gas flow obeys the following five particular laws:

- Law of continuity;
- Law of the balance of inflowing and outflowing fluids at a knot point;

[11] Ends in a combustion cell or in open air (via flue pipes).

- Bernoulli's law (for real fluids);
- Kirchhoff's law of gas-dynamic pressure in a pipe contour;
- Darcy's law (for gas-dynamic losses).

Since these laws and mathematical expressions are used in the engineering analysis and synthesis of gas installations, they will be discussed in brief, in what follows.

2.3.2 Law of Continuity

This is the principle of mass conservation. Its mathematical form is:

$$m^* = \rho_* \dot{V} = \rho_* A_{i*} w_i = \frac{w_{i*} A_i}{v} = \text{const.} \tag{2.6}$$

where:

- m^*, kg/s – mass flow rate;
- $\dot{V}$, m^3/s – volume flow rate ($\dot{V} = m^*/\rho = v_* m^*$)– see also Eq. (2.5);
- $\rho = v^{-1}$, kg/m^3 – gas density;
- A_i, m^2 – area of the ith cross section of the control volume;
- w_i, **m/s** – gas velocity at the ith cross section of the control volume.

Thus presented, the law of continuity (law of conservation of gas mass passing through a control area per unit time) states that the **mass flow rate of a gas** through the ith cross section of a pipe is **constant** per unit time ($m^* = \text{const}$). Consider the entrance and exit of a pipeline transmitting incompressible gas with density ρ ($\rho = \text{const}$). Then, Eq. (2.6) takes the form:

$$w_1 * A_1 = w_2 * A_2. \tag{2.7}$$

and we can find gas velocity w_2 or the cross-sectional area A_2 at the pipe exit, i.e.:

$$A_2 = A_1 \frac{w_1}{w_2} = \frac{\dot{V}}{w_2} \text{ and } w_2 = w_1 \frac{A_1}{A_2} = \frac{\dot{V}}{A_2}, \text{respectively.} \tag{2.8}$$

We can use those relations to calculate pipes with variable cross section (confusers and diffusers) or estimate the gas dynamics losses occurring during gas outflow, knowing velocity (w_1) and cross-sectional area (A_1).

2.3.3 Law of Balance between Intake Flow Rate and Exhaust Flow Rate at a Knot

The law of balance between intake flow rate and exhaust flow rate at a knot point is an analog of Kirchhoff's first law derived for pipelines. It results from the law of fluid continuity in pipe networks transmitting gas to end users. Figure 2.3 shows a

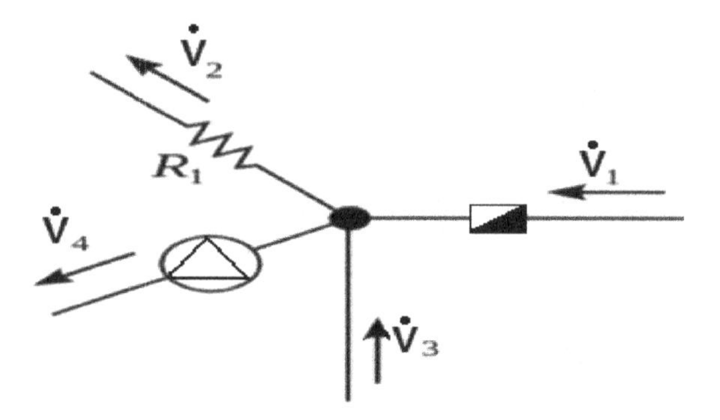

FIGURE 2.3 Model knot of a gas-distributing system.

component of a distributing system, i.e. a junction (knot point) of 4 pipes with inflow rates $\dot{V}_1, \dot{V}_3$ and outflow rates $\dot{V}_2$, and $\dot{V}_4$.

The following law holds at each knot: "**The sum of intake flow rates and exhaust flow rates is zero**".

Its form reads:

$$\sum_{k=1}^{n} \dot{V}_k = 0. \tag{2.9}$$

where (+) denotes inflow and (–) outflow.

Putting Eq. (2.5) into Eq. (2.9) we obtain the following equation of the balance of **"intake and output" power** at the knot:

$$\sum_{k=1}^{n} Q_k = 0. \tag{2.10}$$

where again (+) denotes inflow and (–) outflow.

The design of balance equations of flow rate and power at the knots of a pipe network is the most often and routine activity in solving problems of pipe network synthesis and analysis. The following example treats the input-output gas balance.

Example 2.3

Find the flow rate $\dot{V}_1$ at the knot of the pipe system shown in Figure 2.3 yielding gas balance, if $\dot{V}_2 = 0.004\,m^3/s$, $\dot{V}_3 = 0.0002\,m^3/s$ and $\dot{V}_4 = 0.0065\,m^3/s.$

SOLUTION

Write the balance Eq. (2.9) accounting for flow rates signs:

$$+\dot{V}_1 - \dot{V}_2 + \dot{V}_3 - \dot{V}_4 = 0.$$

Hence:

$$\dot{V}_1 = +\dot{V}_2 - \dot{V}_3 + \dot{V}_4 = +0.004 - 0.0002 + 0.0065 = 0.0103, m^3/s.$$

Flow rate $\dot{V}_1 = 0.0103\,m^3/s$ yields gas balance at the knot.

2.3.4 Bernoulli's Principle (for Real Fluids)

Bernoulli's principle states that an increase in the velocity of a fluid occurs simultaneously with a decrease in pressure or a decrease in the fluid's potential energy. Yet, Leonhard Euler derived **Bernoulli's equation** in its usual form. Bernoulli's principle can be derived from the **principle of conservation of energy**. Consider steady fluid flow in a pipeline. Then, the principle states that the sum of all forms of energy in a fluid along a streamline is the same at all points on that streamline. This means that **the sum of kinetic energy, potential energy and internal energy remains constant**, i.e. $0.5\rho w^2 + p + g\rho\rho = const$. Considering two pipe cross sections, the expression takes the form:

$$0.5\rho w_1^2 + p_1 + g\rho z_1 = 0.5\rho w_2^2 + p_2 + g\rho z_2. \tag{2.11}$$

As for real fluids, one should account for the effect of viscosity and friction forces on fluid flow. These two important characteristics are involved in the theorem of the kinetic energy of a fluid flowing between two arbitrary cross sections of a stream pipe (see Figure 2.4).

Then, we may write the following equation:

$$E_2 - E_1 = A_P + A_G - A_T. \tag{2.12}$$

resulting from the integral form of the kinetic energy theorem [14].

Here,

- E_1 and E_2 – kinetic energy at the first and second cross sections of the stream pipe;
- $A_P = p_1\Delta f_1\Delta l_1 - p_2\Delta f_2\Delta l_2$ work of the surface forces acting on the control surface of the fluid area;
- $A_G = g\rho\left(z_2 - z_1\right)\Delta f$ – work of gravity forces done to transmit a fluid volume;
- $A_T = \lambda N\Delta f\, l_{1\text{-}2}$ negative work of viscous forces opposing fluid flow.

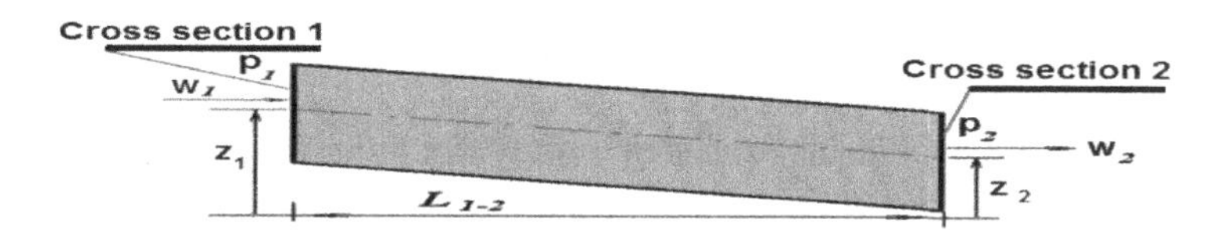

FIGURE 2.4 Stream pipe of gas flow.

Redoing Expr. (2.12), we obtain Bernoulli's equation for real fluid flow:

$$0.5_*\rho\left(w_1^2 - w_2^2\right) + \left(p_1 - p_2\right) + \rho g\left(z_1 - z_2\right) - \lambda\frac{L_{1-2}}{d}\rho\frac{w_2^2}{2} = 0. \qquad (2.13)$$

where:

- $0.5_*\rho\left(w_1^2 - w_2^2\right)$ – change of the dynamic pressure within the pipe;
- $\left(p_1 - p_2\right)$ – change of the static pressure within the pipe;
- $\rho g\left(z_1 - z_2\right)$ – change of the piezometric pressure within the pipe;
- $\lambda\frac{L_{1-2}}{d}\rho\frac{w_2^2}{2}$ – pressure losses due to friction (so-called gas-dynamic pressure losses).

Equation (2.13) differs from Eq. (2.11) containing a term for the pressure drop, and the coefficient of linear friction λ accounts for viscosity. The equation provides a possibility to estimate pressure losses in the pipeline linear section if rewritten in the following form:

$$\Delta p_{GL} = \lambda\frac{L_{1-2}}{d}\rho\frac{w_2^2}{2} = 0.5_*\rho\left(w_1^2 - w_2^2\right) + \left(p_1 - p_2\right) + \rho g\left(z_1 - z_2\right), _{Pa}. \qquad (2.14)$$

It is often used in engineering practice.

2.3.5 Law the of Gas-Dynamic Head in a Closed Loop

The law of gas-dynamic head in a pipe is an analog of Kirchhoff's second law stating that the net electromotive force around a closed circuit loop is equal to the sum of pressure drops around the loop. In the case of a gas-pipe system, it predicts the **needed gas-dynamic head** providing a flow of fuel gas, which would overcome linear and local resistance.

Accordingly, **the algebraic sum of the gas-dynamic losses** $\sum_{1\div n}\Delta p_i$ **in a closed loop is zero i.e.**

$$\sum_{1\div n}\Delta p_i = 0. \qquad (2.15)$$

The effective gas-dynamic pressure losses (pressure drops), e.g. those in linear pipe components, fittings, burners etc. are denoted by (–), while pressure rise in compressors, ventilators, thermo-siphons, ejectors etc. by (+).

The sum takes the following expanded form for the pipe system in Figure 2.5, A), noting that pressure delta Δp_i in components 1, 2 and 3 is negative:

$$-\Delta p_1 - \Delta p_2 - \Delta p_3 + \Delta p_4 = 0.$$

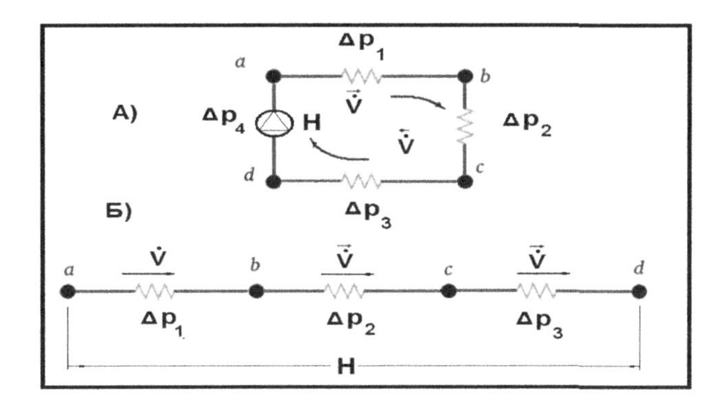

FIGURE 2.5 Gas-dynamic head and losses in pipe systems: A) Closed pipe system; B) Unilaterally open pipe system (point d is the exit to the surroundings).

Gas-dynamic head H is expressed performing the substitution $\Delta p_4 = +H$ and the respective transformation (H is the available pressure rise in mechanical[12], gravity or thermal devices pursuant to their type):

$$H = \Delta p_1 + \Delta p_2 + \Delta p_3. \tag{2.16}$$

The flowchart in Figure 2.5 expresses schematically pipe networks, which are **unilaterally open**. Fuel gas in installations flows under the head provided by the compressor stations of the pipe network.

The required pressure head H, in this case, is the difference between pressures measured at points located after the flowmeter and before the gas device i.e.

$$H = p_{AF} - p_{GD}.$$

Here,

- p_{AF} – pressure measured after the flowmeter (the sought one);
- p_{GD} – pressure measured before the gas device (prescribed by the manufacturer).

Equation (2.16) takes the form:

$$H = p_{AF} - p_{GD} = \Delta p_1 + \Delta p_2,$$

$$\Rightarrow \quad p_{AF} = p_{GD} + \Delta p_1 + \Delta p_2. \tag{2.17}$$

(ΔP_3 is the gas-dynamic pressure drop in the device but it is not present in Eq. (2.17), since it emerges outside the specified ranges).

[12] The use of buster-compressors is admissible only in systems and devices with power higher than 40 kW.

Example 2.4

Calculate pressure at point **a** of the gas pipeline in Figure 2.5, B) if pressure before the gas device $p_{GD.}$ is 1.4 kPa and the gas-dynamic losses in components 1 and 2 are $\Delta p_1 = 5\text{kPa}$, $\Delta p_2 = 3.8\text{kPa}$, respectively.

SOLUTION

Write Eq. (2.16) in the form:

$$(p_a - p_c) = (p_a - p_{GD}) - \Delta p_1 - \Delta p_2 = 0.$$

Then, the pressure at point **a** will be:

$$p_a = p_{GD} + \Delta p_1 + \Delta p_2 = 1.4 + 5.0 + 3.8 = 10.2, \text{kPa}.$$

Gas-dynamic pressure drops $\Delta P_{i=1\div2}$ and the prescribed pressure P_1 are attained when P_a =10.2 kPa.

2.3.6 Darcy's Law[13] (Concerning Gas-Dynamic Losses of Pressure[14] in Linear Pipeline Sections)

As noted in Section 2.2, gas-dynamic pressure loss (pressure drop) depends on six factors, which are present in the Darcy-Weisbach equation[15] assessing pressure losses due to friction:

$$\Delta p_{GL} = \lambda \frac{L_M}{d_p} \frac{w^2}{2} \rho. \tag{2.18}$$

where:

- Δp_{GL},Pa – gas-dynamic pressure loss (pressure drop) along a pipe length;
- λ – coefficient of linear gas-dynamic losses;
- $\frac{L_M}{d_p}$ – dimensionless pipe length (here L_M is the measured physical length of the linear pipe section. Designers use the so-called "calculation length" L_{Cal}. It reduces total (local and linear) losses to linear losses in the pipe section via $L_{Cal} = L_M(1 + b_{eq})$, and b_{eq} is a reduction coefficient);
- d_p, m – internal diameter of the linear pipe section;
- $0.5w^2\rho$, Pa – gas dynamic pressure in the pipe section (w, m/s gas velocity);
- ρ, kg/m^3 – the specific density of natural gas (see Eq. (2.2)).

To asses linear gas-dynamic losses in low-pressure pipes we use the following formulas:

[13] Darcy's law states that pressure gas-dynamic losses are proportional to the square of velocity - $\Delta p \approx w^2$
[14] Fuel gas behaves like an incompressible fluid under low transfer velocity.
[15] The formula is a version of that of Gaspard de Prony – 1855 (J. Weisbach – 1845, A. Darcy – 1857).

Laminar flow ($Re \leq 2000$ and $\lambda = 64/Re$)

$$\Delta p_{GL} = 1.154 \frac{\dot{V}}{d_p^4} * L_M \nu, Pa; \tag{2.19}$$

Transitional flow ($2000 > Re < 4000 \text{ -} \lambda = 0.0025\sqrt[3]{Re}$)

$$\Delta p_{GL} = 0.526 \frac{\dot{V}^{2.333}}{d_p^{5.33} \nu^{0.33}} * L_M * \rho, Pa; \tag{2.20}$$

Turbulent flow $\left(Re > 4000 - \lambda = \left(\frac{\Delta}{d_p} + \frac{68}{Re} \right)^{0.25} \right)$

$$\Delta p_{GL} = 1.62 \left(\frac{\Delta}{d_p} + 53.4 \frac{\nu_* d_p}{\dot{V}} \right)^{0.25} * \frac{\dot{V}^2}{d_p^5} * \rho * L_M, Pa; \tag{2.21}$$

The above expressions include: L_M – measured length of the gas pipe section; Δ – equivalent absolute wall roughness; ν – kinematic viscosity under standard conditions; d_p – pipe diameter; $\dot{V}$– gas flow rate and ρ – gas density under standard conditions (see Eq. (2.2)).

To automate the calculation of linear losses in pipes, we recommend the following formula [12]:

$$\Delta p_{GL} = 0.5 \rho_{NG} w_{NG}^2 \frac{L_{Cal}}{d_p} \left(1.0 + \frac{0.02}{\bar{Re}^{0.94}} \right) * \lambda_{boud}.$$

where:

- $\bar{\lambda} = \frac{\lambda}{\lambda_{boun}}$; $\bar{Re} = \frac{Re}{Re_{boun}}$
- $\lambda_{boun} = 0.048 - 0.0001155 \frac{d_p}{k} + 1.46 * 10^{-7} \left(\frac{d_p}{k} \right)^2$ – limited friction factor;
- $\bar{Re}_{boun} = \left(0.514 \frac{d_p}{k} - 14 \right) 10^3$ – reduced Reynolds number.

The formula guarantees calculation efficiency, due to the simple algebraic transformations and a single use of the standard function. Moreover, one can attain high accuracy of approximation of the experimental data under all three gas-dynamic regimes [12].

We offer solutions to some problems in what follows. They illustrate the application of the method, estimating the pressure of the operating fluid and gas-dynamic losses in the sections of the pipe network.

2.3.7 Calculation of the Parameters of Natural Gas – Examples

Example 2.5

Calculate the linear gas-dynamic losses in a seamless steel pipe of a gas pipeline if gas flow rate $\dot{V}$ is 0.015 m³/s. The pipe length is 150 m and the pipe diameter is 0.15 m.

SOLUTION

Step 1: Determine the hydraulic regime, i.e. find the Reynolds number:

$$Re = \frac{w_{NG*}d_p}{\nu} = \frac{4}{\pi\nu}\frac{\dot{V}}{d_p} = \frac{4}{3.14_*14.3_*10^{-6}}\frac{0.015}{0.15} = 890.8$$

Hence, the gas flow is laminar. The coefficient of linear gas-dynamic losses is:

$$\lambda = \frac{64}{Re} = \frac{64}{890.8} = 0.0718$$

Here, gas kinematic viscosity and density are equal to 14.3×10^{-6} m²/s and 0.73 kg/m³, respectively.

Step 2: Calculation of the gas-dynamic losses via formula (Eq. (2.18)):

$$\Delta p_{GL} = \lambda \frac{L_M}{d}\frac{w_{NG}^2}{2}\rho = 0.0718\frac{150}{0.15}\frac{0.73}{2}\left(\frac{4_*0.015}{3.14_*0.15^2}\right)^2 = 18.9\text{Pa}.$$

If the gas flow rate increases 10 times, the Reynolds number will be 8908.0 and flow is turbulent, while the coefficient of linear gas-dynamic losses is:

$$\lambda = (\Delta/d + 64/Re)^{0.25} =$$

$$= (0.0011/0.15 + 64/8908)^{0.25} = 0.347.^{16}$$

Gas-dynamic losses rise to 10007 Pa.

Gas-dynamic losses emerge not only in linear pipes but also in other pipeline components, such as elbows, T-branches etc., compensators and other fittings (valves, faucets, filters). They are known as **local pressure losses**. Their share is 5–10% of the total pressure losses.

[16] The equivalent absolute roughness of the walls of steel pipes amounts to Δ=0.0011 m after 20 years of exploitation. It is Δ=0.0005 m for new pipes [14].

TABLE 2.4
Coefficient of "local resistance-linear resistance" reduction – b_{Eqv}

1–2 m	2–4 m	5–7 m	8–12 m
4.5	2	1.2	0.5

To simplify the engineering calculation of gas-dynamic pressure losses in pipe networks we introduce a so-called pipeline "**calculation length** – L_{Cal}", which includes the measured length – L_{Cal}, corrected by a so-called "equivalent length" – L_{eqv}. Thus, we indirectly account for local resistance:

$$L_{Cal} = L_M + L_{eqv} = L_M \left(1.0 + \mathbf{b}_{Eqv}\right) \tag{2.22}$$

("$\mathbf{b}_{Eqv}$" – reduction coefficient of local-linear resistance).

The reduction coefficient $\mathbf{b}_{Eqv}$ of pipeline networks under low pressure depends on the length and function of pipes:

- In risers – 20%;
- In pipes connecting risers – 25%;
- In apartment branches – depending on the pipeline arrangement (see Table 2.4).

Example 2.6

Calculate the specific gas-dynamic pressure losses (specific pressure drop) $\Delta\bar{p}_{GL}$ in a pipe network if the length of the pipe connecting the last gas device is 45 m. Local resistance amounts to 25%, and pressure loss of the pipe network after the reducing valve is 1750 Pa.

SOLUTION

The following formula calculates the pipe length to the last end user:

$$L_{Cal} = \left(1 + b_{eq}\right) * L_M = \left(1.0 + 0.25\right) * 45 = 56.25, m$$

The available pressure loss Δp_{GL} is the difference between gas pressure after the valve (1750 Pa) and that before the last gas device (75 Pa). Then, the **specific gas-dynamic pressure losses** will be:

$$\Delta\bar{p}_{GL} = \frac{\Delta\bar{p}_{GL}}{L_{Cal}} = \frac{1750 - 75}{56.25} = \frac{1675}{56.25} = 29.78 \text{Pa/m}.$$

The found value is an important quantity used to calculate the pipe diameter. The **total gas-dynamic pressure loss** Δp_{GL} in the network connecting the end users 1675 Pa, should be equal to the available one.

Example 2.7

The specific gas-dynamic loss of pressure $\Delta\bar{p}_{GL}$ in a network is 3.6 Pa/m. What is the length of the pipeline if losses at local resistance knots (fittings) amount to 20% of the total gas-dynamic loss Δp_{GL} which is 150 Pa.

SOLUTION

The calculated pipeline length is:

$$L_{Cal} = \frac{\Delta p_{GL}}{\Delta\bar{p}_{GL}} = \frac{150}{3.5} = 42.8, m.$$

The actual pipeline length is found using the ratio:

$$L_M = \frac{L_{Cal}}{\left(1+b_{eq}\right)} = \frac{42.8}{1.2} = 35.71, m.$$

That value is used to calculate the distance to the remotest gas device, which could be successfully fed with gas.

Example 2.8

Find the absolute pressure and the gauge pressure of natural gas in the inlet pipe of a network, if the connected U-shaped manometer undergoes pressure equivalent to a 300-mm water column and the atmospheric pressure is 99399.8 Pa. What if the water column is 50 mm high?

SOLUTION

The arms of the U-shape manometer are brought into equilibrium (Figure 2.6). Static pressure along the isobar 3-3 is constant. Hence, "pressure of the natural gas (the left column) is equal to the atmospheric pressure plus that of the water column with height "*h*" (the right column)", i.e.:

$$p_{Abs}^{NG} = p_a + gh\rho_{H_2O}.$$

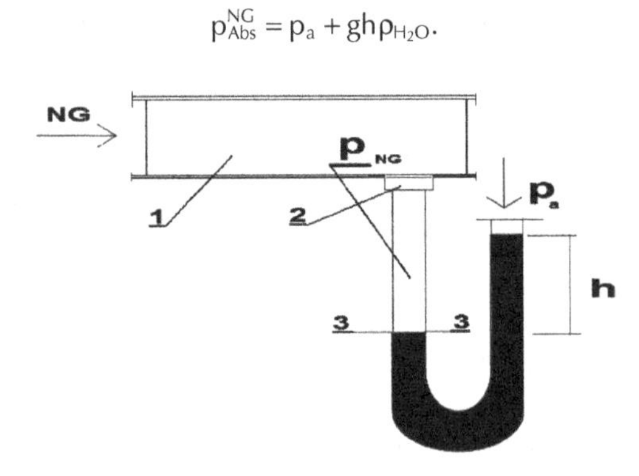

FIGURE 2.6 U-shaped manometer with water: 1-Gas pipe; 2-Connecting orifice; 3-Isobar.

where:

- p_{Abs}^{NG}, Pa – absolute gas pressure;
- P_a, Pa – actual atmospheric pressure;
- g, m/s^2 – earth acceleration;
- h, m – height of the water column;
- ρ_{H_2O}, kg/m^3 – water density (1000 kg/m^3).

Natural gas absolute pressure at *h*=0.3 m will be:
$p_{Abs}^{NG} = 9.81*10132.5 + 9.81*0.3*1000 = 99399.8 + 2943.0 = 102342.4$, Pa. and gas gauge pressure will be:

$$p^{NG} = p_{Abs}^{NG} - p_a = gh\rho_{H_2O} = 9.81*0.3*1000 = 2.943*10^3, \text{Pa}.$$

If the height of the water column changes to 50 mm, the absolute pressure will become 99.89×10^3, Pa and the **gauge pressure** – 0.49×10^3, Pa.

Example 2.9

Calculate the gas flow rate $\dot{V}$ of a gas waterheater with nominal power Q_{WH}=25.0 kW if the efficient combustion of natural gas is 75%.

SOLUTION

The higher calorific value of natural gas is 38.6 MJ/m^3. The flow rate $\dot{V} m^3/s$ of the burner gas providing a power of 25.0 kW is:

$$\dot{V}_{NG} = \frac{Q_{WH}}{\eta * CV} = \frac{25.0*10^3}{0.75*38.6*10^6} = 0.863*10^{-3} m^3/s.$$

Another method of calculating the fuel gas flow rate is to use the modified formula (Eq. (2.5)) containing the Wobbe number N_W and the specific gravity SG, whose values are specified in Table 2.2:

$$\dot{V}_{NG} = \frac{Q_{WH}}{\eta_{\Gamma\kappa} N_W \sqrt{SG}}, m^3/s.$$

It reads as follows for natural gas

$$\dot{V}_{NG} = \frac{25.10^3}{0.75*50.68*10^6\sqrt{0.58}} = 0.863*10^{-3} m^3/s.$$

Hence, the gas device will attain the power of 25 kW, if its consumption is 0.863×10^{-3} m^3/s.

Example 2.10

Regulate the gas flow rate via the regulating valves if the burner needs 0.0012 MPa (1.2 kPa) static gauge pressure measured at the pipe walls.

SOLUTION

The necessary regulation of the fittings should be such as to yield specific pressure acting on the manometer, whose water column height is calculated as:

$$h = \frac{p_{NG}^{Abs} - p_a}{g * \rho_{H_2O}} = \frac{0.1025 * 10^6 - 0.1013 * 10^6}{9.81 * 1000} = \frac{0.0012}{9.81} 10^3 = 0.1223, m.$$

The device regulation should yield pressure on the manometer equivalent to a water column 122 mm high.

2.4 FUEL GAS THERMODYNAMICS

2.4.1 Law of Gas State

According to the assumption of the classic ***kinetic gas-molecular theory***, gas pressure p in a unit control volume is [12]:

$$p = \frac{2}{3} n_A . E_k = \frac{2}{3} n_A . \frac{m.\overline{w}^2}{2}, Pa \tag{2.23}$$

where:

- $n_A = \frac{N_{E_{nv}}}{V_{E_{nv}}}$ is the number of molecules in a unit volume ($N_{E_{nv}}$ is the number of molecules in the control volume);
- $E_k = 0.5 * m\overline{w}^2$, J – kinetic energy of transfer of gas molecules (m – mass and $\overline{\mathbf{w}}^2$ – mean quadratic velocity of molecules).

The following expression for pressure holds within the entire control volume $V_{E_{nv}}$:

$$p = \frac{2}{3} aT \frac{N_{E_{nv}}}{V_{E_{nv}}}, Pa. \tag{2.24}$$

It is assumed that the kinetic energy $\boldsymbol{E_k}$ is proportional to the temperature T of the ideal gas, and one can use the substitution $E_k = a.T$. Then, expression (Eq. (2.23)) takes the form:

$$\frac{pV}{T} = \frac{2}{3} a.N_{E_{nv}} = const.^{17} \tag{2.25}$$

[17] Under standard normal conditions (p_N=101325 Pa and T_N=273.16 K): N_A= 8314 J/kmolK, N_A – number of molecules per unit mass (1 kmol).

Reduced to unit mass, it reads:

$$\frac{pv}{T} = \frac{p}{\rho T} = \text{const.}^{18} \tag{2.26}$$

Equation (2.26) proves that if a thermodynamic system (TDS) is in equilibrium, its basic state parameters (p,T and V) form a complex where the relation between the parameters ***remains constant*** regardless of whether they change their absolute values at a new equilibrium state. Equation (2.26) is known as the Clausius–Clapeyron relation[19], and it **predicts** the values of the TDS state parameters in equilibrium.

Assume that TDS state parameters are pressure (P_1), temperature (T_1) and specific volume (V_1) at a specific moment of time. If a gas, initially in equilibrium, undergoes an external impact (compression, expansion, heating or cooling) and attains a new equilibrium state, its state parameters attain new values – (P_2), (T_2) and (V_2). According to Eq. (2.26), the following relation between the two sets of state parameters holds:

$$\frac{p_1 V_1}{T_1} = \frac{p_2 V_2}{T_2} = \text{const.}$$

The new state parameters are calculated as:

$$V_2 = V_1 \frac{p_1}{p_2} \frac{T_2}{T_1} \text{ or } T_2 = T_1 \frac{p_2}{p_1} \frac{V_2}{V_1}$$

Clausius–Clapeyron relation implies that equal volumes (V) of all gases, at the same temperature (T) and pressure (p), have the same number of molecules – $N_A = 6{,}022.10^{26}$, num/mol=const. This statement is known as Avogadro's law[20].

Under normal conditions (P_N and T_N), an ideal gas with mass 1 kmol[21] will occupy a control volume of 22.4 $\mathbf{m^3}$. This fact follows from Avogadro's law and it is important in finding other operational parameters of the ideal gas under normal conditions, namely:

- Density (ρ_N) – $\rho_N = M/22.4$, kg/m^3;
- Specific volume $v_N = 22{,}4/M$, m^3/kg;
- Gas constant $R = R/M = 8314/M$, $J/k\mu/K\mu$.

[18] The reciprocal of the specific volume v is gas density ρ.

[19] Clapeyron B. (1799–1864).

[20] Avogadro A. (1776–1856).

[21] The molecular weight of a substance, also called the molar mass, M, is the mass of 1 mole of that substance, given in M gram. In the SI system, the unit of M is [kg/kmol]. To calculate the molecular weight of a compound, one should summate the molecular weight of each atom in the molecule – $M = \sum_1^n A_i$, where A_i is Mendeleev's atomic weight. For instance, $M_{CO_2} = A_C + 2A_0 = 12 + 2.16 = 44$, kg, and 1 kmol of CO_2 is equal to 44 kg.

Then, the Clausius–Clapeyron relation takes the form:

$$p.V = R.T. \tag{2.27}$$

Constant *R* has a specific value for different gases (see the Appendix – Table P22).

Another important property of the ideal gas is that the energy it accumulates depends on the variation of its internal energy (*U*) with respect to temperature (*T*), i.e.:

$$\text{Accumulated energy} \approx \frac{dU}{dT}.$$

The kinetic gas-molecular theory of heat states that the energy accumulated per unit mass is constant and the variation of the internal energy depends linearly on the variation of gas temperature:

$$\frac{\text{Accumulated E}}{M} = \frac{du}{dT} = \frac{3N_A}{M} \approx 3R = \text{const.}$$

or

$$du = \text{const} * dT, J. \tag{2.28}$$

Equation (2.28) was derived by Petit and Dulong[22] in 1819. The law of Petit–Dulong states that quantity du/dT is **constant** at gas temperature $T \geq 100K$. Since the constant characterizes **each gas as a medium** accumulating heat, it is called specific heat capacity, specifying the amount of heat necessary to change the temperature of 1 kg of gas by 1°K. It is designated as C_v ($3R = C_v$) and experimentally found for each chemical substance under constant volume.

Equation (2.28) yields the variation of the internal energy per unit mass *du*:

$$du = C_v.dT, J. \tag{2.29}$$

Varying the temperature from T_1 to T_2, the rise of the internal energy Δu regarding two consecutive states of the ideal gas can be written in the form:

$$\Delta u = C_V(T_2 - T_1), J. \tag{2.30}$$

and the resulting internal energy u_2 will read:

$$u_2 = u_1 + C_V(T_2 - T_1), J. \tag{2.31}$$

Besides the case of constant volume (V=const), the specific heat capacity can also be found at constant pressure (p = const), too. There is a difference between the two values, yet in favor of the constant pressure. We denote the specific heat capacity at

[22] A.Petit (1791–1820) and P. Dulong (1785–1838).

(p=const) as C_p, and the specific heat at constant volume for an ideal gas – as C_v. The following relation is derived by **J. von Mayer:**

$$C_p - C_v = R. \tag{2.32}$$

The specific heat capacity C_p is a physical quantity illustrating the capability of an ideal gas to accumulate a specific amount of the total energy (enthalpy: $h = u + pv$) per unit mass:

$$C_p = \frac{dh}{dT}$$

Hence, the increase of the enthalpy dh will be:

$$dh = C_p.dT, J. \tag{2.33}$$

or

$$\Delta h = C_p(T_2 - T_1), J. \tag{2.34}$$

Then, the enthalpy at temperature T_2 will finally read:

$$h_2 = h_1 + C_p(T_2 - T_1), J. \tag{2.35}$$

The Clausius–Clapeyron Eq. (2.26) has three partial solutions where one of the three state parameters – temperature, pressure or volume – remains constant. Three laws express these phenomena (see [41]):

- Charles's law (P=const);
- Boyle's law (T=const)[23];
- Gay Lussac's law (V=const).

We shall analyze the first two of them in what follows.

2.4.2 Charles's Law

The law sets forth the behavior of temperature *T* and volume *V* of a fuel gas under isobaric conditions, i.e. at a fixed pressure $p = const$. Generally, it states that the volume of an ideal gas at constant pressure is directly proportional to the absolute temperature.

Clausius–Clapeyron equation has the following form in this case:

$$\frac{V}{T} = \frac{R}{p} = const \text{ or } V = const * T. \tag{2.36}$$

[23] R. Boyle (1662) and E. Marriot (1676).

Regarding the initial (V_1, T_1) and final (V_2,T_2) equilibrium states it reads:

$$V_2 = V_1 \frac{T_2}{T_1} \tag{2.37}$$

Reducing Eq. (2.37) to unit mass, it takes the form:

$$v_2 = v_1 \frac{T_2}{T_1}$$

where v_1 and v_2 are specific gas volumes at the beginning and at the end of the isobaric process.

Equations (2.36) and (2.37) are mathematical expressions of Charles's law. They show that if a gas system is heated at constant pressure, the change of the specific volume v is directly proportional to the change of temperature *T*, i.e.$\left(v \approx T \right)$. If for instance temperature drops – $dT < 0$, then the system compresses – $dv < 0$ and vice versa, and the gas system undergoes mechanical impacts (work $l_{M_{1-2}}$) due to the surrounding medium.

Example 2.11

Find the change of the fuel gas volume (in %), if the initial volume is 2 m^3, and the temperature rises from 2°C to 24°C.

SOLUTION

The new volume will be: $V_2 = V_1 \frac{T_2}{T_1} = 2.0 \frac{(273+24)}{(273+2)} = 2.0135 m^3$. It is equivalent to a 0.7% increase in the fuel volume.

2.4.3 Boyle's Law

Boyle (1627–1691) formulated a law describing gas behavior at a constant temperature – *T*=const (an "isothermal process"), and the Clausius–Clapeyron equation takes the form:

$$p.V = R.T = \text{const or} \frac{p_1}{p_2} = \frac{V_2}{V_1} \tag{2.38}$$

while gas volume (*V*) is **inversely proportional** to gas pressure (*P*) – if $dp > 0$, then $dV < 0$, $p \uparrow \Rightarrow V \downarrow$ and vice versa.

In other words, if a gas system undergoes an impact keeping its internal energy (*T*=const), the fuel absolute pressure increases and the fuel volume decreases (fuel compresses). Introduce gas into a combustion cell or ejector where pressure is equal to the atmospheric pressure. Then, gas pressure drops and gas volume increases occupying the vessel volume in conformity with Boyle's law.

Example 2.12

Assume that the initial gas volume is 10 m^3 and pressure is 8.0 kPa. How does gas volume change under new pressure P_2 of 2.0 kPa?

SOLUTION

We calculate the new volume using formula (Eq. (2.27)):

$$V_2 = V_1 \frac{p_1}{p_2} = 10 * \frac{(101300 + 8000)}{(101300 + 2000)} = 10.58\,m^3,$$

Applying pressure P_2=800 mBar, we find the new volume as:

$$V_2 = V_1 \frac{p_1}{p_2} = 10 * \frac{(101300 + 8000)}{(101300 + 80000)} = 6.03\,m^3.$$

As seen, the basic analytical relation of the isothermal process predicts the change of gas state parameters in most practical cases.

2.4.4 Graham's Law for Gas Diffusion[24]

Due to the free movement of gas molecules within an occupied volume, gases mix easily. Graham found that the mixing velocity depends on gas density (specific gravity – SG). Light gases occupy a volume faster than heavy gases. Performing a number of experiments, he established the following relation between gas diffusion rate and specific gravity SG:

$$a = (SG)^{-0.5}. \tag{2.39}$$

It shows that if the gas complex $SG^{-0.5}$ is larger, gas diffuses faster in the air. Consider natural gas whose specific gravity is SG_{ng}=0.58. Consider also propane whose specific gravity is SG_{pr}=1.78 ($SG_{ng} < SG_{pr}$). Then, natural gas mixes with the air at a rate exceeding that of propane by 76%. At the same time, butane specific gravity is SG_b=2.0 and $SG_{pr} < SG_b$ (see Table 2.2). Hence, propane mixes with the air at a rate exceeding that of butane by 6.6%.

The fundamental physical laws discussed earlier help specialists in decision-making and prediction of phenomena occurring in gas installations.

2.5 FUEL GAS COMBUSTION

Combustion is a popular word for "non-controlled oxidation" of hydrocarbons, producing as a result of new compounds (water and carbon dioxide) and naturally releasing heat. Methane combustion is schematically presented in Figure 2.7.

---

[24] Thomas Graham (1805–1869)

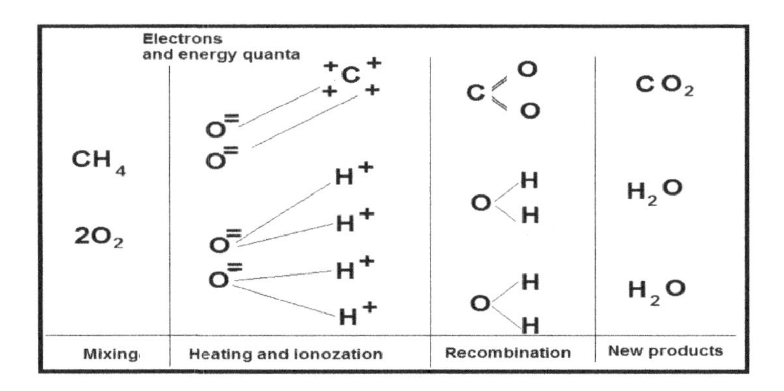

FIGURE 2.7 Phases of methane oxidation.

The combustion of propane and butane is similar, where hydrocarbons do not vanish but transform into other compounds. Since the process is isothermal, various technologies release and utilize heat – cooking, water heating for domestic needs or central heating.

2.5.1 Complete Combustion of Methane

Methane combustion passes through six stages at micro-level (Figure 2.7) under a correct stoichiometric proportion of methane-air mixing (9.81:1) and in the presence of a thermal agent:

- Mixing fuel gas and air (preparing a mix of methane and oxygen molecules in a proportion 1:2);
- Heating of the fuel-oxygen mix (to the flammability temperature – 704°C – Table 2.2);
- Mix ionization (4 free oxygen cations, 4 hydrogen anions and 1 carbon anion emerge and electrons are released at the expense of the introduced energy);
- Spontaneous recombination of oxygen cations and hydrogen/carbon anions (part of the oxygen cations, having release two electrons each, recombine with the hydrogen anions; pairs of the remaining oxygen cations release two electrons each and recombine with one hydrocarbon anion; Free electrons recombine in new molecular structures releasing energy quanta);
- Release of new products (2 molecules of water and 1 molecule of carbon dioxide);
- Release of coproducts (38.5 MJ/m^3 of heat).

$$CH_4 + 2O_2 = CO_2 + 2H_2O + Q.$$

Combustion runs in the area surrounding the burner and the released photons visualize[25] it as a flame. The generated heat is transferred via radiation, convection and thermal conductivity.

[25] The process generates high frequency phonons (with wave length of about 400 nm).

Calculations show that **2.0615 m^3 of oxygen** in 10 m^3 of air are needed in the complete combustion of 1.0 m^3 of natural gas – a mix containing methane as a main component, ethane (5.3%), propane (1.0%) and other gases (butane (0.4%), carbon dioxide (0.6%) and nitrogen (2.4%)). As a result, 2.0 m^3 of water vapor are released equivalent to 0.00125 m^3 of condensate. The amount of oxygen necessary for the combustion of 1.0 m^3 of propane is **4.93 m^3**, i.e. 23.48 m^3 of **air**. Note that propane gas is a mixture of ethane (1.5%), propylene (12.0), propane (85.9%) and butane (0.6%).

Obviously, more fresh air is needed for the combustion of 1.0 m^3 of LPG, and larger volumes of smoke gases are released as compared to natural gas. On the other hand, LPG has a higher flammability limit, and less amount of fuel gas is needed to produce 1 kW. For instance, power generation of 1 kW consumes 0.094 m^3 of natural gas and 0.038 m^3 of LPG. Hence, LPG needs less amount of air.

The released products of combustion are:

- 1 m^3 of natural gas releases 1.058 m^3 of carbon dioxide, 2.019 m^3 of water vapor and 8.777 m^3 of nitrogen (10.854 m^3 of gas products in total). The oxidation of 1 kg of methane results in the release of 2.75 kg of carbon dioxide and 2.25 kg of water;
- This follows from the stoichiometric equation – the combustion of 16 of methane releases 44 of carbon dioxide and 36 of water;
- The combustion of 1 m^3 of LPG releases 2.991 m^3 of carbon dioxide, 3.871 m^3 of water vapor, 18.554 m^3 of nitrogen (25.396 m^3 in total).

Reduced to unit power, natural gas combustion releases 1.034 m^3 gaseous products while the combustion of LPG – 0.965 m^3 gaseous products.

2.5.2 Incomplete Combustion of Methane

The spontaneous recombination of oxygen cations and hydrogen/carbon anions illustrated in Figure 2.7 takes place quickly and results in the formation of various compounds such as alcohol (CH_3OH), aldehyde (HCHO), free carbon (C) and carbon monoxide (CO).

In a medium where free-oxygen cations are present, the recombination proceeds until the end of the process producing water and carbon dioxide. Yet, if the amount of oxygen cations is exhausted or local temperature drops under the flammability limit, the process terminates releasing the newly formed compounds in the surrounding medium.

For instance, if combustion proceeds at a stoichiometric ratio of 1.9:1 (instead of 2:1) the equation of the chemical reaction has the form:

$$CH_4 + 1.9O_2 = 0.05CO + 0.9CO_2 + 9H_2O + 0.05CH_3OH.$$

Shortage of 5% of oxygen is the reason why new products emerge in smoke gases, including carbon monoxide and alcohol.

2.5.3 Controlled Oxidation of Methane

When gas fuel oxidizes in a catalytic medium, the process runs without a direct fuel-oxidizer physical contact. Then, a reaction, type Schoenbein/Grove, takes place:

$$4e^- + 4H^+ + 2O^- \Rightarrow 2H_2O + Q. \tag{2.40}$$

where, despite the basic chemical product (water), free electrons are also released and a flow of electric current takes place.

The construction of a reactor where this interaction develops is different from that of fuel cells[26] (FCs), combustion chambers, furnaces or cylinders of internal combustion engines (ICE).

FCs are complex gas electrothermal devices/reactors where oxidation of inflowing fuel gas takes place in a catalytic environment. Fuel gas may be hydrogen, natural gas, methane, blue gas, propane, hydrocarbons etc. The occurring chemical reaction produces water (H_2O) as a new product and generates new and very useful energy coproducts:

- Heat and
- Electric power[27].

These reactors successfully operated for 30 years in the mid 19th century as the only source of electricity of industrial importance in buildings in England, Germany and France. Since Siemens invented the dynamo in 1867, FCs were replaced by steam turbines and devices employing Rankine's power steam-technology. Their domination proceeds nowadays despite their disadvantages[28].

During the 1970s, the Canadian engineer Paul Ballard (1976) patented a new construction of an FC which differed from W. Grove's cell in that the fuel and the oxidizer were separated by an ion conductive membrane (a pad in a solid or liquid phase – Item 5, Figure 2.8). It prevents the reagents from spontaneous contact and explosive running of the reaction of Eq. (2.40). Yet, the membrane conducts ions charged at the FC anode (Item 2 – Figure 2.8)[29], and it is called an electrolyte.

Figure 2.8 shows an FC scheme and the principle of operation of an FC with a solid-oxide membrane. This is one of the three FC types[30] operating with natural gas. Parallel to the "new" chemical products (water and carbon dioxide), heat is also

[26] The story of FC proceeded in the 20th century after a pause of almost a century, when they were installed in space ships APOLLO to produce drinking water. At the same time, they were experimentally incorporated in busses to operate the internal NASA transport communications as sources of electric power.

[27] FCs are classical co-generators: the chemical work done during oxidation transforms into electric and thermal power. Their real efficiency coefficient in a regime of co-generation reaches 90–95%.

[28] Fatal damage to the environment, total waste of primary energy resources and energy inefficiency resulted in the activation of green house effects.

[29] The anode plated with a catalyst (platinum PL) "splits" the H_2 molecule into two anions H^+ and two electrons e^- [the H_2 molecules inflowing through the fuel supply line (Item 1) dissociate into anions H^+ and electrons e^-]. Water ions diffuse in the electrolyte, whose electric resistance is high.

[30] **Basic types** of FC operating with natural gas – phosphoric-acidic environment (**PAFC**); – solid-oxide environment (**SOFC**) – metal ceramics-molten carbonates (**MCFC**).

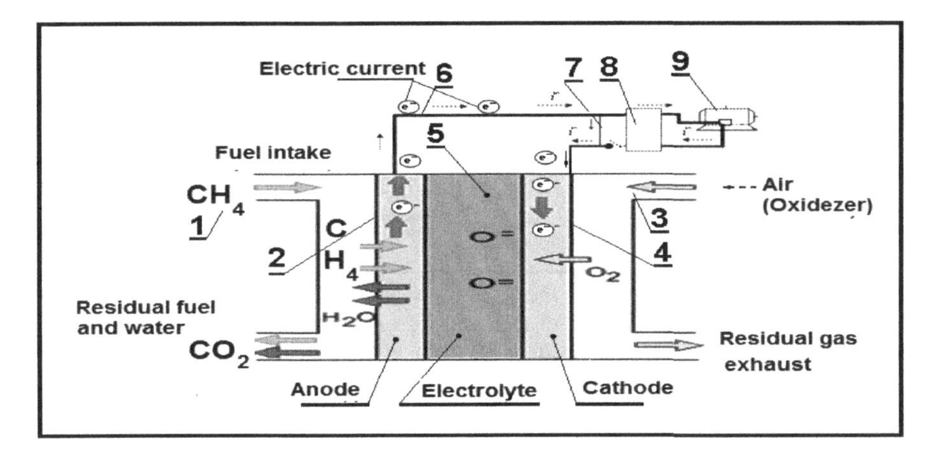

FIGURE 2.8 Scheme of a fuel cell, type "solid-oxide" (SOFC): 1-Fuel intake; 2-Anode; 3-External air intake; 4-Cathode; 5-Solid-oxide membrane; 6-External electric circuit; 7-Bypass; 8-Electric panel; 9-End user.

released (the reaction is exothermic). Moreover, electric charges are generated, and electric current flows in the external circuit and through the electrolyte. The chemical reaction of methane oxidation in an appropriate catalytic medium and in the presence of an electrolyte reads formally:

$$C^{++++} + 4H^{+} + 4O^{=} + 4e^{-} = CO_2 + 2H_2O + Q.$$

FCs using **methane** operate at a specific temperature above 900°C, where the electrolyte is zirconium oxide (porous ceramics) or yttrium. Other cells use molten salts (Na_2CO_3 or $MgCO_3$) as electrolytes. Their operation is stable in the presence of sulfur and carbon monoxide. This extends the spectrum of gas fuels, which may serve as primary energy carriers in energy centers (EC) of buildings. The catalyst facilitates the migration of oxygen atoms through the electrolyte (Figure 2.8).

Oxygen cations diffuse and quickly move to the anode where hydrogen and carbon anions "wait" for them. The released electrons cannot pass through the electrolyte, whose resistance is high. Hence, electricity flows through the external circuit and to the cathode where the electrons associate with the oxygen molecules in a catalytic environment, transforming them into cations.

The oxygen cations diffusing through the electrolyte are carriers of electricity in high-temperature FCs. Linking with hydrogen and carbon anions, they produce water and carbon dioxide, and the reaction is exothermic. An end user closes the external circuit and electrons pass from the anode to the cathode where they associate with oxygen atoms.

The continuous transfer of natural gas in an FC causes a continuous release of electrons, which concentrate at the anode. This results in a continuous action of **electric driving force** through the external wire, the grid and the FC electrolyte. An end user/accumulator closes the external circuit, and electrons pass from the anode to the cathode where they associate with oxygen atoms.

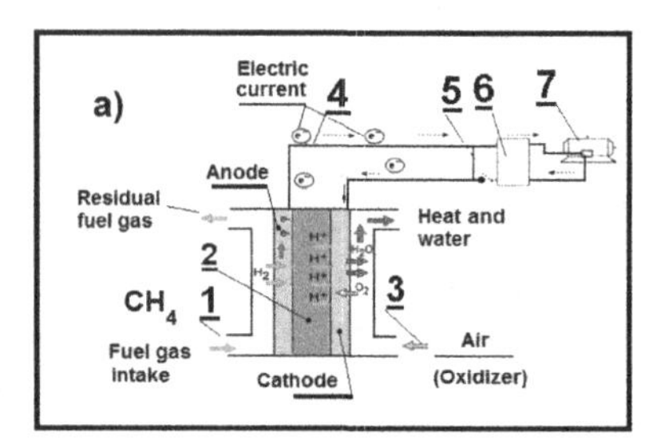

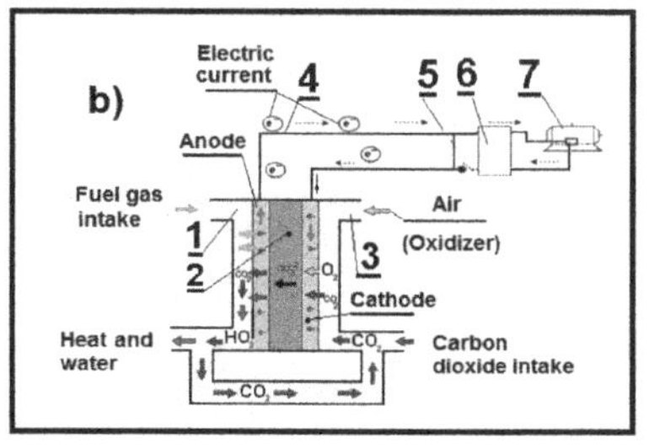

FIGURE 2.9 Operational schemes of fuels cells of various types with respect to the electrolyte used: a) Phosphorus-acidic medium (PA FC); b) Molten carbonates (MC FC): 1-Fuel gas inflow tract; 2-Electrolyte; 3-Oxidizer inflow tract; 4-External electric circuit; 5-Bypass; 6-Distributing panel; 7-End user.

The maximal voltage in an individual FC is **1.23 V:**

$$v = \frac{L_{ce}}{n.N_0 e_0}, \left(v_{max} \Rightarrow 1{,}23\text{Volt}\right)$$

where:

- L_{ce} electric power per unit mass (per 1 g-mol of oxygen). Assessing the variation of Gips's function (L_{ce}=G_2-G_1) we have G_2-G_1=237190.4 kJ/g-mol per 1 q-mol of H_2O;
- N_o is Avogadro's number (6.025.P^{23}) – the number of electrons absorbed by the oxygen molecules per 1 g-mol of H_2O;
- E=1.062.10^{-19} EV – the energy of a single electron.

The theoretical coefficient of energy efficiency of the FC chemical reaction is calculated as:

$$\eta_{th} = \frac{Lce}{(\Delta\Delta^0)} = \frac{n.N_e.e_0.V}{\Delta H^0}$$

Here ΔH^0 is the amount of energy released during the controlled fuel oxidation (fuel enthalpy), ΔH^0=285851.0 kJ/g-mol is needed for hydrogen oxidation. Then, we calculate the theoretical energy efficiency as 85%. **Actually**, it can reach 75%. Besides the excitation of electric current, the release of a significant amount of heat is a useful effect, too. Its release rate however is different for an FC with a different structure. Hence, FCs operate under different temperature regimes.

2.5.4 Fuel Cell Classification and Characteristics. Advantages and Disadvantages

There are several criteria of classification of FCs, adopted by designers of thermal technologies and building energy systems.

FCs are also classified with respect to the used electrolytes. Regarding their application, however, they are:

- FC for steady use including pipe networks;
- Mobile FC, for instance, in cars, submarines and space ships.

Pursuant to the generated power, FCs are:

- Micro-FC (with power of 0.5–10 kW_E);
- Macro-FC (with power > 10 kW_E).

Practically, one can classify FC with respect to the operational temperature, i.e.:

- Low-temperature FC (operating at 200–250°C);
- High-temperature FC (operating at 400–900°C).

Table 2.5 gives the main applications of FC in buildings with respect to their type (column 2). It also shows FC advantages (column 3) and disadvantages (column 4).

TABLE 2.5
A comparison of the fuel cells qualities, used in buildings EC

	Application	Advantages	Disadvantages
Phosphorus-acidic (PAFC)	– Additional supply; – Electrical applications; – Distributed generation of electric power.	– Higher temperature allowing combined heat and electric power generation (CHEPG); – Increased insensitiveness to polluters.	– Platinum catalyst; – Long actuation period; – Weak electric current and low power.
Molten carbonates (MCFC)	– Additional supply; – Electrical applications; – Distributed generation of electric power.	– High efficiency; – Fuel flexibility; – Use of various catalysts; – Appropriate for CHEPG.	– High thermal corrosion and damage; – Long turn-on period; – Low power density.
Solid-oxide type (SOFC)	– Additional supply; – Electrical applications; – Distributed electricity generation.	– High efficiency; – Fuel flexibility; – Use of various catalysts; – Appropriate for CHEPG & KTBE/GT.	– High thermal corrosion and damage of FC components; – Long turn-on period at high temperature.

Table 2.5 proves that fuel cells should be employed in power centers of buildings as:

- FC for main supply;
- FC for additional supply;
- FC as components of systems for distributed (decentralized) electric power generation.

Phosphorus-acidic fuel cells (PAFC), FC with molten carbonates (MCFC) and solid-oxide fuel cells (SOFC) can be elements of systems for additional power supply of buildings, including the use of wind power in so-called "passive buildings", since those devices operate successfully in micro-systems for combined generation of thermal and electric power (mHPS). Except for natural gas, FCs allow the use of other fuel gases such as propane, biogas or sin gas, if available nearby.

The focus here is on phosphorus-acidic (PAFC), solid-oxide (SOFC) and molten carbonate (MCFC) FCs as components of systems for distributed (decentralized) power generation. This is due to their high efficiency under cogeneration and high insensitiveness to fuel gas pollutants.

Table 2.6 gives some additional technical characteristics of FC, facilitating their selection and incorporation in building power systems.

TABLE 2.6
Basic technical fuel cells characteristics

	PAFC	MCFC	SOFC
Temperature of operation	150–200°C	600–700°C	700–1000°C
Oxidizer	Air per depleted air	Air	Air
Electricity carriers	H^+ ions (solution of H_3PO_4)	$CO_3^=$ ions (typical molten carbonate $LiKaCO_3$)	$O^=$ ions (stabilized ceramic matrices with free $O^=$ ions)
Common electrolyte	Liquid phosphoric acid having penetrated a matrix of lithium aluminum oxide	A solution of lithium, sodium and/or potassium carbonate having penetrated a ceramic matrix	Solid ceramic zirconium stabilized with yttrium (YSZ)
Typical constructions	Carbon, porous ceramics	High-temperature materials, porous ceramics	High-temperature materials
External reforming	No	Yes, good manufacturing practice	Yes, good manufacturing practice
Sensitivity to pollutants	CO < 1%, S	S	S
Efficiency of electric power generation	35–45%	40–50%	45–55%

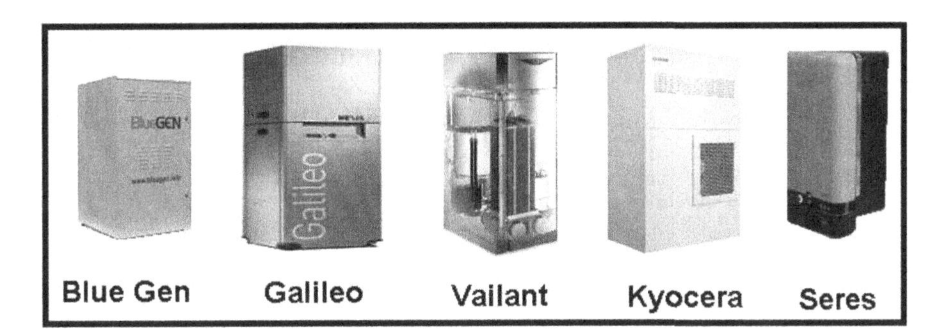

FIGURE 2.10 Micro-cogeneration systems (mHPSs) of fuel cells.

2.5.5 Fuel Gas Cells in Household Micro-Cogeneration Systems (mHPS)

The use of FCs in household energy systems[31] (in residential buildings, flats or low-rise blocks of flats), simultaneously generating electric and thermal power, seems to turn into a technological "epidemy", similar to the smartphone epidemy[32] – in Japan[33], Korea and Germany.

Those states display an urge to change the paradigm of generation and distribution of electric power in the last 10–15 years. Opposite to the world centralized power supply, new household power systems emerge.

These compact systems are called micro-cogeneration systems, or micro heat and power systems (mHPSs). They are "solid oxide (SOFC)" FCs, "phosphorus-acidic (PAFC)" FCs and "molten-carbonate (MCFC)[34]" FCs[35] with the size of a home fridge. The market offers more than twenty different brands at a growing rate – see Figure 2.10. As seen in Table 2.6, various electrolytes are used in FCs – liquid phosphoric acid in a lithium aluminum oxide matrix, solutions of lithium, sodium and/or potassium carbonate in a ceramic matrix or in solid zirconium stabilized with yttrium (YSZ).

The three electrolytes are not sensitive to admixtures and CO_2, eventually present in the fuel.

[31] Known as local, decentralized micro-cogeneration systems or mCHP systems.

[32] At present, the Ene Farm program operates in Japan, and programs NIP1 (2006–2016), NIP2 (2016–2025) and CRFCG (kfw.de/433) operate in Germany. Similar programs are in operation in (4th Plan for NRE Technologies), Italy (SEN-PER), Sweden (HIT-NIP-SE), Austria (energiestrategie.at, 2010), Denmark (Hyrogennet.dk) and the USA (EPA2005, sec.805; EISA2007, 110-140) and even in the UK (Air quality 2016–2017).

[33] The program of the Japanese government to commercialize the introduction of home fuel cells successfully started in 2005 where 3300 fuel cells were installed. Some 120,000 cells were installed in 2009, and their number was doubled in 2012. The Japanese government plans to install 1.4×106 fuel cells by 2020 and 5.3×106 cells by 2030 (there are over 50×106 households in Japan).

[34] Brands of domestic mCHP systems, included in the program ENE-FARM, are AISIN (type SOFC), JX ENERGY (type SOFC), PANASONIC (type PECF) and Toshiba (type PECF) with durability of 90 000, 90 000, 60 000 and 90 000 hours, respectively.

[35] Another possibility is to employ internal combustion engines that use gas as fuel (see Figure 2.13,b). Yet, their operational cycle is significantly shorter than that of fuel cells.

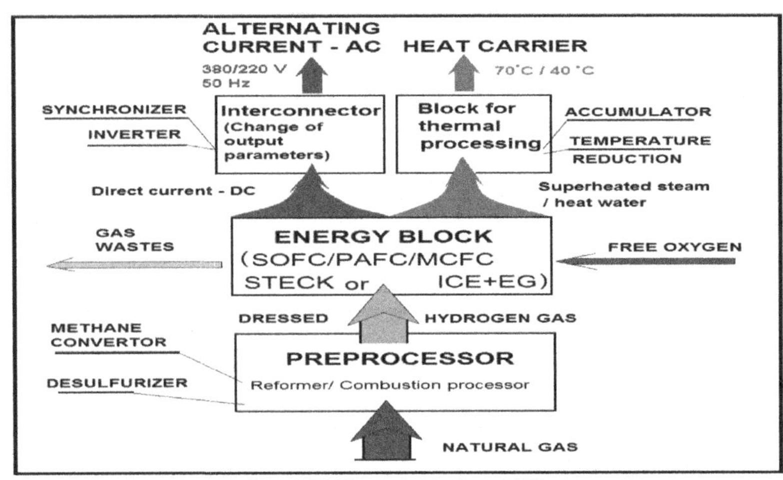

FIGURE 2.11 Flow-chart of an mHPS.

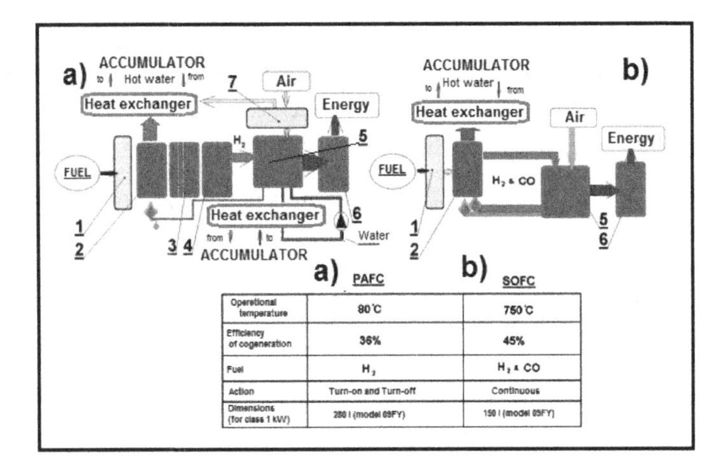

	PAFC	SOFC
Operational temperature	80°C	750°C
Efficiency of cogeneration	36%	45%
Fuel	H_2	H_2 ± CO
Action	Turn-on and Turn-off	Continuous
Dimensions (for class 1 kW)	280 l (model 09FY)	150 l (model 05FY)

FIGURE 2.12 Structure of an mHPS [40]: a) Phosphoric-acidic medium (PAFC); b) Solid oxide fuel cell (SOFC). 1-Desulfurizer; 2-Reformer; 3-Switcher; 4-Separator of CO; 5-Fuel cell package; 6-Interconnector; 7-De-moisturizer.

The general structure of an mHPS corresponds to the running energy conversion technologies.

Consider the second component of the preprocessor – the reformer (Item 2) in Figure 2.12,a) and b). **Methane conversion** into a rich hydrogen gas mix takes place there under the treatment with water vapors, and the conversion runs in a catalytic environment. The technological vapor[36] is supplied by the FC itself, when it is ready

[36] The incorporated burner is used to turn on the fuel cell – see Figure 2.12.

to operate. Carbon oxide generated in the reformer contaminates the catalyst in the energy block and it should be removed from the fuel. This takes place in the separator (Figure 2.12 – Item 4).

It is graphically presented by a functional flow chart – see Figure 2.11, and its components are:

- Preprocessor[37];
- Energy block;
- Block adapting the parameters of the output current and synchronizing them with the parameters of the electric grid (interconnector[38]);
- Block for thermal treatment.

The function of the preprocessor is to prepare the fuel for the next usage, if necessary, by removing sulfur and its compounds in a desulfurizer (Figure 2.12 – Item 1). Besides, the fuel gas composition stabilizes there via a low-temperature conversion of methane homologs into methane. The basic function of the desulfurizer, however, is to transform the inflowing hydrocarbon fuel into an air mix where hydrogen prevails.

The main module of an mHPS is the "**power block**" transforming the fuel chemical energy into a final product – electric and thermal power – Figure 2.11. The transformation runs following:

- The described technology of controlled oxidation in the FC – see Figure 2.13, a);
- The Siemens cascade technology, consisting of the successive transformation "combustion→ mechanical motion → electromagnetic induction → electric power" in a machine aggregate "dynamo-machine – ICE" (following the direct cycle of Miller, Brayton or Sterling[39] – see Figure 2.13, b).

To apply the technology of controlled oxidation, FCs are aggregated in series of 195 pieces, since the maximal voltage generated by an individual FC is 1.23 V. To attain standard values of the grid voltage at the exit of the power block, cells should be connected consecutively – see Item (1) in Figure 2.13.

The structural flowchart of an mHPS includes two additional modules (Figure 2.11):

- **Block harmonizing voltage and amperage** of the electric power generated by the mHPS with the parameters of the internal electric installation and with those of the external grid, called interconnector (Item 6 – Figure 2.12).

[37] For micro-systems with fuel cells and with Engines with Internal Combustion (EIC), using hydrogen as fuel.

[38] ASHRAE2001; ASME Standard PTC 50.

[39] The efficiency coefficients amount to 40–42%, 56% and 12%, respectively. An ICE operating under Miller's direct cycle uses directly natural gas or propane. The number of revolutions per minute of an ICE operating under Brayton's direct cycle is large and the engine is noisy. An ICE having adopted Sterling's direct cycle uses waste technological heat, which scatters in the surrounding medium.

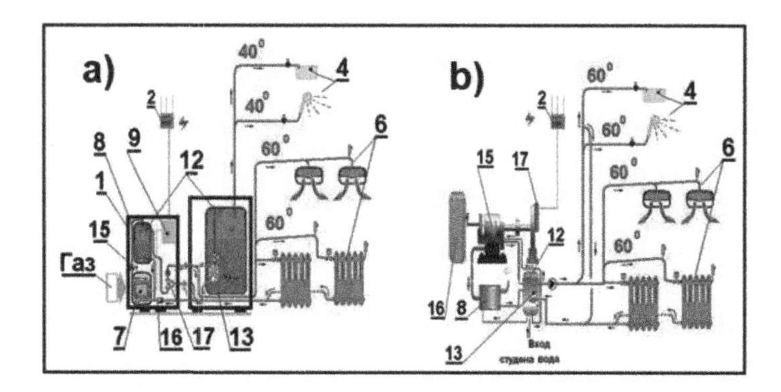

FIGURE 2.13 Structure of an mHPS: a) With a fuel cell; b) With an EIC: 1-Fuel cell body; 2-Distributing board with synchronizer; 3-Block for thermal processing; 4-Domestic end user of hot water; 5-End user of heat; 6-End user of heat for air heating for HVAC systems; 7-Fuel cell package; 8-High temperature heat exchanger; 9-Interconnector; 10-Separator; 11-Heat exchanger; 12-Heat accumulator; 13-Heat exchanger for end users; 14-Tank; 15-Gas ICE; 16-Accumulator; 17-Electric generator.

The interconnector includes two modules:

- **Inverter**, transforming the direct current (DC), generated by the micro-system into alternating current (AC) and
- **Synchronizer**, harmonizing the phase of the internal voltage with that of the external grid voltage, when the mHPS turns on to operate in parallel with the grid.

The interconnector operates under the control of a system for automatic control (SAC).

- **Utilizing block (utilizer) thermally treating the heat carrier**. Its operation is similar to that of the interconnector.

It controls the parameters of the thermal energy (pressure and temperature of the heat carrier), guaranteeing the performance of the pipework components.

The structure of the utilizer of an mHPS is more complex, since the system comprises **high-temperature** FCs, type PEMFC or SOFC for instance, where the temperature of operation is within ranges 250÷750°C. This is illustrated in Figure 2.13, a), where the temperature is utilized to 40°C and 60°C by means of a multi-contour system (utilizing, heat exchanging and mixing valves and heat accumulators). For comparison, Figure 2.13, b) shows the block thermally treating the heat carrier which runs in an mHPS with an ICE, while the temperature is kept at 60°C only.

Despite the FC equipment with industrial utilizers, sometimes individual utilizers should be needed in specific building systems. They should operate in conformity with the needs of the end users. The selection of a scheme/structure of a utilizing block (with respect to type and characteristics) depends on the power consumption.

Chapter 4 of this book discusses how one can save energy carriers, facilitating the operation of the National Central Energy System and operator's performance by installing an **mHPS**, i.e. a domestic mHPS. This saves power and decreases emissions of greenhouse gases. An analysis of the volume of primary energy carriers consumed by the basic household appliances, permanently connected to the pipe network, proves this effect. The **basic**[40] power needed by households is a stochastic quantity depending on a number of factors – occupation, number of occupants and their habits and activities, climate, season etc. Consider a four-person family per household. To meet the requirements of the present analysis, we assume with admissible practical accuracy the mean daily **basic** power to be 1.5 kW_e and 0.6 kW_{th}. It is the total power of continuously switched-on electrical **home appliances** (lighting system, refrigerators, laptops, routers, radios, chargers, TV, audio systems) not exceeding 1.0 kW_E, plus power for water heating.

Consider two types of a household power supply:

- Central supply;
- Supply by a local mHPS equipped with the FCs (see Table P24 and Figure 4.73).

Assume that the two systems have equivalent power. Then, calculations prove that the local system consumes an amount of fuel gas, which is 2.8 times smaller than that consumed by the central system. As far as the central system needs 179.7×10^{-6} m^3/s of fuel gas, the local system needs only 64.2×10^{-6} m^3/s of primary energy carrier. Thus, the micro-cogeneration technology saves 64.3% of fuel.

Chapter 4 considers several schemes of household installations of fuel cells consuming natural gas, as well as their integration into other building installations.

2.5.6 Technical Characteristics of mHPS[41]

Generalizing the material in Section 2.5.2, we give the technical characteristics of mHPSs aiming to help the correct choice:

- Main characteristics: – used fuel; – nominal electric power, kW_E; – nominal thermal power, kW_{th}; – total coefficient of power efficiency, %; – overall dimensions, WXLxH, mm; – weight, kg (see Table P24).

[40] Power needed for the operation of those appliances is called "basic power". It is the sum of the power of the simultaneously operating electric home appliances.

[41] "BlueGen" and "ENGEN 2500" of SOLID POWER GmbH(Gm) company; "Hexis Galileo 1000 N" of HEXIS (Sw) company; "G5" of VAILLANT(Gm) company; "Kyocera" of KYOCERA (Korea) company; "Steel Gen" of CERES POWER (Ho) company; "Gamma PREMIO" and "DashInnoGen" of BAXIINNOTRCH, BAXI (UK) company; "Deluxe Unit" of HONDA (Jp) company; "Aisin fuel cell system" of AISIN TOYOTA (Jp) company; "Vitovalor 300-P" of VIESSMANN (Gm) company; "Panasonic" of PANASONIC (Jp) company; "Toshiba PEFC" of CHOFU SEISAKUSHO Co Ltd (Jp) company; "JX Eneos fuel cell system" of JX ENEOS–SANYO (Jp) company; "Tepsoe fuel cell" of TEPSOE (Cz) company; "Phillips66 Fuel Cell system" of PHILLIPS66 (US) company; "OxEon Energy Fuel Cell system" of OXEON ENERGY() company; "Dachs InnoGen" of SenerTec (Gm) company; "POSCO Energy fuel cell system" of POSCO Energy() company; "S1-10C, S1-20C, … S1-60C" of POWER CELL (Sw) company; "Energy server" of BLOOM ENERGY (US) company; "Cerapower FC10" and "Logapower FC10" of THERMOTECHNIK–BOOSH (Gm) company; "Inhouse5000+" of RBZ (Gm) company; "LIPEM" of DANTHERM POWER (Dm) company; "Elcore 2400" of ELCORE company.

- Additional characteristics: – coefficient of electric power efficiency, %; – coefficient of thermal power efficiency, %; – fuel flow rate, m^3/s; – necessary gauge pressure before the gas regulating station, kPa; – temperature of the outflowing water, °C; – temperature of the inflowing water, °C; – parameters of the electric power of switch-on (number of electric phases and wires; voltage, V; frequency, Hz); – emission of greenhouse gases, kg/MW-hr (see Table P24).

Consider a local mHPS with an FC with power 1.5 kW_e and 0.6 kW_{th}. Consider also an equivalent centralized system. The comparison and analysis of both systems yields:

1. The power generation of the local system is 2.8 times more efficient than that of the centralized system.
2. The annual economy of the operation of **only one system** with an FC amounts to 3592.5 m^3.
3. Another "1.8" devices may operate using the saved fuel, and the amount of generated power will be 1.8 times greater.

The comparative analysis of data (annual energy savings in a local gas micro-system and emissions by the equivalent central system) shows that the local system:

- Reduces the emissions of $\mathbf{CO_2}$ by 64.3% annually;
- Reduces the emission of **water vapor** by 64.3% annually;
- Reduces the emissions of $\mathbf{NO_x}$ by 64.3% annually.

As for ecology protection, conclusions are:

- The local gas mHPS, even without a special catalytic tip installed on the output terminal, is an environment-sparing technology **3 times more efficient** than the equivalent central system;
- Extrapolate the data and assume that 200.103 [42] FCs operate in a series. Let the individual power of each cell be 1.5 kW_E and 0.6 kW_{th}, respectively. Then, cells generate power equivalent to 420 MW (3.0×10^5 kW_E, + 1.2×10^5 kW_{th}), i.e.:
- The amount of primary energy for inefficient centralized manufacture and transfer technologies drops (41.39 + 4.97 PJ/annually in this case);
- The consumption of natural gas by a decentralized power generation system decreases 3 times (46.39PJ: 15.58 PJ) supplying the end user with the same amount of energy (electricity and heat). In general, the efficiency of decentralized systems is 3 times greater than the efficiency of centralized systems.

[42] The figure is prognostic.

The economic aspects of the introduction of a micro-cogenerating system are discussed at the end of Chapter 4.

The appendix – Tables P24 and P25 – gives the technical characteristics and actual costs of some FCs on natural gas as of 2018. They are offered on the market and are commercial and practical success.

2.6. GAS FLAME AND COMBUSTION DEVICES

Gas flame visualizes combustion. The flame pattern generated by the photon emission at the flame front shows the structure of the oxidation area of the fuel gas. Practically, there is a vast variety of shapes, dimensions and colors of flames, where flames are:

- Post-aerated;
- Pre-aerated.

Flames of the first type emerge without a preliminary supply of air (i.e. without aeration). The outflowing gas sucks the surrounding air during the disintegration of the gas/air interface into large vortices, and it ignites and burns, mixing with the ejected air (see Figure 2.14, a) and [39]. That type of mixing is quite rare in modern gas systems.

Modern systems generate flames of the second type (see Figure 2.14, b). Part of the air needed (about 40%) is sucked from an area where its purity and oxygen contents are guaranteed. This "primary" air transferred by gravity or centrifugal forces mixes with the fuel in the ejection space of the burner (Item 4). The gas mixture enters the combustion chamber where it ignites. Additional air (called "secondary") comes from the flame area.

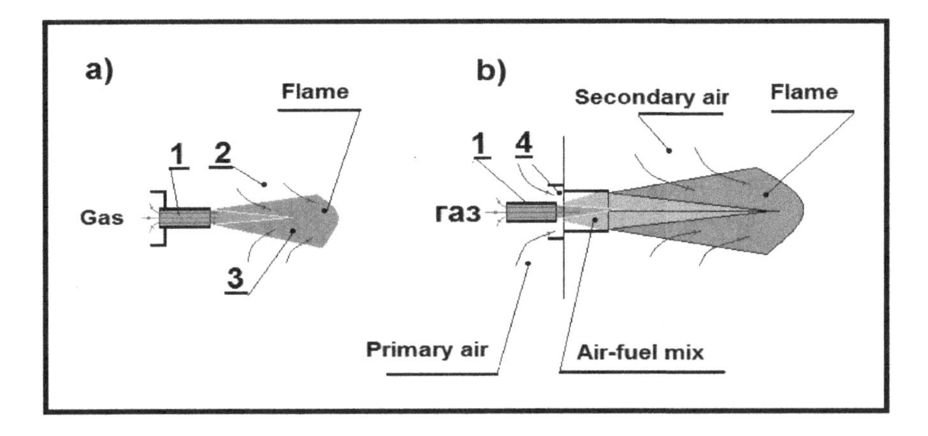

FIGURE 2.14 Basic schemes of flame formation and flame nozzles: a) Post-aerated flame; b) Pre-aerated flame: 1-Gas nozzle; 2-Air surrounding the flame; 3-Combustion air-fuel mix; 4-Ejection mixing chamber.

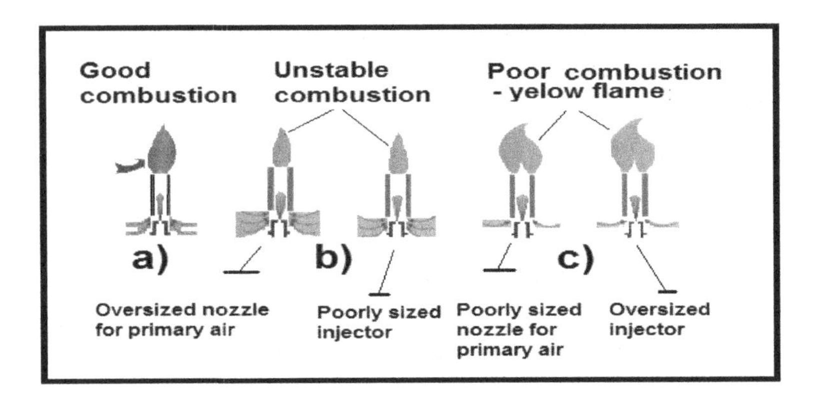

FIGURE 2.15 Flame color and shape are symptoms of the combustion quality: a) Good flame mixing and combustion (blue flame); b) Unstable flame mixing (air excess) and combustion (thin yellow flame); c) Poor flame mixing (air shortage) and combustion (thick yellow flame) [19].

The real combustion takes place along a complex stereometric surface[43] called flame "front" where the current fluid velocity equals that of flame propagation.

To generate a stable flame, the draught of the gas-air mix should agree with that prescribed by the manufacturer. Moreover, the operator should control the shape and color of the flame.

In the case of an outflow velocity higher than the prescribed one, the flame elevates above the nozzle, breaks off and extinguishes. If the outflow velocity is lower than the prescribed one, the flame sinks into the nozzle, while incomplete combustion takes place in both cases.

Experienced gas specialists successfully regulate gas combustion with respect to flame color, which is an indicator of combustion quality. This is illustrated in Figure 2.15. A blue flame with a green core and light-blue periphery is an indication of good combustion. A yellow flame is an indication of poor combustion owing to the incorrect dimensioning of: (i) the primary air nozzle or (ii) the ejector. Flame cooling below the flammability limit owing to the low temperature of the surrounding medium or the presence of cold 3D objects in the flame area (i.e. cooling below 500–800°C) also results in unstable combustion and flame extinguishment. Then, intensive soot generation and carbon release and concentration take place.

Flames generated by lattice, rod-shaped, package and frame burners belong to the class of pre-aerated flames. The single large nozzle is replaced there by a number of

[43] Flame "front" is a 3D non-steady surface in the area of instability of large-scale vortices of the jet draught. Average and small-scale turbulences emerge there owing to the development of a cascade mechanism of disintegration (the degree of the absolute turbulent intensity reaches $\varepsilon_u^0 = 80 \sim 90\%$). Turbulences are responsible for the quality and intensity of oxygen cations merging with carbon and hydrogen anions. The shape of the flame front depends on the conditions at the interface, including nozzle structure, shape and dimensions of its outlet cross section, gas input and burner power. The flame shape is conical in the special case of axisymmetric outflow, only (gas velocity along the axis is the highest one and the flame is pushed far away from the outlet).

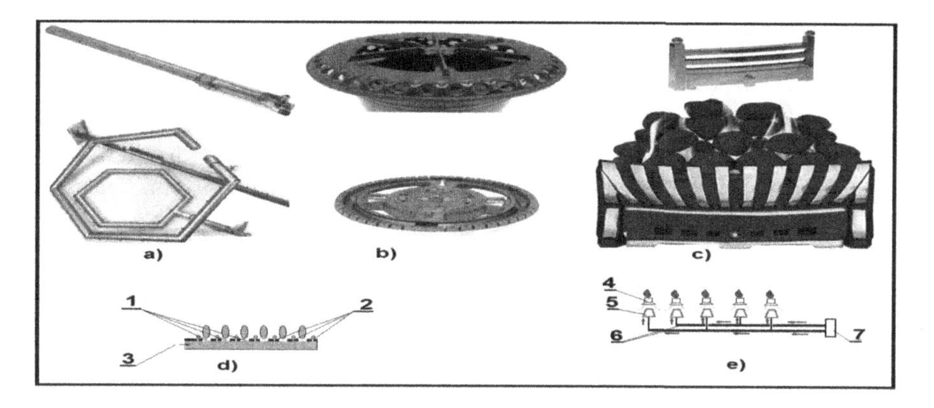

FIGURE 2.16 Systems of flames and burners: a) Rod-shaped burner; b) Package burner; c) Frame-shaped burner; d) Lattice burner with main and auxiliary flames; e) Duplex burner: 1-Main nozzles; 2-Auxiliary nozzles; 3-Gas-air mixture; 4-Ejector; 5-Gas confusor; 6-Duplex gas pipelines; 7-Gas collector.

miniature nozzles generating a system of flames (see Figure 2.16, d), which stabilize each other. Modern heating gas appliances often use a so-called "duplex burner" in order to increase the reliability of the combustion devices (see Figure 2.16, e). Opposite to a lattice burner (Figure 2.16, d), mix preparation in a duplex burner is individual in each nozzle and takes place into an individual injector (Item 4). Burners of that type are shown in Figure 2.16. Their vast variety results from the variety of gas devices, which they are incorporated in. For instance, cooking stoves use circular package burners while water heaters and heaters – long box or rod-shaped burners. Under identical parameters (input and power) a pre-aerated flame produces higher temperature as compared to a post-aerated one. This is due to the better mechanical mixing of the chemical agents prior to their interaction with each other. Hence, burners with a pre-aerated flame are more popular.

However, those burners may be classified into two groups employing another criterion (the method of primary air supply):

- Atmospheric;
- Forced draught.

Primary air enters an atmospheric burner through the inflow nozzle under atmospheric pressure. A burner with forced draught sucks air by means of a fan installed in its smoke flue.

Burners belonging to the two groups create a pre-aerated flame and their main components along the route of fuel mixing and distribution are:

- Primary air nozzle;
- Ejector;
- Mixing chamber (pipe);
- Head.

The first two of them arrange in advance the flow of fresh air and fuel gas prior to mixing. The primary air nozzle is of "confusor" type where gas fuel accelerates, since the decrease of the cross section dA > 0 yields increase of the outflow velocity w_2 (see Eq. 2.7) – the result is $w_2 > w_1$. Pressure P_2 drops below P_1 at the inlet of the horizontal ejector after fuel flows out of the nozzle-confusor (1), which follows from Eq. 2.11. A successful selection of the confusor yields a drop of pressure P_2 as seen in the following expression:

$$p_2 = p_1 - 0.5\rho\left(w_2^2 - w_1^2\right), \quad \text{Pa} \quad \Rightarrow p_2 \leq p_1 .$$

Then, an area of negative pressure emerges at the ejector inlet creating conditions of fresh air ejection.

The expertise in the field shows that the burner power depends on the ejector's dimensions and structure. Fuel gas flow rate also depends on the diameter of the ejector aperture (on the contraction number, respectively). For instance, if the diameter increases by 20% (from ø5 to ø6), the flow rate will rise by about 40%.

The selection of an appropriate ejector is of crucial importance to the efficient operation of any device. Hence, some fuel devices are designed together with their "controlling" mechanisms – these are iris diaphragms of valves, for instance, which guarantee adequate mixing and control of the output power.

The third burner component – the mixing chamber (pipe) provides the quality of the gas-primary air mix. This technological requirement sometimes contradicts to the designer's pursuit of compact and cheap mixers. To minimize the overall dimensions of the mixing chamber some devices use tangential units or grids[45].

The fourth component in some burners is the head whose function is to supply uniformly **secondary air** and terminate combustion. Structurally, this is achieved by an appropriate location of the outflow nozzles within the burner head.

We shall consider in what follows some typical burners installed in *gas devices and heat generators. The structure of a gas device as an* element of various building systems is discussed in Chapter 4 together with its function.

Figure 2.17 shows the structure of an atmospheric burner widely used in cookers. As seen, the ejector (2) sucks primary air (2) under atmospheric pressure.

Thus, a parallel flow emerges at the ejector entrance, consisting of a core and a periphery, i.e. a gas jet blown through the (Item 1), and fresh air blown into the mixing pipe (Item 3) under atmospheric pressure.

To provide good mixing of the two parallel flows, the pipe (Item 3) has a significantly large volume and mixing route. The prepared fuel-air mix is rich in fuel but poor in oxidizer, since it receives only about 40–45% of the air needed. The rest 55–60% come from the area around the burner head. The head generates a system of radial flames aerated by secondary air to attain the necessary stoichiometric ratio.

A burner blowing a preliminarily prepared gas-air mix belongs to the group of devices generating a pre-aerated flame (Figure 2.18), and gas/air ratio is close to the stoichiometric one. The mix is prepared within the burner and its ignition takes place at the burner outlet.

As a result, no burner head with nozzles for better aeration is present, but there are only outlet apertures drilled through a corrugated metal sheet and creating a system

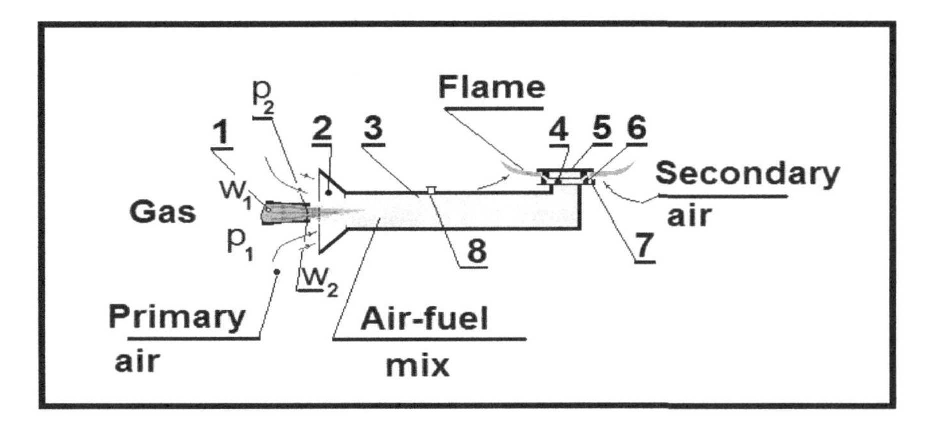

FIGURE 2.17 Structure of an atmospheric burner of a cooker blowing pre-aerated flame: 1-Fuel gas nozzle (confusor); 2-Ejector; 3-Mixing chamber (pipe); 4-Head; 5-Cap; 6-Crown; 7-Electrode; 8-Cleaning aperture.

of flames. This proves to be a successful technique preventing flame break off and extinguishment. Flames are compact and the combustion temperature is higher. The combustion chamber operates at a higher temperature, it is compact and there is no risk of flame cooling down and extinguishment by the chamber walls.

Figure 2.18 shows two schemes of preliminary mixing of fuel gas and air. Both burners are of a forced type and create a set of flames. They use air blown by a fan (Item 1) installed before the chamber.

Consider the first scheme (Figure 2.18, a). Two synchronized valves (Item 2) automatically control the process of mixing, depending on the input of the fan (Item 1). Change of fan's input and the number of revolutions yields change of valve position in order to secure the correct stoichiometric ratio of combustion.

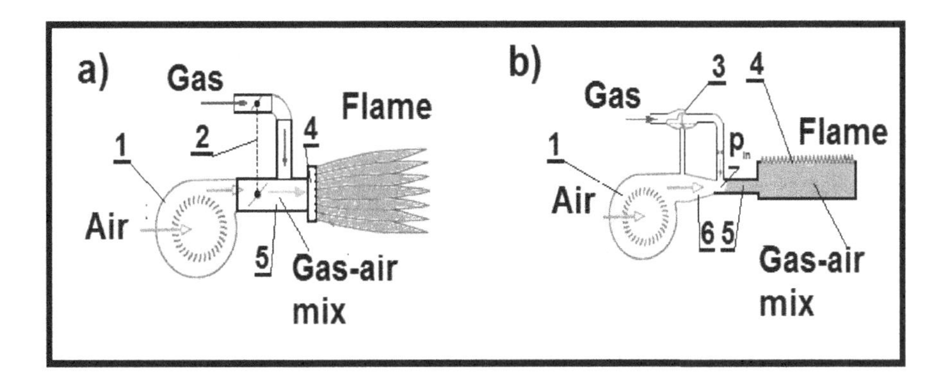

FIGURE 2.18 Preliminary preparation of the air-gas mixture: a) Active control of mixing; b) Passive control of mixing. 1-Fan; 2-Choke valves; 3-Regulator; 4-Burner; 5-Mixing pipe; 6-Nozzle.

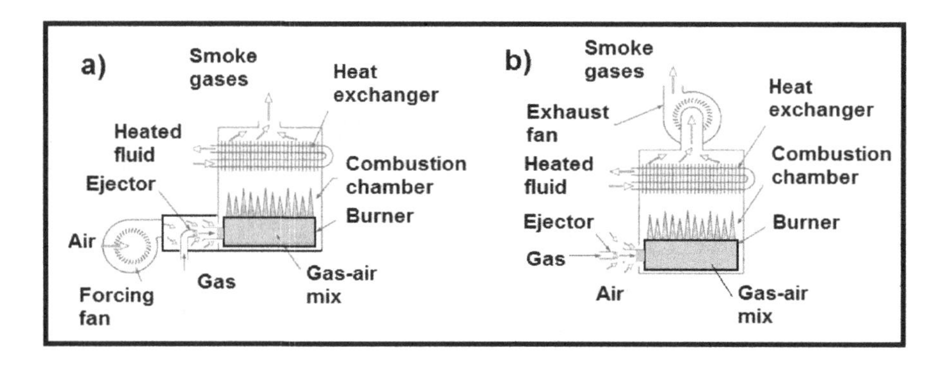

FIGURE 2.19 A burner operating jointly with a fan: a) Fan before the burner; b) Fan after the burner. 1-Heat generator; 2-Fan; 3-Burner; 4-Heat exchanger; 5-Tract of combustion products.

Consider the second scheme (Figure 2.18, b). Mixing takes place in the ejector after the (Item 6) installed along the route of the fresh air. The change of the air input yields change of pressure in the mixing pipe of the ejector (Item 5) – see Section 2.1.3.3. Change of the pressure gradient results in a change of gas pressure after the regulator (Item 3) and a change of the gas flow rate.

The characteristic feature of both schemes is that burners operate thanks to the **forced draught** generated by the fan, whose revolutions conform to the power providing the needed air flow rate. Designers of fuel devices offer two positions of the fan – along the air route and along the route of combustion products:

- Before the burner (the burner is called forced draught burner or swelled burner)[46] – see Figure 2.19, a);
- After the burner (this is a de-forced burner or an induced draught burner)[47] –see Figure 2.19, b).

In forced draught burners, the fan is installed before the burner generating pressure on the entire air-fuel duct. If the sealing of the pipe network/smoke duct is poor, smoke may enter inhabited areas.

Hence, this scheme is employed in industrial installations and large boilers installed in premises where the requirements to the air quality are lower. The second fuel scheme (Figure 2.19) is a priority of the design of heating installations for residential and public buildings.

Note that the installations can operate in accordance with the ventilating system (gas burners suck air from the interior and the entire building undergoes negative pressure). Hence, fresh air infiltrates from outside removing pollutants on its route to the sucking inlet of the gas devices – see Section 4.4.2 for more details.

Finally, we consider a so-called "package" burner, which completes the wide variety of combustion devices – Figure 2.20. It is an autonomous aggregate consisting of a fan, a motor, a gas pipe with a controlling valve and a mixing pipe ending with a turbulizing head. The aggregate is assembled at the factory.

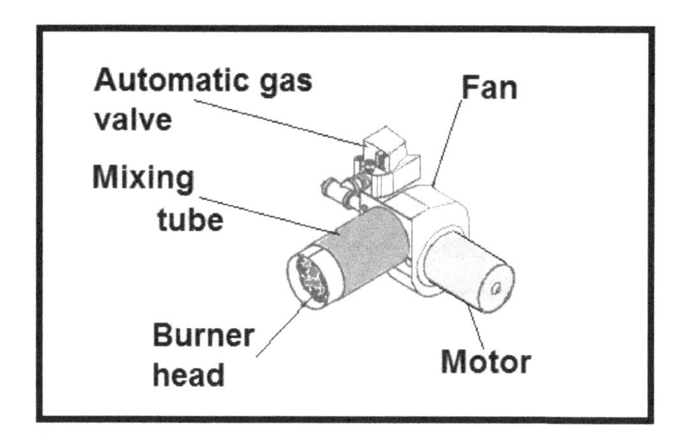

FIGURE 2.20 Package gas burner (a monoblock).

Package burners can regulate gas and air inputs, setting up the number of fan's revolutions, operative input and gas valve regime of operation. The devices are used in the combustion chambers and they should agree with the overall dimensions and thermal loading of the chamber. Note that the burner type should correspond to the type of systems leading away the combustion products. The nature of that correspondence is discussed in Section 3.4.

2.7. A FLAMELESS COMBUSTION OF FUEL GAS AND COMBUSTION DEVICES

The pursuit of increase of the power efficiency of gas combustion devices combined with the high fire safety and ecology requirements results in the implementation of new combustion technologies. Another, practically important, technology emerged in the last 20–30 years parallel to gas FCs. The innovation is called a **flameless gas radiator**, and its application as a basic component of "radiation heating systems" intensifies.

The survey of specialized literature shows that two gas combustion flameless technologies exist at present, and **flameless radiators** with a different structure are designed on that basis (see Chapter 4 for more details), namely:

- Flameless combustion of fuel gas (Figure 2.21);
- Catalytic combustion (Figure 2.22).

The first one is a variant of the classical injection technology (see Figure 2.17), where fuel gas is prepared such as to adopt an appropriate stoichiometric ratio but with a small excess of air (α=1.05). The gas-fuel mix is introduced and distributed within a system of a number of ceramic nozzles. The distance between them is 2–3 mm, whereas the outflowing jets are very close to each other. They cover a small distance prior to merging. Then, the mix ignites and burns, forming a very thin (0.5–1.5 mm) flameless layer over the nozzles and generating high temperature and a white combustion layer.

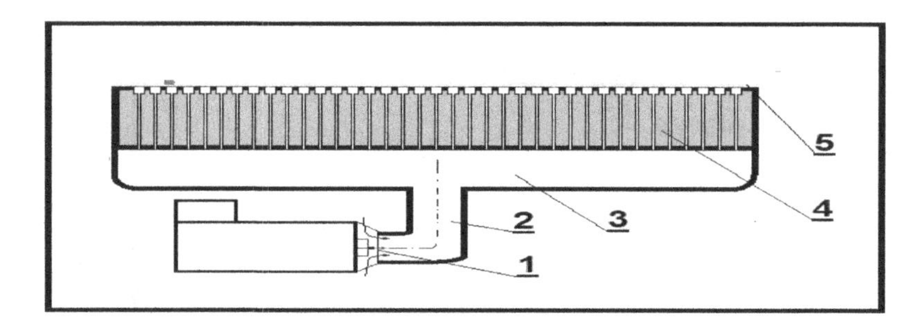

FIGURE 2.21 Flameless combustion device: 1-Ejector; 2-Mixing pipe; 3-Distributor; 4-Ceramic nozzle channels; 5-Combustion layer.

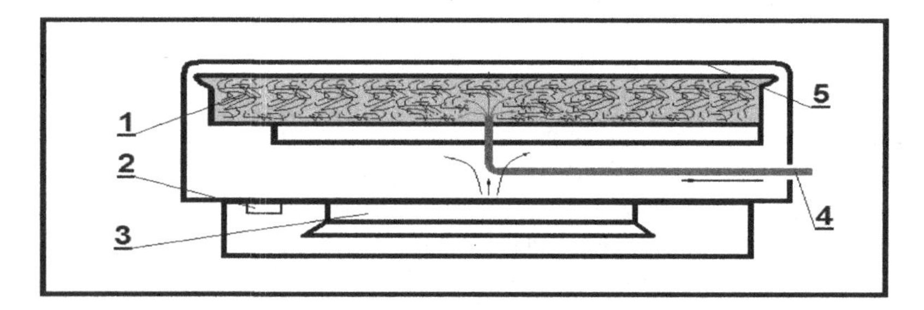

FIGURE 2.22 A combustion device for catalytic flameless combustion: 1-Reactor filled with ceramic fibers; 2-Electronic block; 3-Fan; 4-Gas pipe; 5-Box.

The second technology of flameless combustion takes place in the catalytic environment of a chemical reactor (Item 1 – Figure 2.22), similar to that of an FC (see Section 2.5 – Figure 2.8). The formal similarity consists in the use of a catalyst (platinum, solid oxides, meltable carbonates etc.) which controls the bonding between methane and oxygen. Another formal similarity is the running of exothermic reactions in both devices.

The difference between the FCs and the catalytic radiators consists in the management of the chemical reactions running in the reagent chambers. As far as there is no direct contact between the two reagents in the FC (they are physically divided by the electrolyte layer), the fuel gas of the catalytic radiators is introduced in a chamber which is filled with a platinum catalyst, layered on ceramic or silicon fibers.

During that contact the fuel consisting of hydrocarbons (methane, propane etc.) dissociates into hydrogen and carbon ions, which interact with the oxygen of the ejected air. Hence, heat is released increasing the temperature to 600–800°C. Catalytic combustion takes place without an open flame, which makes the radiators fire safe.

The difference between the FCs and the catalytic radiators consists also in the nature of final coproducts – FCs generate electricity, heat and water, while gas radiators generate a thermal flux of medium intensity – see additional data on catalytic radiators in Chapter 4 (4.1.2.3.1). Combustion takes place on a specific surface (ceramic wadding).

3 Gas Supply of Urbanized Regions

In urban areas, natural gas (NG) distributes through networks in a gaseous or liquid state. The method of transport depends on a number of factors – distance to the deposit and available transport structure, availability of gas depots, economic optimization of the transport considering also the political stability and supply safety.

For almost half a century, the basic methods of distant transport have been:

- Through pipelines: under super pressure of 7.5–20.0 MPa (GTP);
- By tanks filled with compressed natural gas (CNG);
- By tanks filled with liquefied natural gas (LNG);
- By tanks filled with liquefied hydrocarbons extracted from gas deposits (GTL).

3.1 TERRITORIAL (STATE) PIPE NETWORK OF GAS TRANSPORT

Natural gas in Southeast Europe, including our country, is transported via a pipe network with high pressure (HP) and average pressure (AP). The networks are state-owned. In our country, its system manager is the joint-stock company "Bulgartransgas" Ltd. The entire network[1] includes:

- Intake and exhaust gas terminals and measuring stations;
- High and average pressure pipeline networks;
- Compressor stations (3 HP with a total power of 49 MW);
- Regulating stations (68 in number);
- Measuring stations (10 in number);
- Transit pipelines transporting gas to other Balkan states: Turkey, Greece, Sirbia and Macedonia;
- Gas deposits Chiren and Galata. They serve as gas buffers and emergency reserve, since the available gas amounts can satisfy local needs for 2–3 weeks under daily consumption of $4{,}0–4{,}5 \times 10^9$ m^3. They are equipped with compressor stations of their own.

Figure 3.1 shows schematically the state gas-transport map of Southeast Europe, where you can see the strategical position of Bulgaria. The scheme resembles a rhomboid with a diagonal oriented "Northeast – South-West", which is in a process of

[1] Re-export oriented to the other Balkan states.

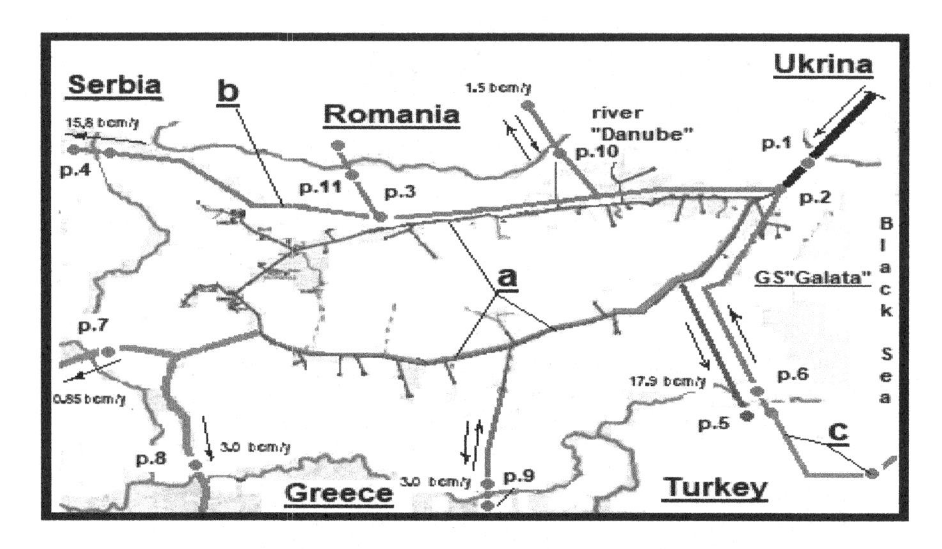

FIGURE 3.1 Scheme of the gas hub "Balkan": a) Bulgarian NG distributor ring; b) New pipeline to Serbia; c) Black sea pipeline interconnector: p.1 – Negru Voda; p.2 – Providia; p.3 – Polski Senovez; p.4 – Zaichar; p.5 – Malkolchan; p.6 – Strandja; p.7 – Kystendil; p.8 – Kulata; p.9 – Komotini; p.10 – Ruse; p.11 – Oriahovo.

modernization to be converted in a gas hub "Balkan". It should satisfy the Directive N93 of EC for NG.

The basic components of the national network are the **main pipelines**[2], conveying fuel gas from the terminals to the distributing networks of end or privileged users (see Figure 3.1,a). The pipelines consist of sections, whose length depends on the pipe diameter, i.e. on the gas maximal intake, and on the seismic conditions of the terrain. Yet, they should not exceed 30 km in length, while shutoff valves are additionally installed:

a. at branches;
b. before and after the compressor stations;
c. before and after water barriers.

Three key segments of the gas hub "Balkan" are currently under construction:

a. Modernization of the system pipeline segment (from Provadia-p.2 to Polski Senovez-p.3 with length 166 km and d_p=1200 mm);
b. New system pipeline interconnector to Serbia, which can be continued to Hungary and Slovakia – Figure 3.1,b) (from Polski-p.3 Senovez to Zaichar-p.4 with length 308 km and d_p=1200mm);

[2] Conforming to their location, they are classified as: – class "1" "– in a region with less than 10 buildings"; – class "2" "– in a region with 10 – 45 buildings"; – class "3" "– in a region with more than 45 buildings"; – class "4" " – among buildings higher than 4 stories".

c. New pipeline interconnections with Black sea pipeline or Turkish corridor (from Strandja-p.6 with length 11 km) and with new liquefied gas terminal in Alexandropoulos (from p.9).

The total transition is projected to be 16 bcm-y (3bcm-y for Serbia, 9bcm-y for Hungary and 4bcm-y for Slovakia).

Except for shutoff valves, gas pipelines are equipped with branches, controlling-measuring equipment, and automatic equipment controlling the switch-on/off of devices for separation and diagnostics, a system for corrosion and electrochemical protection, as well as a telecommunication system for remote control. Pipe drainage devices and candles for pipe scavenge under gas transfer termination are installed between the shutoff valves. Instruments measuring pressure and transmitted gas volumes are installed in the branches. The pipeline route is marked by metal or steel reinforced concrete poles, 1 m high, indicating each pipe bend. The calculated pipeline strength is a complex quantity reflecting the complex loading of the facility – internal pressure combined with temperature and seismic impacts (including effects of soil freezing). Pipe thickness is prescribed pursuant to BDS EN 1594, Par. 14, BDS EN 12007 and BDS EN 10208-3. Each pipeline section, longer than 1600 m and under pressure higher than 1.6 MPa, is fenced with a servitude strip, 400 m wide, meeting the requirements of Par. 10 of the Territorial, Urban and Local Development Act (TULDA).

Usually, main pipelines are laid under the ground[3]. The crossing of other logistic systems such as railways, highways or communication and power lines should be in conformity with the normative acts, not violating the operation of the structure[4] [1, 2]. Above-ground installation of pipelines is performed in mountain and swampy areas or when obstacles are to be overcome.

Gas-compressor stations are installed at each 120–150 km of a pipeline consisting of long pipes under high pressure. They compensate gas-dynamic losses raising gas pressure to the designed value- 5.4 MPa, usually. Gas multi-step piston units or turbo-compressors are installed in a station to facilitate their operation. They are equipped with cooling and lubricating lines. Energy centers – ECs (including depots for consumables, heating system etc.) should supply them with electricity, if the area is not centrally electrified. Additional units also participate in gas systems – equipment removing solid particles and condensates, gas-odorant mixing installations, measurement and control equipment (MCE) (incl. pressure regulators, flow meters and remote control lines).

Distributing and measuring stations are also part of the gas complex. Their function is to "reduce and stabilize gas pressure" and measure amounts of diverted gas before connections to big users or regional networks serving so-called "privileged users".

[3] The typical depth of pipe embedding is 0.8 m. Additional trench digging is needed in a number of cases: pipe diameter exceeding 1.0 m; risk of freezing; pipes imbedded in areas undergoing erosion/corrosion impacts; meeting specific technological requirements.

[4] The pipeline is installed in a protective ventilated casing (steel pipe).

3.2 URBAN AND REGIONAL GAS NETWORKS[5]. CLASSIFICATION. STRUCTURE

Once supplied or extracted, NG reaches end users (building with home or industrial pipe networks) via the urban and regional networks.

Conditions of access, distribution and activities of the operators of gas-transport and gas-distributing networks are formulated in conformity with the regulations known as "Regulations of access to gas-transport/gas-distributing networks", Par. 192 of the Energy Act.

In fact, the public gas supplier[6] sells imported or extracted amounts to companies called **privileged gas consumers**. The Commission of Energy and Water Regulation (CEWR) enters them in a special "**Registry of licensed privileged users**"[7]. At present, CEWR has licensed 43 companies distributing NG. Four of them have regional licenses. The rest of them have local licenses only.

For now, some of the main "privileged users" develop investment projects. They build and exploit the distribution networks as gas dealers. The principle of cascade gas distribution applies to a non-privileged user in towns and regions, where privileged users retail fuel gas supplied by public distributing companies. The end users are juridical persons or individual economic subjects (budget, industrial or domestic subjects). They are owners of gas installations, but may join the network and purchase gas under gas supply contracts only.

There is a vast variety of gas-distributing networks with respect to intake, pipeline length and density. Networks are designed and built accounting for the following factors: intake prognosis including future consumption and connection; design pressure and gas-dynamic losses; map of the area; prognosis of urban development and operation under failures.

3.2.1 Classification of Urban Gas-Distributing Networks

We use here three criteria to classify urban gas-distributing networks:

- Operational pressure;
- Pipeline topology;
- Method of construction.

[5] The extension of a gas-transport infrastructure in areas with an existing transport network is an actual task nowadays. Yet, urban areas lacking blue fuel also need a respective infrastructure. Designers aim at the optimization and precision of mutual operability, since the gas route is a crucial issue in the construction of pipelines. Note here the importance of route passability, hydrology and geology. There were over 2100 km of pipelines by 2013 supplying end users all over the country.

[6] Pursuant to Par. 43, Sub-. par. 1, p. 3 of the Bulgarian Energy Act, only one license for public territorial gas supply is issued. Following CEWR's Decision No P-046 from November 29 2006, a license with 35 years validity was issued to ",,Bulgargas" Ltd (No Л-214-14/29.11.2006).;

[7] Annual consumption not less than 20×10^6 m^3.

Conforming to the first criterion, gas-distributing networks are divided into four groups:

- High-pressure pipe networks (HP) – from 0.5 MPa to 1.6 MPa;
- Medium-pressure pipe networks (MP) – from 0.2 MPa to 0.5 MPa;
- Intermediate-pressure pipe networks (IP) – from 0.01 MPa to 0.2 MPa;
- Low-pressure pipeline networks (LP) – below 0.01 MPa.

The total length of gas-distributing networks is over 870 km, and pipe diameters are within the range 27–40". Internal branches are about 950 km long with pipe diameter 4–20".

HP and MP networks supply big users, as well as MP and LP networks. LP pipelines connect residential buildings and small trading companies.

Regarding topology, the distributing networks are:

- Radial (Figure 3.2,a);
- Ring-wise (Figure 3.2,b).

Ring-wise schemes are built in compact populated areas with large construction density. The main pipeline HP (Item 1) is designed such as to form a closed contour surrounding the area.

Supply of residential micro-areas and counties can be through gas-holder stations, which are buffers and balance the unsteady supply and consumption of NG. Ring-wise schemes are flexible under repair, failure and supply diversification, but their cost is higher.

The radial schemes are very popular in large inhabited areas as well as in every newly gasified settlement and quarter. Their distributing pipelines (Item 4) form gas-conveying rays reaching all radial users. All ray branches should have shut-off fittings to block gas transfer during a repair.

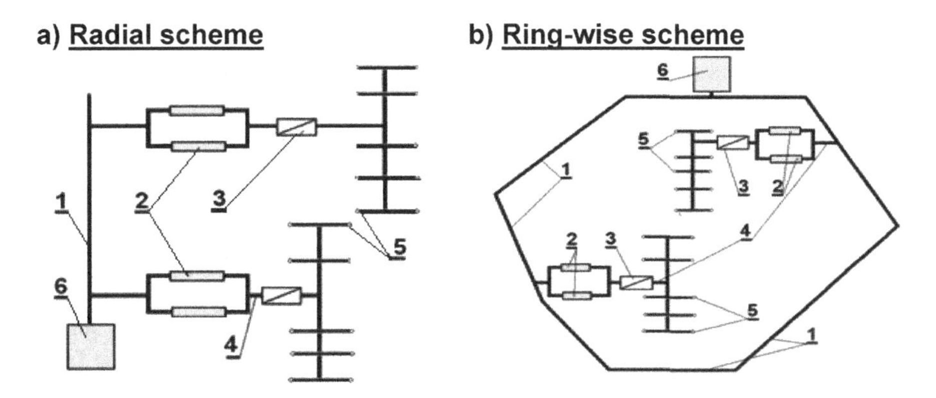

FIGURE 3.2 Types of distributing networks with respect to their topology: 1 – Main pipeline HP (0.6 MPa); 2 – Gas holder station (GHS); 3 – Gas-regulating station (MP); 4 – Distributing duct (main pipeline) – MP; 5 – Gas-regulating unit (LP); 6 – Automatic gas-distributing station (AGDS) reducing the HP from 5.5 to LP 0.6 MPa.

Each scheme shown in Figure 3.2 has gas-regulating units (**GRU**-Item 5) at the end of the gas route where gas parameters are regulated such as to correspond to the parameters of the gas pipeline systems.

Briefly, we must note that **GRU are classified** with respect to the installations they feed. They are individual and combined units. Pursuant to their operation, GRU are regulating, measuring and regulating-measuring units. Besides, GRU may be single and integrated into so-called building EC (see Section. 3.3 and 3.3.1).

According to the methods and techniques of installation, the external distributing networks are underground and above-ground ones. The depth of pipeline embedding depends on the moisture contents of the transported fuel gas.

- If soil freezes and pipes convey moist gas they should be laid under the freezing soil layer;
- The embedding depth should be not less than 0.8 m, if dry gas is conveyed. It may be reduced to 0.6 m in some special cases.

The underground pipelines are laid far from buildings, facilities and communication lines[8]. Pipes should be installed in ventilated containers under street pavement, green areas or agricultural land, near roads and under river beds or other water reservoirs.

The horizontal pipe sections are laid at a slope of $2^0/_0$, guaranteeing condensate outflow in tanks and protection against water stoppers and hydraulic shocks.

When crossing communication lines, the installation height should be not less than 0.15 m above the ground. If electric cables pass over the pipeline, the installation height should be not less than 0.5 m above the ground. When crossing a railway, the pipeline should be in a steel casing. Special regulations apply to the pipeline protection from stray current.

Gas pipelines should rise above the ground when passing through industrial and residential areas. The TULDA prescribes the distance to facilities and buildings. Pipeline components should be fabricated from metal or other certified material, and they should be inserted in a protective casing. The pipes should be installed on:

- Inflammable separate columns, supports or trestles higher than 0.35 m and above the ground;
- Massive walls (under the explicit permission of the owners);
- Inflammable bridges and overpasses.

Gas pipelines are built using strengthened[9] and insulated steel pipes (BDS-EN-ISO-3183:2013) or high-density polyethylene pipes (BDS-EN 1555-1:2010). They are welded[10] or soldered to each other. Flanges and threads connect pipes to fittings and devices – shut-off valves and taps, insulating flanges, pressure regulators, measuring and control instruments. Connecting and fashion components are fabricated from

[8] Mainly those with an open cross-section such as channels, as well as water ducts, telephone wires or tramways

[9] Pipes of underground pipelines are with a thickness exceeding 3 mm, while the thickness of pipes of above-ground pipelines exceeds 2 mm.;

[10] The regulating standards are BDS EN 12732 and 12007-2.

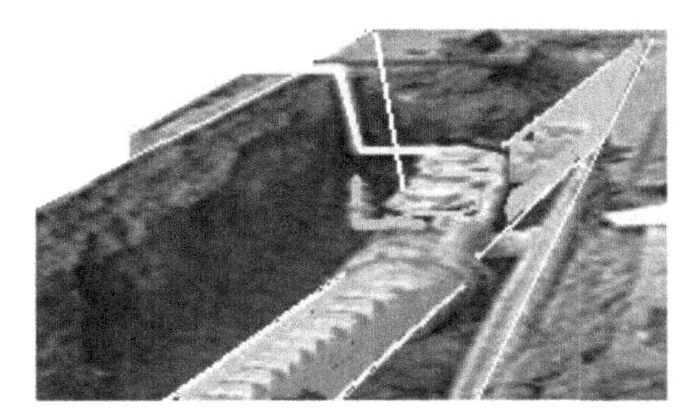

FIGURE 3.3 Underground gas pipeline in a ditch, 0.8 m deep.

cast iron. Shut-off valves are installed to provide operability, safety and continuous gas supply through sections not undergoing repair (planned, prophylactic or emergency) or failure.

The location of the shut-off fittings depends on:

- Operational necessity;
- Connection of potential users or expansion of the network;
- Operational pressure and pipe diameter.

Shut-off valves are also installed at specific points, for instance before and after protective casings, along drainage and blowing pipes etc.

Due to an eventual change of the operational temperature resulting from the change of the environmental conditions, pipelines are supplied with compensators of temperature deformations. Lenticular compensators are usually used for that purpose, installed in steel reinforced concrete wells together with a slide valve – see Figure 3.4. The materials used in distributing networks should satisfy the requirements of NSIOSSN (Chapter 3, Section 3. 1) and BDS EN ISO 3183:2013[11].

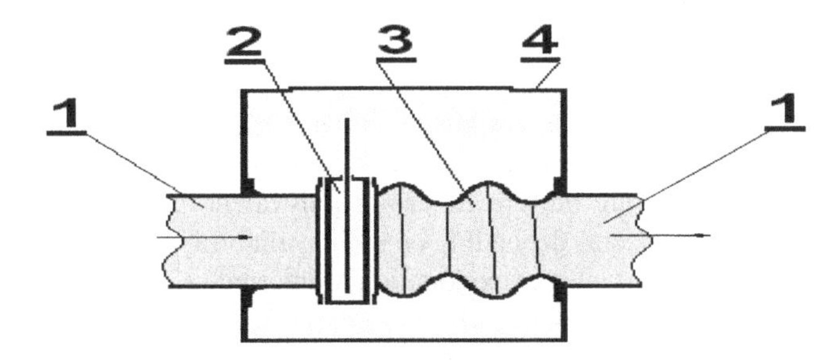

FIGURE 3.4 Lenticular compensator, installed in a steel reinforced concrete well: 1 – Pipeline; 2 – Slide valve; 3 – Lenticular compensator; 4 – Steel reinforced concrete well.

[11] High density polyethylene pipes should meet the requirements of BDS EN 1555-1:2010.; Copper pipes – pursuant to BDS EN 1057:2006 and A1:2010.

3.3 DECENTRALIZATION OF BUILDING POWER SUPPLY. ENERGY CENTERS

Providing buildings with the necessary amount of primary energy is a main logistic issue, which is a component of the general task of building logistics. Hence, modern investment projects solve this problem by giving priority to safety, stability and diversification of the supply of primary energy carriers. Note here the national and regional macro-economic conditions and policies such as taxes and funds for environmental protection, costs of supply, suppliers' rating, waste storage etc. Several types of energy carriers are supplied to high security and reliable buildings. For instance, there are various technologies of energy supply – central supply of thermal and electric power, central gas supply and the implementation of on-site energy generators of heat and electricity. The use of wind power electric generators also contributes to the increase of energy safety and reliability, if buildings are equipped with so-called "backup systems" operating with traditional energy carriers. Thus, investment projects should have a special section called "Building logistics". It should be imposed upon the duties of the chief designer or architect.

3.3.1 Gas Supply of Buildings. Types of Gas-Supplied Energy Centers (GEC) of Buildings. Classification

The design of an ***energy center*** (EC) attributes to the technological treatment of the problem of energy safety and reliability, and EC is part of the entire investment project. Conforming to modern ideas, EC is the unit where preparation of the primary energy carriers takes place in view of their further use in building subsystems. Energy fluxes enter the EC and distribute among users, while flow temperature should be below 90°C, pressure – below 2.2 kPa and voltage-below 110/220 V. The ECs are located in common or special premises such as boiler rooms, user's stations and electric board or transformer rooms. Depots for primary energy carries (fuels) are often needed.

EC classification is shown in Figure 3.5 and it reflects the existing status quo in building energy supply, i.e. the building is a net user of central supply energy. However, if a building generates useful energy, it is only for local needs. Depending on the energy supply, the ECs operate either as **terminal stations** of central supply (as substations, traffic posts or user stations) – Figure 3.5,a) or as **autonomous units** (having local heat or power sources) – Figure 3.5,c) [9].

ECs of terminal type are built as mono- or polyvalent (electro/heat or electro/gas – see Figure 3.5,a) systems depending on their availability or closeness to central heating units. The safe and reliable exploitation of internal systems and installations requires reduction of the parameters of energy carriers to a suitable level. Pursuant to state regulations, energy users are non-privileged buyers.

Autonomous EC (see Figure 3.5,c) are located in buildings far from distributing networks. They are also built if the energy cost is inconceivable. Their function is to transform the primary energy into useful energy satisfying the needs of the end users. However, different energy generators operate: electric or thermal ones, as well as mHPSs (micro heat and power systems) installed in special premises. Storehouses

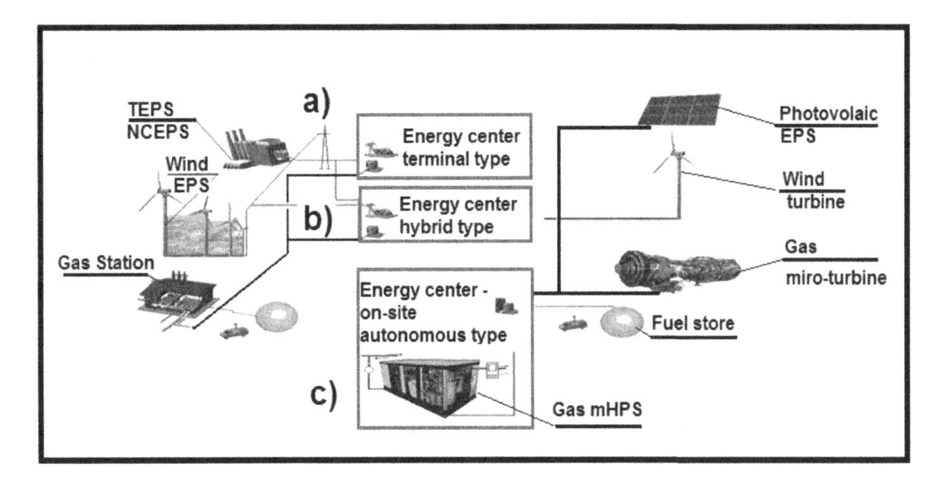

FIGURE 3.5 Classification of building energy centers.

for primary energy carriers are designed and built as part of those EC. Their capacity depends either on the intensity of energy consumption or on the EC capability of regular supply of primary energy carriers.

There are buildings with hybrid EC where either central or on-site energy supply runs (Figure 3.5,b) [9]. They generate on-site energy for alternative supply. Buildings with hybrid EC are more expensive. Yet, they operate more reliably, are capable to operate in a co-energizing regime with central and regional systems and can send back the generated extra energy. ECs are also built as part of wind electric power systems, increasing the reliability and stability of the supply of electricity and decreasing the operative costs.

Each EC discussed herein may adopt a **mono-variant** (considering a single energy carrier), **double-variant or poly-variant** power supply – see Figures 3.6, 3.7 and 3.8, respectively. This depends on the type of the energy carrier. Moreover, reliability, efficiency and diversification of supply may thus improve.

This book stresses upon the possibility of diversification of the energy supply using various fuel gases. We shall discuss in what follows four schemes of an EC, where fuel gas is the principal energy carrier [9], including:

- Mono-valent, autonomous EC – Figure 3.6;
- Mono-valent, hybrid EC – Figure 3.7;
- Bi-valent, hybrid EC – Figure 3.8;
- Bi-valent energy supply of buildings by an EC of terminal type – Figure 3.9.

The first scheme (Figure 3.6) is an autonomous EC, which is widely used in all types of buildings, which are far from central logistic units – electric power substations, traffic posts, gas substations, regional heating stations or group user's stations. The only logistic requirement to the building is its equipment with a specialized room to house the fuel gas batteries (Item 1).

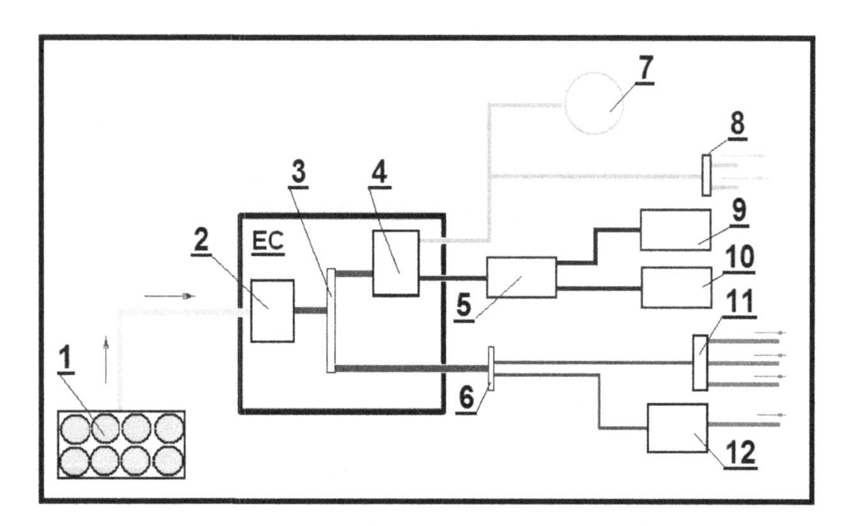

FIGURE 3.6 Mono-valent, autonomous EC, supplied from a local fuel store and consisting of bottle batteries with compressed/liquefied gas: 1 – Fuel store; 2 – Gas Regulating Station (GRS); 3 – Gas distributor; 4 – Gas mHPS; 5 – Main eletro-distributing panel; 6 – Gas pipe distributor; 7 – Hot water accumulator; 8 – Hot water pipe distributor; 9 – Sub-panel-users; 10 – Batteries of electric accumulators; 11 – To home gas appliances; 12 – Gas cooker.

The proposed scheme includes a "technological novelty" consisting in the use of a gas mHPS (Item 3) generating the necessary electrical and thermal power. The balance of energy consumption is attained via a series of appropriately calculated heat and electricity accumulators – (Item 7) and (Item 10). Gas supply by interchangeable bottle batteries takes place following a scheme designed by the operator. The EC of that type (Item 2) adapts gas parameters to the technological parameters of the appliances as stated by their manufacturer. It has the potential of satisfying the gas needs

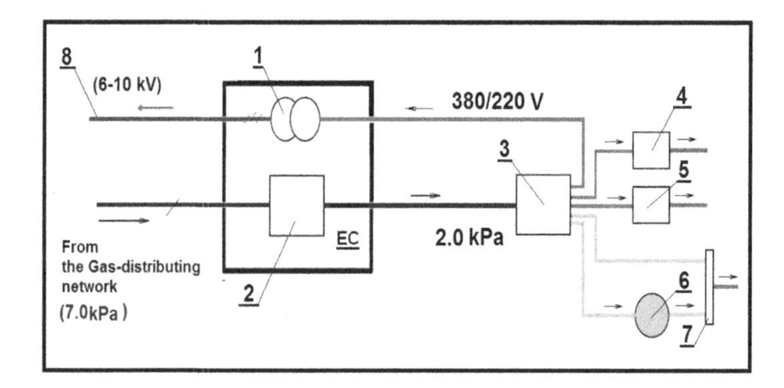

FIGURE 3.7 Mono-valent hybrid energy center supplied via a low-pressure gas-distributing network: 1 – Upgrading transformer; 2 – Gas-regulation station (GRS); 3 – Gas mHPS; 4 – Sub-panel lighting; 5 – Sub-panel power switches; 6 – Hot water accumulator; 7 – Hot water distributor tube; 8 – Returning HP cable to the regional substation.

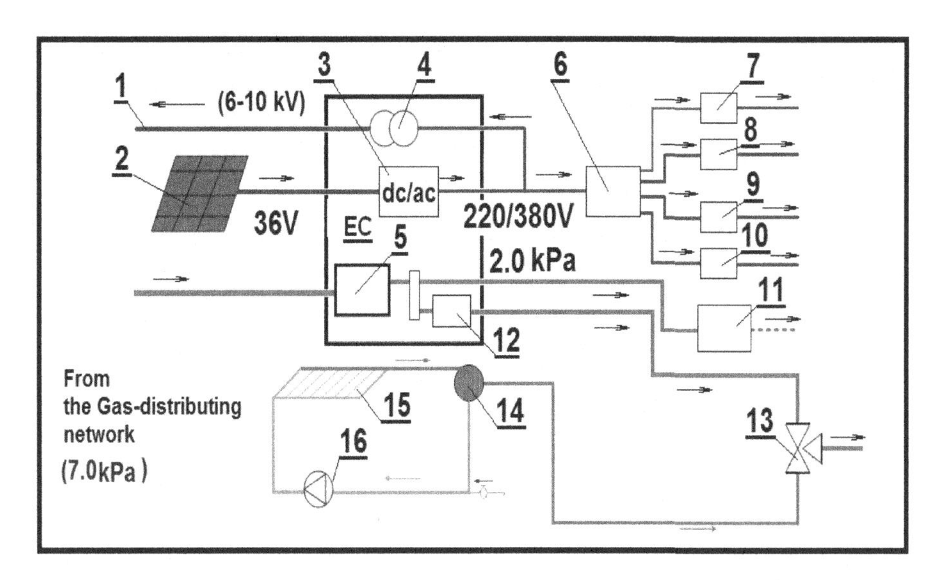

FIGURE 3.8 Bi-valent hybrid energy center connected to a high-pressure distributing network and to two solar generators: 1 – Reversible high-voltage connection to/from a regional substations; 2 – Photovoltaic generator of electricity; 3 – Inverter – DC/AC; 4 – Upgrading transformer; 5 – Gas-regulating and measuring system – **GRMS**; 6 – Main distributing electrical panel; 7 – Lighting sub-panel; 8 – Sub-panel of power contacts; 9 – HVAC; 10 – Lifts; 11 – Gas cooker; 12 – Gas instantaneous water heater; 13 – Three-way valve – SAC; 14 – Hot water accumulator; 15 – Solar heat generator; 16 – Circulation pump.

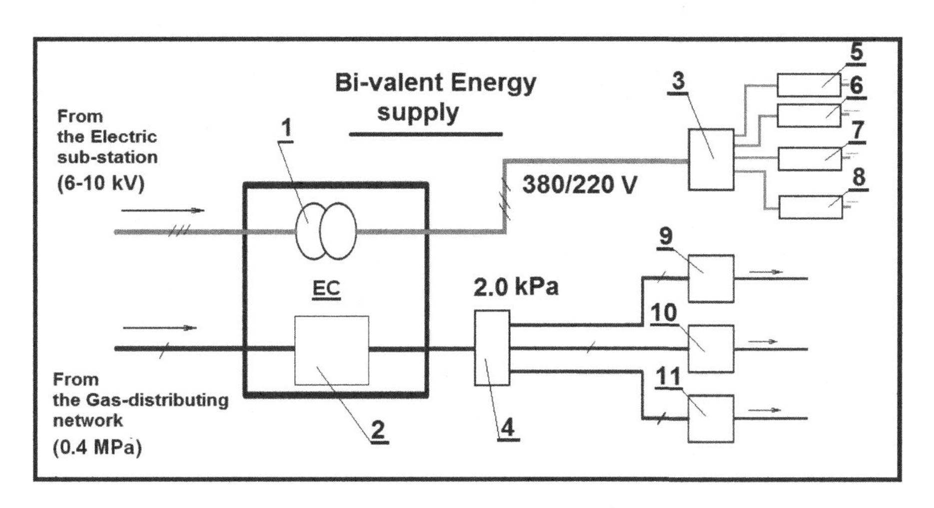

FIGURE 3.9 A typical central bi-variant energy supply of buildings via a terminal EC: – Supply by an electric power substation 6 or 10 kV; – Gas supply by a gas-distributing network 0.4 MPa: 1 – Lowering transformer; 2 – Gas-regulating unit – GRU; 3 – Main electric panel; 4 – Gas distributing tube; 5 – Sub-panel "lighting"; 6 – Sub-panel "power contacts"; 7 – Sub-panel "lift"; 8 – Sub-panel "common sections"; 9 – Gas instantaneous water heater; 10 – Gas kitchen stove; 11 – Gas heating devices-convectors.

of the building. Yet, it demands serious investments, but less expensive variants are also possible.

Consider for instance an individual one-family house. It needs an mHPS[12]. Fuel gas should be supplied in replaceable gas bottles. Then, one should solve the inverse problem, selecting an appropriate mHPS and installing an appropriate building envelope – the power of the cogenerator should satisfy occupants' needs, while the peak demand for power should not exceed the installed and limited power. Gas appliances should be selected and connected in a rational scheme, and they should operate according to a specific schedule. The availability of electric substations or a traffic post near the building facilitates the exploitation of the EC as shown in Figure 3.7. It operates under a co-energizing regime jointly with the central electro-supplying system. A reversible connection with minimal electric or thermal power, for instance 2.5–3.0 kW or 5.0–7.0 kW is used for that purpose consisting of an mHPS (Item 3), an upgrading transformer (Item 1) and a reversing cable (Item 8).

Together with the satisfaction of power needs, the gas mHPS can "send back" energy to the operator of the electric system. Following [4], **the use of a cogenerator** provides the EC operator with the legal status of a "**privileged buyer**" on the low-end gas market. This creates favorable economic conditions for the fast reimbursement of the initial investments[13].

Gasified energy centers (**GEC**) [14] successfully participate in the supply of solar energy. The scheme is shown in Figure 3.8 [9]. The main energy suppliers are two solar generators – of electricity (Item 2) and heat (Item 15), respectively. EC gas supply in this case takes place by means of a gas-regulating and measuring system – GRMS (Item 5). Gas supply takes place when the amount of **on-site** solar energy (assessed via the temperature of water heated in the accumulator – Item 14) is insufficient to satisfy the needs of the internal users.

To increase the reliability of the system, one should include an additional gas instantaneous water heater (Item 12) operating in compliance with the solar heat generator (Item 15) through the three-way mixing valve. When the temperature of water supplied by the accumulator (Item 14) is lower than the set point, the instantaneous water heater (Item 12) automatically activates. It heats the water to the needed temperature, i.e. fuel gas is used only upon necessity.

The EC scheme provides for the installation of an electric reversible connection to the system of central supply of electricity (Item 1). When there is no need for photovoltaic energy, it is directed to the regional substation via the reversible connection. It "returns" there and is sold to the electro-system operator (ESO). Capital flows help to the fast reimbursement of the initial investments. At night or in gloomy weather, when the photovoltaic electric power is insufficient, the building would receive and

[12] See the Table P24.

[13] The problem of reimbursing the initial investments is solved by comparing the project variant to the general issue where no energy returns to the EC system and the status of a privileged gas buyer is not introduced.;

[14] GRS/GRMS –- facilitiyfacility regulating pressure and measuring gas consumption at the inlet of the internal network installed in the ventilated secure room.; GRS/GRMS – facility regulating pressure and measuring gas consumption at the inlet of the internal network installed in the ventilated secure cabinet.

"buy" energy at a nominal price from the system of central electricity supply. The scheme under consideration may be used in **zero energy houses (ZEH)**, which grow popular in the USA and are expected to become a priority of the EU by 2020. In that case, the power of photovoltaic and other on-site generators is arranged such as to satisfy the annual energy needs of the building. At the same time, ZEH designers apply all passive and active architectural approaches with a task of attaining energy economy of about 70%.

3.3.2 Structure of the Gas Regulating Station in an EC

A typical gas EC (Figure 3.9) is connected to the central heating systems and operates as a terminal station. A system for automatic regulation controls the GEC operation, where high-quality management of energy is needed to provide economy and operability of the entire energy system.

In this case, the GEC function is to adapt the parameters of the supplied energy carriers (fuel gas and electricity) to the parameters of the internal systems and installations, to measure the amount of the supplied energy carriers and to distribute the energy fluxes between the end users. The EC scheme includes a lowering transformer (Item 3), gas-regulating and measuring panel (Item 2) and respective distributing devices – panels and tube collectors (Items 3 and 4) – Figure 3.9.

The gas-regulating and measuring unit (panel) is the most important component of a bi- and poly-variant EC. Its basic element is the pressure regulator, whose functions are as follows:

- To reduce pressure in the distributing network (0.4 MPa, for instance) to the operational pressure of the internal installation (about 2.0 kPa, for instance);
- To maintain constant gas pressure at the installation intake.

An example of a GRU is shown in Figure 3.10. The scheme shows a filter (Item 3), as well as a pressure reducing valve with an incorporated protective-cutting valve

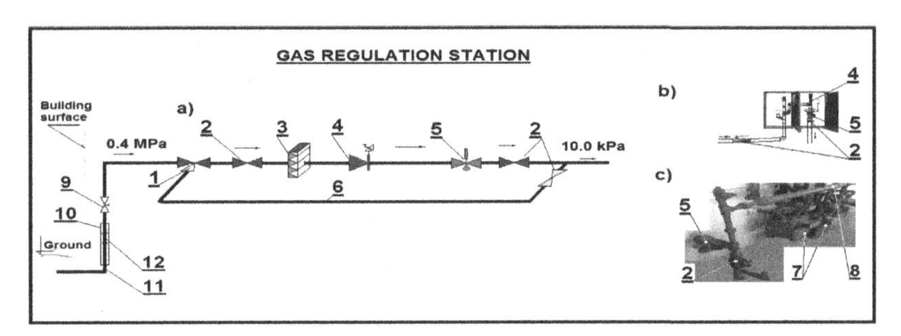

FIGURE 3.10 Gas-regulating and measuring station (GRMS) as a component of an EC: a) Principal scheme; b) View of a GRMS; c) View of a group GRMS: 1 – Triple mixing valve; 2 – Valve; 3 – Filter; 4 – Reverse-protecting valve; 5 – Pressure reducing valve with incorporated protective-cutting valve; 6 – Bypass joint; 7 – Gas flowmeter; 8 – Gas tube distributor; 9 – Shut-off valve with a dismountable joint; 10 – Pipe sleeve; 11 – Underground gas pipeline; 12 – Pipe passage PE-HD/steel.

(Item 5). These protective fittings are needed, since the fuel gas contains mechanical inclusions and drops of condensed water causing material deposition on the pipe walls and decrease of its cross-section. The reverse-protective valve (Item 4) serves to deflect the inverse gas flux from the internal installation when the supply through the bypass (Item 6) terminates.

In an emergency or a prophylactic check, a bypass connection (Item 6) can be additionally installed.

The pipeline connecting residential and public buildings intakes the room of the GRU[15] on a support, which is an element of the building structure. Thus, eventual axial displacements may be limited if a thermal expansion of the linear section takes place. A shut-off valve and a dismountable joint are installed along that section – Item 9.

When gas supply takes place through an underground polyethylene pipeline, PE-HD for instance, one should use a pipe sleeve – see Item 12 (Figure 3.10). A scavenge valve is installed after the gas regulation station (GRS) to facilitate an eventual repair.

Yet, the design of pipe mechanical compensators should be meticulous in order to avoid future violation of the system density, i.e.:

- The initial 2m-section of the gas pipework should not lie on fixed supports. Knees are installed along that section to compensate the thermal expansion (pipe axes bend at an angle of 90°);
- Thread joins should be installed in Z-shaped segments of the network;
- Lenticular compensators should be fabricated from alloyed steel pursuant to DIN 30 681;
- Soft joints should be made pursuant to DIN 30 663.

Conforming to Par. 160 [3], the adopted scheme of gas supply foresees measurement and registration of the fuel gas amount supplied after the GRS.

As far as the measurement of gas consumption is traditionally performed via a GMS, fixed (and accessible) to one of the side facades of the building, a number of panels measure the fuel gas consumption in multistorey residential and business buildings occupied by a particular number of individual users. GMS location is determined as dependent on the user's location and the building volume. Typical sites of GMS installation are:

- Walls of installation shafts integrated into the stairwells – see Figure 3.11,d);
- Stairwells and belonging terraces and balconies – see Figure 3.11,b) and c);
- The external facade wall of a stairwell – see Figure 3.11,d);
- Common ventilated rooms located on inter-storey platforms – see Figure 3.11,e).

[15] Besides GRU, type RM 10.0 kPa and RM 22.0 kPa, gas regulating and measuring units (GRMS), type RM 22 kPa and RM 10.0 kPa, as well as gas measuring units (GMU), type M20 kPa or GMU, type SM 10 and 20 kPa[25] are incorporated in an EC. GRS may be outside the building yard, fixed to the building or in a special ventilated room in the building common cellar.

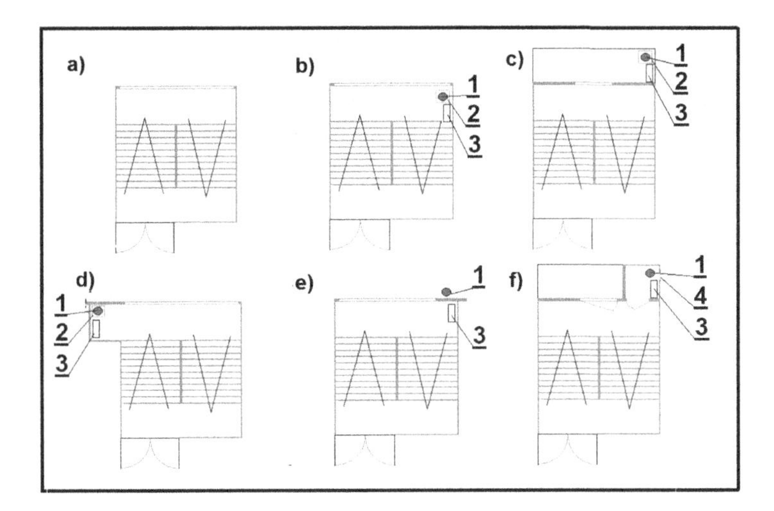

FIGURE 3.11 Variants of the location of the vertical pipe and the individual GMS in multi-story residential buildings: a) Zero variant; b) In a stairwell; c) In adjacent terrace or balcony; d) On a wall of an installation shaft; e) On the external facade wall of the stairwell; f) In a common ventilated room on an inter-story platform: 1 – Vertical pipe; 2 – Ventilation aperture in the stairwell plate (≥0.01 m^2); 3 – Apartment GMS; 4 – Upper and lower vent on the room vertical wall.

The transversal dimensions of the installation shaft (Figure 3.11,d) and the ventilated room (Figure 3.11,e) are such as to facilitate the installation of gas-measuring devices and make them service-friendly. A wide variety of instruments are employed to record fuel gas consumption. The methods of record are:

- Record of the gas amount consumed by the EC;
- Indication of the gas output.

There are two classes of recording devices: volume devices (measuring gas volume in liters) and analog devices. Volume devices measure the amount of gas:

- Via a diaphragm or a chamber (using a synthetic fiber called "reinforced nitro-rubber"). The range of devices, type U4 and U160, is 4–160 m^3/h, respectively;
- Via a rotor: an inpeller or a propeller. The output range is 45–9100 m^3/h. All rotor devices operate under a nominal pressure of 0.2 MPa, but there are also devices that operate under 10.0 MPa.

Similar devices are:

- Turbines – measuring the rotor angular velocity (range 250–6500 m^3/h);
- Diaphragms measuring the gas pressure losses (range up to 200 m^3/h);
- Thermal anemometers – measuring current amperage via the active arm of a Wheatstone bridge.

(Specialized literature provides additional information on those devices and their principle of operation).

Consumption-measuring devices are installed in dry and underground rooms of a building[16] that are plastered with inflammable putty. Their volume should not be less than 3 m^3 and they should be naturally ventilated. They should be separated from the first and closest user by an inflammable screen with a limit time of fire resistance amounting to 60 min. Conforming to normative documents, the measuring devices should be installed in an inflammable niche, cabin or cabinet supplied with a locked door and natural ventilation. This should be done when the devices are to be attached to an external wall, staircase or in a common room. Niche walls should be plastered with inflammable putty. Devices should not be installed in premises with A- and B-fire safety degree.

Niches should be guarded against other EC communication lines (electric cables, water ducts, drainage pipes etc.) by means of inflammable screens – natural ventilation should be provided.

3.3.3 Gas Boiler/mHPS Room

3.3.3.1 Structure of EC with Gas Water Heaters/mHPS Station (Boiler Room)

Rooms for gas water heaters [with gas boiler and mHPS (EIC or FC)] are located in the building basements and above-ground premises[17], whose construction type and operational space allow safe exploitation, repair and dismantle. The units combine pipelines, facilities and devices for fuel gas combustion, including a system conveying combustion waste. A gas pipeline under pressure of up to 0.5 MPa may be installed in boiler rooms, if operational and emergency ventilation is provided together with gas detection instruments and explosion proof illumination.

The heat capacity of an installed gas boiler is a function of the prognostic power, which in its turn depends on a number of factors, including:

- Climatic conditions and landscape of the construction terrain;
- Function and overall dimension of the building;
- Building envelope;
- Regime of occupation, number of occupants and building exploitation.

There are two types of NG boilers: conventional and condensing. They are also classified as wall installation and floor installation devices. Boilers for floor installation[18] are widely applied in gas-supplying systems due to their compactness, energy efficiency and integration with generator batteries for simultaneous operation (Figure 3.12). Their installation is cheap and technology-friendly. Construction practice shows that they are appropriate either for internal (Figure 3.12,f and g) or for external (facade)

---

[16] Measuring devices of gas consumption in small buildings are installed on external walls. Thus, fuel gas consumption can be charged.

[17] Autonomous thermal boxes are installed in internal yards or at the buildings.

[18] The power of instantaneous gas water heaters is within wide ranges - (2– ~ 10 kW).

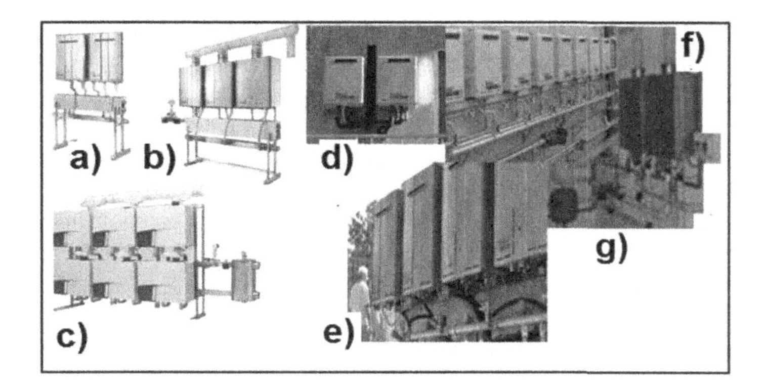

FIGURE 3.12 Variants of the location of instantaneous water heaters for wall assembly: a) Aggregate of two IWH on a metal frame; b) Aggregate of three IWH; c) Aggregate with six IWH; d) On a floor platform; e) On a yard platform; f) and g) Indoor assembly.

installation (Figure 3.12,d and e). Floor-installation boilers (drum-type, firetube type, module type) use a significant volume of water as a heat-transferring medium and thermal capacity (Figure 3.13).

The choice of the water heater type depends on the schedule of hot water usage and the scheme of energy flux distribution. When the energy demand is steady throughout the year and with low intensity, it is advisable to use non-capacitive and instantaneous water heaters for wall assembly. In case of a large, periodic and spontaneous demand for thermal energy, it is advisable to use capacitive, thermal generators with a large volume of accumulated hot water, while water temperature should be automatically maintained within specified limits. If a multi-variant system with renewable energy sources is integrated into an EC, a good engineering approach is the use of a combination of instantaneous water heaters and volume heat accumulators.

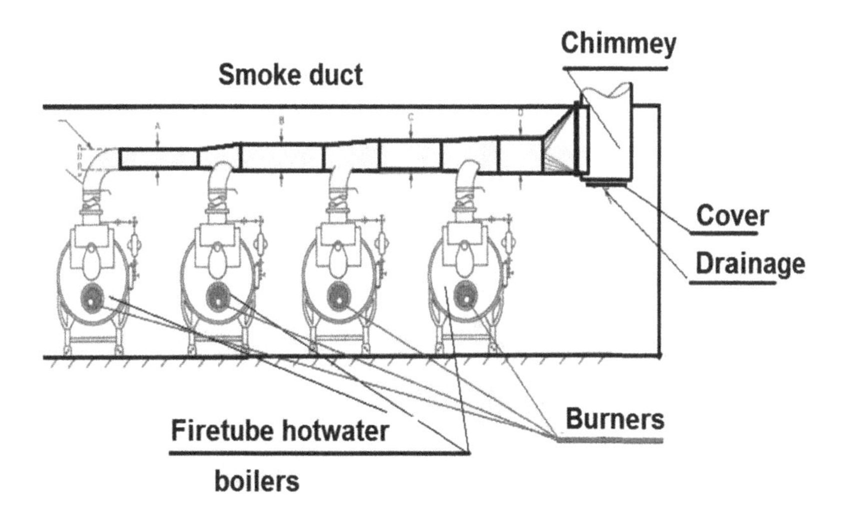

FIGURE 3.13 Boiler room with firetube hot water boilers floor installation.

The boiler room should allow repair of the equipment, conservation, technical control, installation of additional control-measuring instruments, easy control, easy technical manipulations, including air blowing of gas pipes and exhaustion of combustion product.

The installation of heat generators, mHPS, burners, gas lines and control panels should allow access to the combustion system and smoke ducts. Servitude distance between the walls, the heat-technological equipment and its components should be such as to guarantee safe work, keeping the sanitary requirements (see Figure 3.14). Neutral earthing of the electrical devices is mandatory.

Actuators and relays should secure all motors. Boilers, FC and EIC under pressure should be equipped with the necessary mechanical protective devices: protective valves, balance valves and expansion vessels (open and closed). Boards with appropriate explaining information should be attached to heat generators and smoke ducts, gas and electric lines.

3.3.3.2 General Requirements. Operative and Indoor Environment of the Boiler Rooms. Calculation of Warehouse Gas Amount

The boiler site should be supplied with:

- Resting room for the operational staff;
- Fire protection system;
- Natural and mechanical ventilation;
- Drainage system (special shafts should be available – with increased output in order to drain water and remove sediments);
- Protection against local explosions in gas boilers (roof and fencing structures);
- Emergency light and low-voltage illuminators;
- Alarming system (telephone, ring or siren, clock);
- Doors opening out.

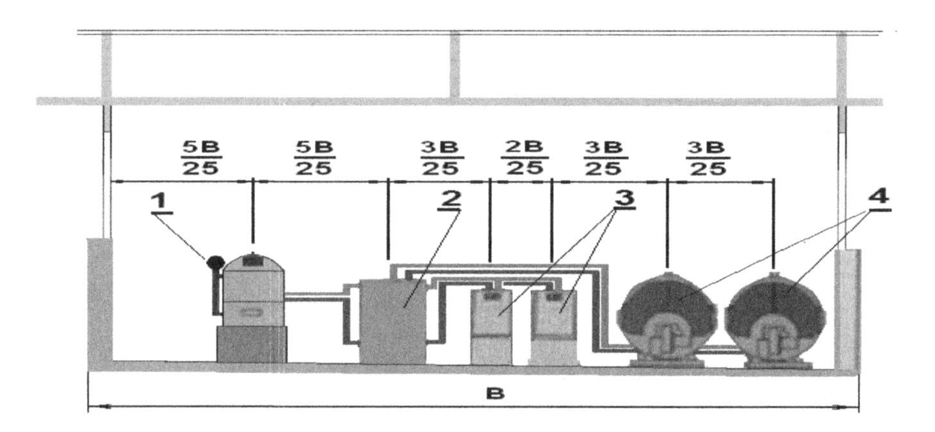

FIGURE 3.14 Vertical cross-section of the engine department of the building energy center: 1 – Gas mHPS; 2 – Thermal accumulator; 3 – Heat pump; 4 – Heat generator.

The structure should be manufactured from materials whose time of fire resistance is at least 90 min.

The design of the general installation, exhausting combustion products from gas devices of one and the same type – “B1” and “B2”[19], should obey the following requirements:

1. Each fuel gas installations should have an individual smoke flue;
2. Gas exhaustion should take place at different levels;
3. The distance between the inlets of the highest and the lowest smoke flues should not exceed 6.5 m, where larger distance is allowed as an exception if fire safety is provided.

Operative indoor conditions:

The work environment should guarantee healthy and safe work of the operational staff, obeying all requirements of the Labor inspection and the Hygiene-epidemiology inspection.

If there are no specific requirements, the designers should satisfy the following normative requirements:

- Norm of illumination – 100lx;
- Thermal comfort – air temperature t_a, relative air humidity φ_a and air velocity w_a accounting for operators who do medium-heavy work with clothes 2Klo. Typical temperatures are: 18°C at the floor, 25°C above the floor and 35°C under the ceiling);
- Noise level – 50 dB;
- To ventilate[20] the cauldron room, one should provide air output in conformity with the regulations of Table 3.1.

TABLE 3.1
Minimal flow rates of the mechanical ventilators of boiler rooms

Gas cauldron type	Pressurizing ventilation m^3/h per kW	Suction ventilation m^3/h per kW
With flow inverter	2.8	0.73
Without inverter but capable of stabilization	2.6	1.25
Indirect air aggregate with flow inverter	4.3	5~10 % Less than the inflowing one
Indirect air aggregate without flow inverter	4.14	5~10 % Less than the inflowing one

[19] See Sub-par. 3.4.3.

[20] In case it is not specified in the table, one should adopt the norm of 2 m^3/s per kW of the installed gas boilers.

TABLE 3.2
Dimensions (live section) of ventilation grids

Type and power of the gas boilers kW	Inflow of outdoor air-minimum m^2/kW	Discharge of polluted air to the surrounding atmosphere -maximum m^2/kW
Open smoke flue	0.0004	50% of the inlet grid
> 70 ~ 1800 <		
> 1800	0.0004 plus 0.927 m^2	50% of the inlet grid
In a box		0.0002

Note: The grid is installed 0.1m above the floor and under the ceiling

When the boiler room is with natural ventilation, the dimensions (net section) of the ventilating grids should be determined on the basis of the installed power and norms indicated in Table 3.2.

Calculation data, not specified in Tables 3.1 and 3.2, are enclosed in the Appendix (Tables P10–P14).

Necessary gas amount in the storehouse

Gas storehouse amount is calculated as a reserve necessary for 25–30 days of operation of the heat generator, depending on the average winter temperature (in January):

1. Calculation of the energy needed:
   - For building heating:

$$E_{Buil}^{Heating} = 0.418_* \frac{V_{Buil}^{0.666}}{R_{Ref}} {}_* IEE_{Env*}\ HDD_{January*} 10^6 \ \text{J/january}; \tag{3.1}$$

   - For heating and ventilation:

$$E_{Jan} = E_{Jan}^{Heating} + E_{Jan}^{Vent} =$$
$$= \left(1 + a_{Vent}\right) * 0.418_* \frac{V_{Buil}^{0.666}}{R_{Ref}} {}_* IEE_{Env}\ HDD_{Jan*} 10^6 \quad \text{J / Jan.}$$

Here V_{Buil} is the heated volume, R_{Ref} – reference value of the thermal resistance m^2K/W, IEE_{Env} – index of energy efficiency of the building envelope [37], HDD_{Jan} – heating day/degree ratio in January valid in the building location area and a_{Vent} – coefficient of infiltration considering the energy share needed to heat the entering outdoor air.

Calculation of the storehouse gas amount:

$$V = \frac{E_{Jan}}{\eta_* N_{W*} \sqrt{SG}}, \quad m^3,$$

where SG is the specific gravity, N_W, J/m^3 is the Wobbe number, while η is the efficiency coefficient of the gas device.

Example 3.1

Consider the town of Montana, North-West Bulgaria. Estimate the January gas amount needed by the building of the regional administration. Its overall dimensions are 46/35/20 m, and the class of energy efficiency is "CEE=B" (IEE_{Env}=0.875). Propane heats the building. The reference value of the thermal resistance is R_{Ref}=2.5 m^2K/W. The building undergoes natural ventilation where the filtration coefficient is a_{Vent}=0.3, and 75% of the building volume is heated. The efficiency coefficient of the gas generator is η=0.85. Find the propane amount needed by the EC fuel gas unit.

SOLUTION

January heating day-degree numbers amounts to HDD_{Jan}=840. Since 75% of the building is bound to heating in January, the volume adopted in the calculations is 24150.0, m^3.

1. Calculation of the needed energy:
   - Heating of the building:

$$E_{Buil}^{Heating} = 0.418_* \frac{V_{Buil}^{0.666}}{R_{Ref}} {}_*IEE_{Env} \; {}_*HDD_{Jan*}10^6 =$$

$$= 0.418 \frac{24150^{0.666}}{2.5} 0.875.840.10^6 = 101.99, \text{ GJ/January };$$

   - Heating and ventilation:

$$E_{Jan} = E_{Jan}^{Heating} + E_{Jan}^{Vent} =$$

$$= \left(1 + 0.3\right) * 101.99 = 132.58, \text{GJ/January.}$$

2. Calculation of the warehouse gas amount:
   The specific gravity of propane-butane is SG=1.78 and the Wobbe number is N_W=64.47×10^6 J/m^3.
   The January gas volume is:

$$V_{Jan} = \frac{E_{Jan}}{\eta_* N_{W*} \sqrt{SG}} = \frac{132.58_* 10^9}{0.85_* 64.47_* 10^6 \sqrt{1.78}} = 1819.1, m^3.$$

   The propane-butane transportation pressure is 20.0×10^6 Pa. The transportation volume is found using Boyle's law:

$$V_2 = V_1 \frac{p_1}{p_2} = 1819.1 \frac{1.10^5}{200.10^5} = 9.096, m^3.$$

Here the organization of the storehouse and the logistics of the adjacent area should be accounted for.

Together with fuel gas supply and storage, the problem of exhaustion of combustion products also needs specific treatment. The next Sub-par. 3.4.3 tackles this problem.

Example 3.2

To illustrate the material in Sub-par. 2.5.3, we discussed in next table the technical characteristics of a fuel cell, operating as an energy server. The calculation of the structural components of the cell is also set forth.

Used fuel	*Natural Gas,*	*Directed Biogas*
Necessary pressure	103422.0 Pa	
Nominal electric power (AC)	200 kW	
Coefficient of electric efficiency	45.0 %	
Parameters of the electric power of switch-on:	480V; 60 Hz; 4-wires; 3 phases	
Weight, kg	4, 540	
LxBxH	5689.6 × 2133.6 × 2057.4	
Emissions	NO_x < 0.07 lbs/MW-hr	
	S_{ox} neglectful	
	CO < 0.10 lbs/MW-hr	
	VOC_s < 0.02 lbs/MW-hr	

Calculation of the infrastructural components of the fuel cells

We consider here a numerical example using catalogue data. We must calculate the necessary infrastructural components connecting the fuel cell to the EC of an office building (Figure 3.15), where: η_e=0.45; η_{tot}=0.9; w_{NG}=5m/s; w_{HW}=3 m/s, and the electric current density is 5 A/m^2.

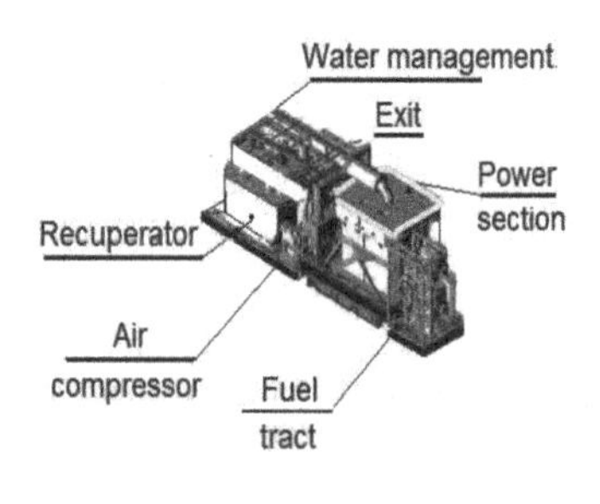

FIGURE 3.15 Micro HPS.

SOLUTION

Estimation of the total power and fuel cell consumption:

Total power: $Q_{tot}=P_e/\eta_e=200000/0.45=444\ 444.4$ w

Gas consumption:

Total power: J/s	444 444.4
Gas consumption: m^3/s	0.01279
Gas consumption: m^3/day	1105
Gas consumption: $m^3/month$	33151
Gas consumption: $m^3/year$	397 820

Electric parameters:

Electric power: J/s	200 000
Amperage: A	525.5

Thermal parameters:

Thermal power: J/s	244 444.4
Output of the circulating fluid: m^3/s	0.003

1. **Calculation of mains**

   Gas mains: $d_{NG}=(4V_{NG}/\pi/w_{NG})^{0.5}=(4\times12.79\times10^{-3}/3.1415/5.0)^{0.5}=0.057$ m

   Supply main cable: $d_E=(4A_e/\pi)^{0.5}=(4\times525.5/3.1415/5.0)^{0.5}=11.56$ mm=0.012 m

   Hot water mains: $d_{HW}=(4V_{HW}/\pi/w_{HW})^{0.5}=(4\times0.003/3.1415/3)^{0.5}=0.035$ m.

   Gas reducing valve - / 103422 Pa

2. **Calculation of the components of the electric installation:**

Contactors:	(for 525 A);
Fuses:	(for 525 A);
Dividers:	(for 525 A);
Returning traffic post:	(360 V×11 kV for 200 kW_e);
Returning main cable:	(11 kV for 200 kW_e I=10.5 A);

3. **Calculation of the components of the heating installation:**

Circulation pumps:	pursuant to the circulation contours;
Mixing valves:	pursuant to the circulation contours;
Intermediate heat exchangers:	pursuant to the load.

3.4 SYSTEMS EXHAUSTING GAS WASTE

Methane combustion is described in Par. 2 (Sub. Par. 2.5). Useful heat is released during gas combustion, consumed by home and industrial technologies. Yet, it is accompanied by the release of coproducts, such as nitrogen (70%) and three-atomic gases including carbon dioxide (9.5%), water vapors (20.5%), sulfur and nitrogen dioxides (minimal amounts), carbon monoxide and alcohol. Released gas waste yields ventilation discomfort to the occupants and damages the equipment (home appliances, gas heat pumps, water heaters, heat generators or mHPS). To neutralize wastes, they should be exhausted outside the building. Hence, a **specialized system**

for the exhaustion of combustion products or smoke gases is needed as part of each gas device.

The type and structure of that system depend on a number of factors: device type, i.e. its operation as part of the building installation, device power (flow rate of released gases[21]) and the arrangement of the building.

Modern engineering practice adopts the following methods of exhaustion of combustion products:

- Self-scattering waste gases (without the use of flues – class A);
- Forced exhaustion of waste gases (open-class **B** or contour-closed – class **C**).

According to the exhaustion mechanism, the respective systems are:

- With natural convection (with index "**1**");
- With forced convection (the device undergoes sub-pressure – with index "**2**", or the device undergoes overpressure – with index "**3**").

Figure 3.16 illustrates the systems of smoke exhaustion **A, B** and **C**. The system, class **C**, is used in low-power devices (up to 7 kW), located in premises with large span and active total ventilation/central air conditioning, type "air-air", where the air exchange is over 10 times larger (in kitchens, offices and conference halls). As illustrated in Figure 3.16,a) specific feature of the system is that the generated combustion products scatter within the same control volume, which supplied fresh air needed for combustion.

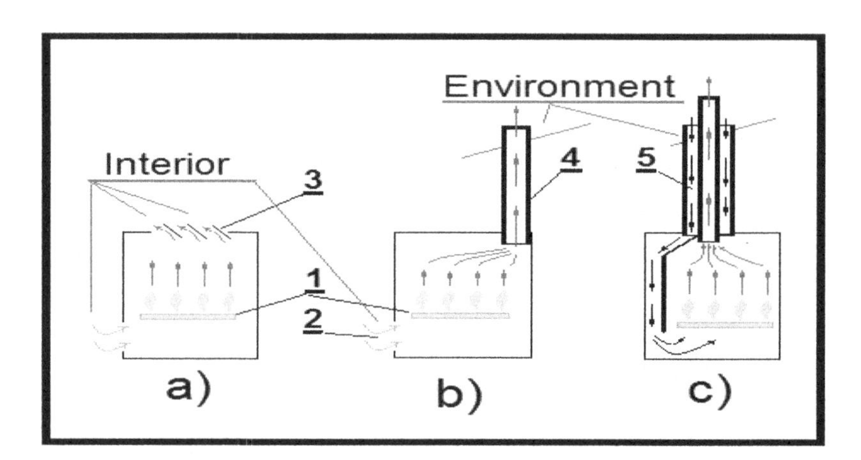

FIGURE 3.16 Types of systems for the exhaustion of combustion products: a) Class A – without smoke ducts; b) Class B – open; c) Class C – closed contour: 1 – Burner; 2 – Fresh air; 3 – Exhausting register; 4 – Outlet smoke duct; 5 – Channel supplying fresh ventilated air.

[21] Typical, combustion of 1 m^3 natural gas releases 1.058 m^3 carbon dioxide.

Waste gases mix with the available indoor air and with the outdoor cleaning air inflowing through the ventilation installation or infiltrated through the voids of the building envelope. Thus, the generated carbon dioxide[22] pollutes the air in the occupied area. It is treated as a "load" applied to the ventilation installation that provides air comfort.

Adopting a particular flow rate of the gas devices and the generated amount of carbon dioxide CO_2, one can predict the outdoor air needed to remove carbon dioxide[23]. If the input of the cleaning air is somehow limited, the capacity of the gas device will be also limited. The satisfaction of inequality (Eq. (3.2))[24] yields an estimation of the power (Q, kW) of a device, **type A,** that uses a system releasing combustion products:

$$Q \leq \frac{60}{1.058} * \dot{V}_a * (C_{Ad} - C_a) * \eta_{GD} * N_W \sqrt{SG}, \quad kW. \tag{3.2}$$

- $\dot{V}_a$ – flow rate of outdoor air supplied by the entire ventilation system (for instance, 7.1 l/s);
- $C_{Ad}^{CO_2}$ – admissible indoor concentration of CO_2 (for instance, $C_{Ad}^{CO_2} = 0.001$ or 1000 ppm);
- $C_a^{CO_2}$ – concentration of CO_2 introduced together with the outdoor air ($C_a^{CO_2} = 0{,}0003$ or 300 ppm).

If the power of the device does not satisfy the above inequality[25], this means that the released combustion products will contaminate the ambient air, thus exceeding the acceptable indoor level. Hence, outdoor air supply should intensify prior to the installation of the device.

The second type of systems exhausting smoke gases, called **open** systems, constitute a separate class – **B**. They have two modifications:

- Class **B1** (with a switch[26] or draught breaker);
- Class **B2** (without a switch).

Most often, they are components of heat generators installed on the floor (see Figures 3.13 and 3.14). They "collect" the fresh air from the building interior (boiler room)

[22] Carbon dioxide is harmless for humans. Yet, its presence in premises is unwanted, since it decreases the amount of available oxygen. It is assumed that the admissible concentration of CO_2 should not exceed 1000 ppm.

[23] See "Indoor air quality procedure of Standard 62-1999".

[24] The combustion of 1 m^3 of natural gas releases 1.058 m^3 of carbon dioxide, 2.019 m^3 of water vapors and 8.777 m^3 of nitrogen, i.e. 10.854 m^3 of gas products in total.

[25] Practice shows that if thermal power is less than **7 kW**, the system conveying and exhausting waste gases operates using indoor air supplied via natural ventilation (infiltration through the windows and doors). When power is greater fresh air supply should be forced.

[26] If pressure at the terminals is high, the draught valve should turn on automatically to avoid return of gases to the fuel chamber. The opposite flux unloads into the boiler room. Then, the emergency ventilation turns on and conveys the returned gases outside the building.

or from the gas device room, thus generating depressurization in the area. Gas devices operate as "suction intakes" of air from the interior. Combustion products are exhausted outside the control volume by means of systems, class **B**, and subsequently conveyed outside the building.

As a whole, air infiltrating through the voids of the building envelope (window and door looseness) restores the air balance. The described operation of systems, class **B**, maintains the air freshness in the building and in the boiler room. When a system, class **B**, is applicable to gas devices of one-family homes, the thermal power necessary to heat the infiltrating air should be considered in the calculation of the thermal loads.

The open systems exhausting smoke of gas combustion products, type **B**, were employed by the oldest combustion technologies introduced at the dawn of the industrial revolution in England and continental Europe. The enclosed Examples 3.3 and 3.4 give details of the design and calculation of systems, class **B**.

The ventilation effects arising during the operation of systems, class **B**, in premises with gas devices subside or are negligible in systems of smoke conveyance, class **C**, with a closed contour (see Figure 3.16,c).

The air necessary for combustion comes from the exterior space through a specialized duct for fresh combustion air (Item 5). Thus, one can avoid the occurrence of sub-pressure in the surrounding space. A flue (Item 4) conveys the released combustion products outside the building.

The channel of fresh combustion air, the combustion chamber, together with the ducts for combustion products, forms a sealed and closed fluid contour. The intaken-exhausted gas fluxes balance each other and satisfy the continuity law. Thus, air rarefaction around the device is eliminated, without violating the running combustion.

The practice has recognized a wide variety of systems, class **C**, operating by:

	Characteristics	**Notation**
Fig.3.16, a	A free diffusing non-controlled system, the device operates under natural convection	**A1**
Фиг.3.16, b	An open-class system, without a head switch, the device operates under natural convection	**B21**
Фиг.3.16, c	A system class with a closed contour intake/exhaust nozzles installed on the roof, the device operates under natural convection	**C31**

- Change of the ***location*** of the terminals (intake/exhaust diffusers): on the facades (**C1**), on the roof (**C3**) or on the roof and at the ceiling (**C7**);
- Change of the structure of intake/exhaust channels: connection to a multi-passage flue (**C2**), use of U-shaped channels (**C4**) or unbalanced gas pipelines (**C5**) and (**C8**);
- Employment of various ***principles of control*** of smoke gas ventilation (see the basic control systems in Figure 3.17).

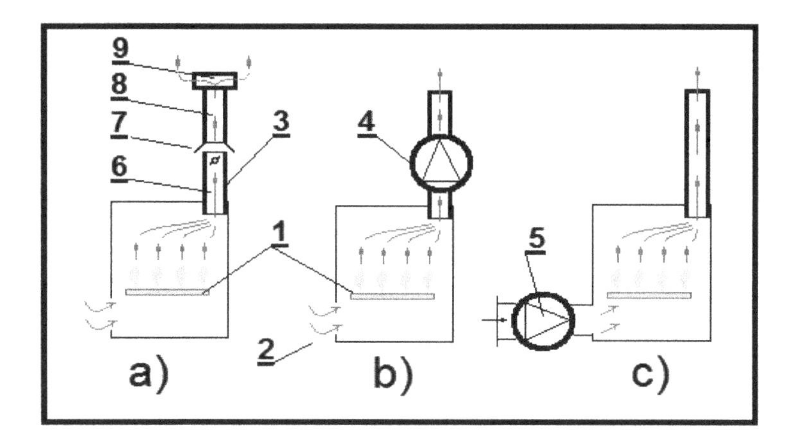

FIGURE 3.17 Systems controlling the exhaustion of the gas combusting products: a) Index "1" – natural exhaustion; b) Index "2" – forced exhaustion, the device undergoes negative gauge pressure; c) Index "3" – forced exhausting, the device undergoes positive gauge pressure: 1 – Burner; 2 – Intake of fresh air for burning; 3 – Outlet flue; 4 – Smoke suction fan; 5 – Pressurizing fan; 6 – Primary flue; 7 – Head switch; 8 – Secondary flue; 9 – Exhausting nozzle (terminal).

	Characteristics 2	Notation
Fig.3.17, a	An open-class system, with a head switch, the device operates under natural convection	**B11**
Fig.3.17, b	An open-class system, without a head switch, forced convection, the device operates under depressurization	**B22**
Fig.3.17, c	An open-class system, without head switch, forced convection, the device operates under pressurization	**B23**

The designation of systems exhausting the combustion products is a combination of letters and digits, consisting of **three symbols**, for instance **B21: – B** and **2** denote the class of the control systems while **1** denotes the system type. Here, B2 means that the system is open (Figure 3.16,b) and without a head switch. Digit "**1**" means that the system exhausts the gas combustion products via natural convection (Figure 3.16,a). The above tables specify the designation of systems shown in Figures 3.16 and 3.17.

Another example is the designation **B12** of an open-class system (Figure 3.17,b) with head switch, which exhausts the gas combustion products via forced convection, and the device undergoes depressurization.

The third example is a system, type **C13**, with a closed contour (Figure 3.16,c) and with a horizontal facade exhausting the combustion products via forced convection, while the device undergoes pressurization.

Specification of materials and flues

A system exhausting the combustion products consists of individual components with different function and location. They are pipes, terminals, diffusers, supports and insulations. The documentation specifies flues[27] by a formal letter-digit combination providing the following information:

- Temperature class (**T** – the operational temperature of the smoke gases);
- Pressure class (**P** – pressurization or positive gauge-pressure, **N** – depressurization or negative gauge-pressure, **H** – deep depressurization or deep negative gauge-pressure);
- Resistance to soot ignition (with **S** and without **O**);
- Resistance to condensation (**D** – dry and **W** – wet);
- Resistance to corrosion (class **1**, class **2**, class **3** for liquid and solid fuel);
- Thermal resistance ($m^2K/W \times 10^2$);
- Minimal distance to inflammable materials – **C**.

For instance, the specification of a gas cooker "**T200 P1 S W 1 R28 C0.1**" means a structure operating at maximal temperature of 200°C and under positive gauge-pressure (pressurized). It should be resistant to soot ignition, water vapor condensation and corrosion (class **1**). The thermal resistance should be 2800 W/Km^2. Do not place flammable materials close to the device – the distance should be longer than 0.1 m.

3.4.1 Open System Exhausting Gas Combustion Products via Natural Convection

The system is the most advantageous option to a designer of systems exhausting gas combustion products, considering funds needed, i.e. there is no need for operative and maintenance investments. Lastly, its ventilation performance provides indoor comfort in a space with gas appliances, and its operation is noiseless.

The system consists of four components (see Figure 3.17,a):

- Primary flue;
- Flux switch;
- Secondary flue;
- Flue with an outflow diffuser (terminal).

Since the flue route is the most important issue of the building design, the first stage of the overall design consists in drawing the flue lines in plan view drawings and in vertical cross-sections. The flue intake should be vertical. This is the optimal solution if the architectural plan and function of the premises allow such an arrangement (see Figure 3.18,a). It is advisable to avoid deflection from the vertical axis. If this is not possible, the deflection angle with respect to the vertical axis should exceed 135° in order to avoid significant violations of the natural convection (see Figure 3.18,b)[28]. The distance between the first bend (knee) and the gas device or flux switch should be at least 0.6 m.

[27] BS EN 1443.

[28] An eventual horizontal flue section should not be longer than 2 m.

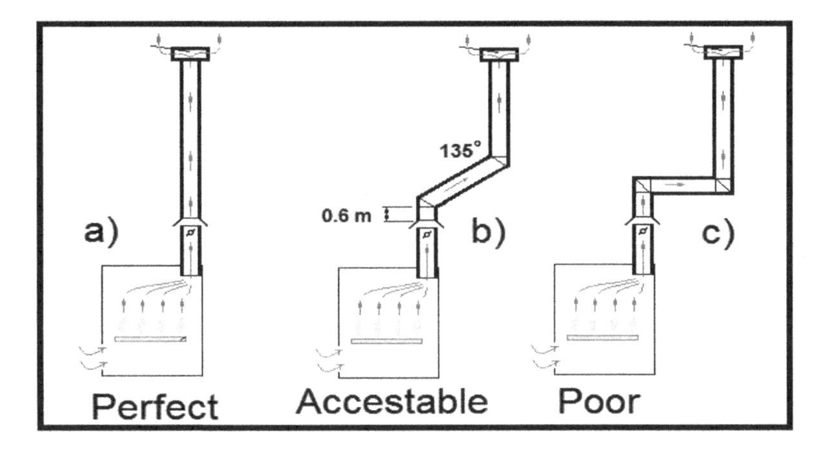

Split, dismountable	2.0	0.125	0.012
Other gas systems	2.4		
Power < 0.07MW	1.0	As that of the Devices	
Power 0.07÷3.5 MW	3.0		
Device-minimal dimensions	H_{min}, m	D_{min}, m	A_{min}, m^2

FIGURE 3.18 Assembly types and minimal dimensions of an open system exhausting gas combustion products.

The next step of the project is to calculate the flues. There are different methods of calculation. We illustrate here a method applicable to pipes with a diameter of 0.15 m[29]. It involves so-called "flue efficient height"[30] as a criterion of sufficiency. The "efficient height H_E" is found as:

$$H_E = H_A \frac{1}{1 - \dfrac{H_A K_E - \sum K}{K_I + K_0}}, \tag{3.3}$$

where:

- H_A – vertical flue height, m – found via in situ measurements and specified in the documentation;
- K_0 – factor of gas-dynamic resistance at the exhaust of gas combustion products – Table P1;

[29] IGE/UO/10.

[30] The efficient height of a flue is smaller than flue physical height measured in situ.

- K_I – factor of internal gas-dynamic resistance of the device – Table P1;
- K_E – factor of gas-dynamic resistance – Table P2;
- ΣK – sum of gas-dynamic pressures, m, included in the pipe total length (see Example 3.3).

The results for $\boldsymbol{H_E}$ found using Eq. (3.3) are compared to those for $\boldsymbol{H_{\min}}$ taken from Figure 3.18. If $\boldsymbol{H_E} < \boldsymbol{H_{\min}}$, the diameter of the vertical flue $\boldsymbol{D}$ needs no correction and height $\boldsymbol{H_A}$ should not be increased. To illustrate the method discussed so far, we present the following example.

Example 3.3

Calculate the flues of a condensing gas boiler with power 30 kW – Figure 3.19. The boiler is installed in the basement of a two-storey building with a denivelation of 6.1 m between the roof and the ridge-exhaust stack damper. The chimney is integrated with the building structure and composed of steel pipes. Its signature is "**T70 N1 O D 1 R22 C0.05**". Select the cross-section of the vertical flue and calculate its height.

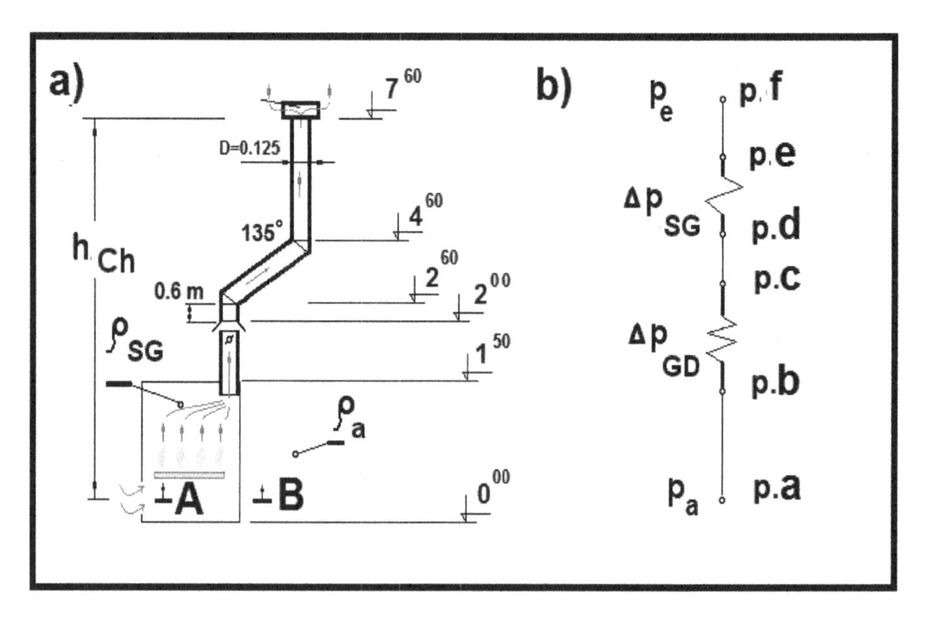

FIGURE 3.19 Vertical scheme of an open system for the exhaustion of gas combustion products: a) Physical scheme; b) Gas-dynamic analog:

- p_a, p_a – pressure in the combustion chamber and in the surrounding area;
- $\gamma_{SG.}$ и γ_a – specific weights of smoke gas and air;
- h_{Ch} – height of the flue system;
- Δp_{GD}, Δp_{SG} – gas-dynamic losses in the gas device and flue.

SOLUTION

The system class is **B11**. The planned solution implies a 2-m deflection rightward with respect to the straight vertical axis. Two 135° knees are used. They are 0.6 m above the switch of the smoke gas flux. The total flue length amounts to 0.5+0.6+2.82+3.0=6.97 m (Figure 3.19).

We find the minimal diameter and height of the vertical flue using data in Figure 3.18 – $\boldsymbol{D}_{min}$ and $\boldsymbol{H}_{min}$, respectively. We chose $\boldsymbol{D}_{min}$=0.125 m and a minimal height $\boldsymbol{H}_{min}$ of at least **2.4 m**. Using Table P1 we find the factors of internal and external gas-dynamic resistance of the device – $\boldsymbol{K}_I$ and $\boldsymbol{K}_0$, K_I=1.0, K_0=1.0, corresponding to a flue with diameter $\boldsymbol{D}_{min}$=0.125 m. The factor of gas-dynamic resistance of the pipes $\boldsymbol{K}_E$ is $\boldsymbol{K}_E$=0.25 per linear meter, given in Table P2. Using the same table, we find that the factors of gas-dynamic resistance of the two knees and the flue are 2×0.25 and 0.25, respectively, corresponding to a flue diameter D=0.125 m.

Next, we calculate the sum of the gas-dynamic resistances along the total pipe length – ΣK, m:

- Pipe section: 6.97 m×0.25=1.74
- Two knees: 2×0.25=0.5
- Flue: 0.25

Total **ΣK**=2.49 m

The input data are given in the following table:

H_A	K_I	K_0	K_E pipes	K_E 2 knees 135^0	K_E terminal	ΣK
6.1	1.0	1.0	0.25/ per linear meter	2*0.25	0.25	2.49

Substitute the above data in formula (Eq. (3.3)):

$$H_E = 6.1 \frac{1}{1 - \frac{6.1 * 0.25 - 2.9}{1.0 + 1.0}} = 3.61\text{m}.$$

It is found that the effective flue height is 3.61 m. It is by c 50% larger than the value recommended in Figure 3.18 – $\boldsymbol{H}_{min}$=2.4 m, backing up the operation of system **B11**.

Despite these advantages, the discussed open systems exhausting gas combustion products by means of natural convection (class **B1** and **B2**) have an identical disadvantage. They are not reliable if exploited throughout the year, since they have no stable gas lift. The systems employing natural convection are more successful in winter when the outdoor temperature is low. Yet, their performance in the summer and in the transition periods is problematic. The analysis of factors affecting the gas-dynamic lift in a flue confirms this phenomenon.

The prognostic value of the gas-dynamic lift of the system – H_{GLift} is determined, applying the law of distribution of the gas-dynamic head (Par. 2.3.4) to an open smoke-exhausting system.

It is convenient to design a gas-dynamic scheme (Figure 3.19,b) equivalent to the system shown in Figure 3.19,a). Then, Eq. (2.15) reads:

$$\sum_{1\div n} \Delta p_i = \left(p_a - p_e \right) - \Delta p_{GD} - \Delta p_{SD} = 0 \text{ or}$$

$$H_{GLift} = \left(p_a - p_e \right) = \Delta p_{GD} + \Delta p_{SD}. \quad (3.4)$$

On the other hand, considering that the outside and inside gas – dynamic pressure is one and the same along the isobar line A-B (see Figure 3.19,a), i.e. $p_a = p_B$ (see Example 2.6), we have:

$$\left(p_a + h_{Ch} * \gamma_{SG} \right)_A = \left(p_e + h_{Ch} * \gamma_a \right)_B.$$

Here, p_a and p_e are pressure values in the fuel cell and in the surrounding area while $\gamma_{SG.}$ and γ_a are specific weights of smoke gas and air. Parameter h_{Ch} is the height of the flue system. We have an expression for the gas-dynamic lift occurring in the gas device due to gravity, i.e. $H_{GLift} = p_a - p_e$:

$$H_{GLift} = \left(p_a - p_e \right) = h_{Ch} * g * \left(\rho_e - \rho_{SG} \right). \quad (3.5)$$

It follows from Eq. (3.4) and Eq. (3.5) that the gas-dynamic lift depends on the difference between smoke gas density and the ambient air density. It is directly proportional to the flue height h_{Ch} (i.e. the denivelation between the fuel device and the flue terminal):

$$H_{GLift} = h_{Ch} * g * \left(\rho_a - \rho_{SG} \right). \quad (3.6)$$

Eq. (3.6) proves implicitly that the gas-dynamic lift depends on the ambient temperature, which determines the density of the ambient air. In other words, the lift in a system with natural convection depends on the local climate, on the current temperature and on the wind velocity and direction[31] – note that the remaining conditions are kept constant. Due to the stochastic character of those factors, the operation of the entire system is unstable[32] and dependent on the current outdoor temperature and wind.

The engineering approach to the improvement of the stability and reliability of an open exhausting system foresees the incorporation of a mechanical draught that would dominate the gravity draught. This is attained in open systems, class **B12, B13, B14** and **B23**.

---

[31] It specifies the so-called "aerodynamic trace" of the building, whose characteristics are the total pressure at blocking points nearby the terminals (intake and outflow diffusers) and the emergence of opposite fluxes in the entire system.

[32] Eq. (3.6) shows that H_{GLift} inverts, when density of the ambient air becomes commensurable with or less than the density of the smoke gases. Then, natural convection damps or inverts its direction.

3.4.2 Open Systems with Forced Convection

An additional fifth component – a fan, is included in that class of systems (see Figure 3.20). It can be mounted before and after the gas device.

In the first case, the system exhausting gas combustion products operates under pressurization (positive gauge-pressure). There is risk, however, of waste gas release. In the second case, the system operates under depressurization (negative gauge-pressure). Then, suction of "extra" air can take place cooling down smoke gases in non-dense flues before the fan. The smoke fan itself must be capable of operating at high temperatures.

The prognostic value of the gas-dynamic lift in case of forced convection is found applying the law of distribution of the gas-dynamic draught (Sub-par. 2.3.4) to the scheme illustrated in Figure 3.20,e). Its form is:

$$\left(p_a - p_e \right) + \Delta p_{Fan} - \Delta p_{GD} - \Delta p_{SG} = 0,$$

and

$$H_{GLift} = \Delta p_{Fan} + \left(p_a - p_e \right) = \Delta p_{GD} + \Delta p_{SG}.$$

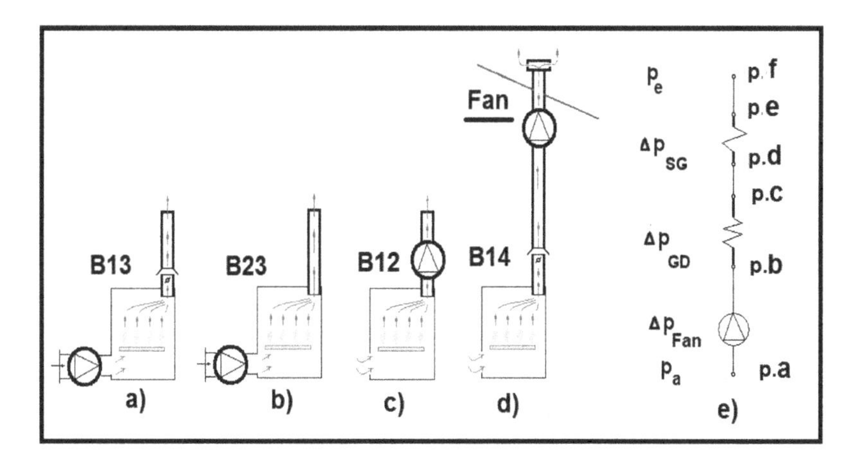

FIGURE 3.20 Variants of incorporation of the mechanical draught in a system exhausting gas combustion products: a) Combustion chamber (CC), equipped with a flux switch, is under the draught of a pressurizing fan; b) CC, without a flux switch, is under the draught of a pressurizing fan; c) CC, without a flux switch, is depressurized by a suction fan; d) CC, equipped with a flux switch, is depressurized by a suction fan; e) Similar scheme of an open system exhausting combustion products via forced convection:

- p_a, p_e – pressure in the combustion chamber and in the ambient atmosphere;
- γ_{SG} and γ_B – specific gravity of smoke gases and air;
- h_{Ch} – height of the flue system;
- Δp_{Fan} – gas-dynamic head of the fan;
- Δp_{GD}, Δp_{SD} – gas-dynamic drops in the device and flue.

The intensive gas-dynamic lift generated by the fan will overcome the gas-dynamic losses Δp_{GD} in the combustion chamber and in the flues, as proved by the equation.

Considering Eq. (3.5), the gas-dynamic lift consists of two components: lift generated by the fan and lift generated by gravity (i.e. a constant and a variable part):

$$H_{GLift} = \Delta p_{Fan} + h_{Ch*} g_{*} \left(\rho_a - \rho_{SG} \right).$$

This is the first advantage of a system with forced convection.

Other useful effects of the ventilation are:

- The location of the gas terminal (flue exhausting diffuser) is freely chosen;
- The flue route is specified based on design and intuition, but condensation should be avoided. Pipe dimensions are smaller. Horizontal flue sections are allowed;
- Smaller smoke – exhausting devices are used;
- Smoke – exhausting devices are installed on the roof, but they may be also mounted on the building facade;
- Better cleaning and neutralization of smoke gases is guaranteed prior to release;
- One may use plastic pipes and more efficient devices;
- The system operates under an automatic regulation of the flow rate of gas combustion products, coordinating the fan revolutions with the flue pressure.

As said, the incorporation of a smoke fan (Fan) facilitates the operation of the system, but it yields an increase of the initial investments and operational and maintenance costs. The use of a smoke fan however is necessary if two or more devices join the system exhausting gas combustion products[33]. They have a common flue connected to the flues of the gas devices.

3.4.3 Centralized Systems Exhausting Gas Combustion Products by Means of Forced Convection

Formally, those systems constitute two representative groups:

- Modular systems;
- Apartment systems.

The first group is popular in industrial buildings and EC of public buildings with a massive concentration of gas devices. The second group concern systems installed in residential buildings, hotels, offices, holiday homes etc. They have identical premises

[33] The basic requirement to the multiple linking of several devices in a common system is that their combustion system should be of one and the same type, i.e. burners should be identical –- with atmospheric or mechanical draught (see Sub. -par. 1.2.6). No devices consuming other fuels (coal, petrol or fuel oil) should be connected to the system without permission by the manufacturer.;

such as apartments, office complexes etc. grouped with respect to the equipment and the installation logistics – kitchens, bathrooms and toilets, for instance.

The construction of systems of smoke gas exhaustion with a common flue requires the observation of the following principles:

- Safety of each device provided by a switch or flux cutter;
- Access to the combustion system of each device;
- Access to the main flue;
- All devices in the boiler room should be connected to a common system for automatic control of the combustion activation/deactivation and the exhaustion of combustion products;
- Optimization of the dimensions of the flue such as to handle the total gas output under minimal initial investments[34].

Finally, manufacturers should consult designers on the technological compatibility between the used devices and the exhausting system.

3.4.3.1 Modular Systems

The assembly of those systems is based on the idea that they should operate parallel to or jointly with the exhausting systems of a group of gas devices[35], while flues should be composed of factory-made modules.

The choice of an exhausting system depends on the number and power of gas devices. In the ideal case, flues of two gas devices with identical power and combustion systems, connected to the main flue, should be parallel. If the number of devices is even, the connection is pursuant to the scheme shown in Figure 3.21,a). Then, a device with frequent and prolonged operation should be located close to the vertical flue. When all devices operate under an identical regime, flues are to be dynamically regulated to avoid device blocking. An alternative approach is to assemble a flue with a variable cross-section so that pressure drops in flue branches equalize.

If the number of gas devices is uneven or they have different power, the most popular scheme is to connect flues in a successive series[36] (see Figure 3.21,b). Here again the most exploited device should be located close to the vertical flue. Yet, there is a disadvantage, which is hard to remove despite the regulation and design, i.e. different gas-dynamic losses in the primary flues require different methods of cleaning the combustion chambers.

The scheme in Figure 3.21,c) shows how to eliminate this disadvantage. As seen, the primary flues are connected in a horizontal series where a suction fan creates a parallel flux of fresh air. The idea is to make all devices equivalent to each other with respect to the gas-dynamic lift of the vertical flue.

---

[34] The cross-sectional area of the main flue should be at least 0.04 m^2.

[35] In the absence of mechanical draught, systems with common central flue allow the connection of not more than eight gas devices with identical combustion systems. If a flue segment is rectilinear, the number of devices should be reduced to six.

[36] The horizontal sections of the systems with common central draught and without ventilation should not be longer than 2 m.

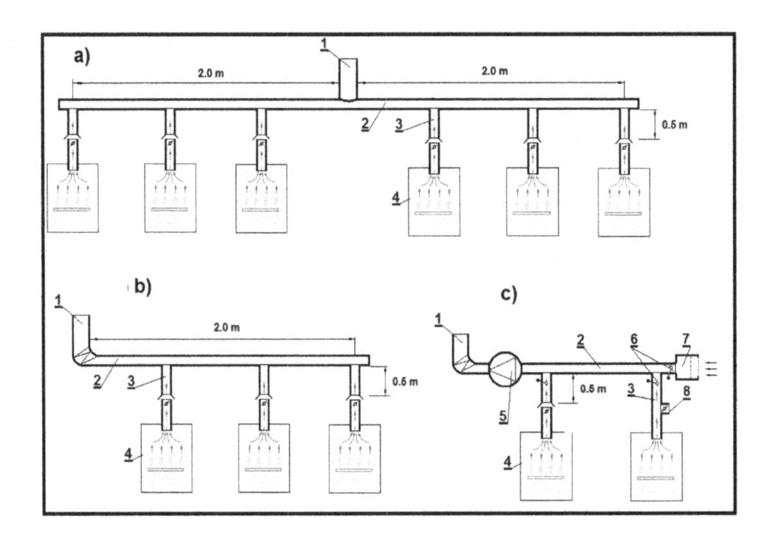

FIGURE 3.21 Modular systems: a) Parallel connection of the exhausting system of gas combustion products (ESGCP); b) Serial connection of the ESGCP; c) Serial connection but with a cleaning fan: 1 – Main vertical flue; 2 – Flue horizontal section; 3 – Primary flue; 4 – Gas boiler; 5 – Cleaning fun; 6 – Control damper with a lock; 7 – Confusor intaking fresh air from the surroundings; 8 – Flux switch.

Another advantage is the opportunity to decrease the mix concentration to acceptable limits[37] by an appropriate choice of the flow rate of the ventilating air mixed with smoke gases. Thus, the air-smoke mix may be released through a device covered by a jalousie grid[38] and mounted on the building facade. The angle of gas release with respect to the horizon is 30°. Thus, it does not cause discomfort in the area.

To attain better mixing, fan and flue designers recommend the velocity of the cleaning air to be within the ranges $6 \leq w_a \leq 8$ m/s, while combustion air should also be supplied. The minimal flow rate of the fan $\dot{V}_{BeHT}$ should be selected such as dependent on the installed power Q_{GD}, kW, pursuant to the formula:

$$\dot{V}_{Fan} = \frac{Const * Q_{GD}}{3600}, \ m^3/s \tag{3.7}$$

where Const=10.8 for NG and Const=12.8 for LPG.

Gas appliances should not be turned on if the cleaning fan does not operate. Besides, a locking control damper[39] should be installed in the suction air duct. Such damper should also be installed along the primary flue (Figure 3.21,c) before the flux switch.

[37] The acceptable levels are: – $CO_2 < 1\%$, – $CO < 50$ ppm, $NO_x < 5$ ppm.

[38] Considering gas devices with power of up to 1 MW, the height is 2 m above the ground. If the power is greater, the height rises to 3 m.

[39] The damper is installed during a planned regulation of the system controlling the mix of fresh air and smoke gases.

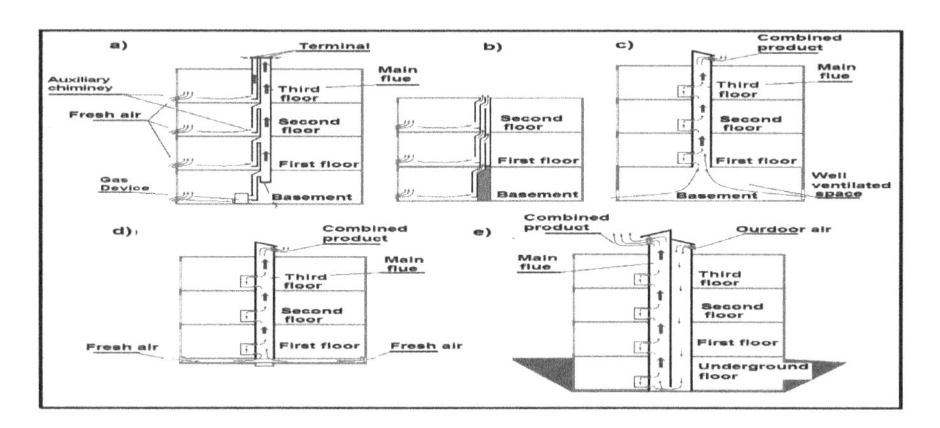

FIGURE 3.22 Schemes of open apartment systems for the conveyance of combustion products: a) Branched flues – for water heaters; b) Branched flues – for gas stoves; c) Ventilated sub-terrain; d) SE – flue; e) U – flue.

3.4.3.2 Apartment Systems

The construction of apartment systems exhausting gas combustion products requires explicit consent between owners and investors.

In contrast to modular flue installations, gas devices have low power (cookers, water heaters, convectors etc.) and they are connected to the main flue integrated into the structure. The flue passes through the building core. Yet, it does not contact the envelope, thus avoiding water vapor condensation on its walls.

Figure 3.22 illustrates some of the popular apartment systems. The five schemes have an identical feature – the connection of the individual devices to the system is not direct but via an auxiliary flue, whose task is to create an independent draught. The height of the auxiliary flue of water heaters should exceed 1.2 m. It should exceed 3 m in gas cookers. Another feature is that the fuel systems of gas devices connected to the flue should be identical. These are for instance gas cookers, water heaters with systems, class **A**, exhausting the combustion products etc.

Fresh air is supplied through the envelope voids or fan apertures (Figure 3.22,a and b), through a ventilated basement (Figure 3.22,c and d) through coupled channels on the roof (Figure 3.22,e). The choice of a method of outdoor air supply should be made considering the building envelope and the HVAC project.

3.4.4 Releasing Devices (Terminals) Installed on the Building Envelope[40]

A terminal is a diffusive nozzle fixed to the end segment of the main flue and preventing it from rain inflow, wastes or small birds. Its second function is to block the formation of an inverse air flux in the flue, when the static or entire pressure in the area is excessively high.

[40] Flue terminal (releasing device) is the mandatory component of a flue with $D<0.17$ m.

There are two types of terminals, which are widely used:

- With a jalousie grid;
- With a guarding network (the network dimensions are 0.006 × 0.006 m – 0.016 × 0.016 m).

The velocity of smoke gas intake is in the of range 4–7 m/s. Practice shows that if the net section of the releasing device is twice larger than the cross-section of the main vertical flue, the latter operates successfully and jets do not stick to the external flue surface.

The position of the terminal on the building envelope should be well selected to avoid inconveniences. The terminal should be outside the aerodynamic shadow of the building, accounting for the orientation of the wind "rose". Its traditional location is on the roof. The exhausting diffusers of the flues can be installed on the building facade, too, but they should be appropriately integrated with the other envelope components, keeping fire safety and not violating the architectural style and esthetics.

3.4.4.1 Terminals Installed on Roof Structures

The installation of terminals on the building roof is the simplest solution, since it effectively aids the natural convection of combustion products and their subsequent scatter in the surrounding atmosphere.

The choice of the location of a terminal on the building roof depends on a set of factors, and the roof structure is the most important one, i.e. whether it is flat or sloping. The recommended distance between the terminal and the building structure is given in Figure 3.23, pursuant to the roof type.

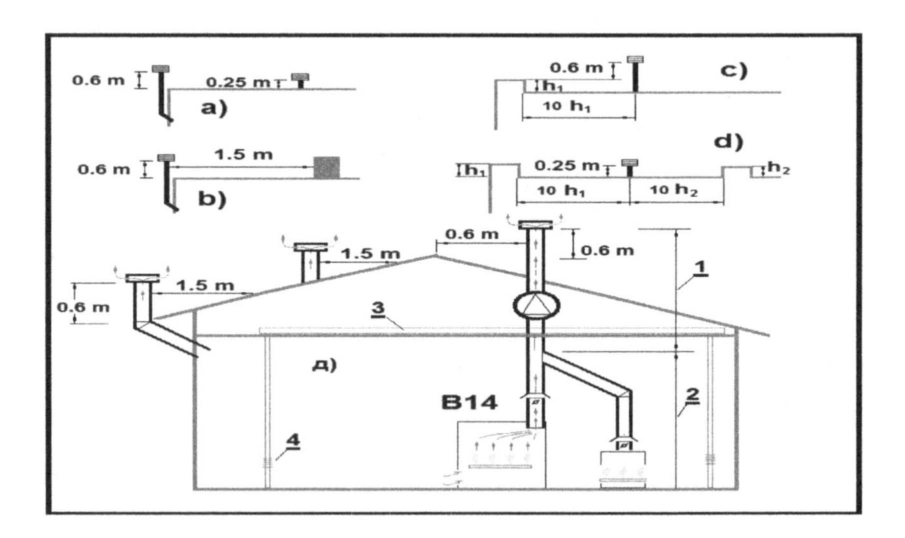

FIGURE 3.23 Location of the terminal: a) Flat roof; b) Flat roof with a structure; c) Flat roof with a lateral board; d) Flat roof with a bilateral board; e) Sloping roof. 1 – Main flue; 2 – Flue branches; 3 – Air duct; 4 – Diffuser of outdoor air.

TABLE 3.3
The minimal installation distance of terminals

Power	Roof type		Location of the terminal
		Sloping roof	• At the ridge or 0.6 m away from the roof line
< 70 kW	Flat roof	With breastwork or external flue	• 0. 6 m above the roof line;
		With breastwork or external flue	• 0. 25 m above the roof line;
> 70 kW	Sloping or flat roof		• 1. 0 m above the roof

Table 3.3 gives the minimal installation distance of terminals, depending on the gas device power and roof type.

The general rule is that the terminals should release combustion products beyond the so-called aerodynamic shadow of the building.

Standstill circulation areas emerge there with a concentration of noxious emissions, exceeding the admissible limit. This makes the areas inappropriate for the installation of ventilation diffusers and fresh air ducts.

The number of releasing devices on the roof is also an essential factor.

When there are three or more terminals, they should be installed at one and the same level, unless their diameters exceed 0.3 m.

All of these recommendations concern the location of terminal devices on the roof, only. When the building "concentration" in an area is large, most buildings are grouped in clusters and have different height. Then, the gas combustion products should be exhausted beyond the combined aerodynamic shadow.

3.4.4.2 Terminals Installed on Facades

We discussed in Par. 3 (Sub-par. 3.4.3.3.1) the advantages of the scheme in Figure 3.21,b) and explained that the intake/exhausting diffusers of open modular systems are installed on the building facades and at the level of flue's horizontal segment.

To avoid air contamination by the released smoke gases, an additional cleaning fan should participate in the scheme. This however cannot solve the main problem arising in open systems of smoke gas exhaustion, i.e. the depressurization of the interior (vacuum generation).

The gas industry has produced a special class of gas devices capable of operating on building facades. They attribute to the elimination of the disadvantages of the open systems and mainly the disadvantages of systems creating internal ventilation effects. Figure 3.24 shows some of their modifications.

Note that all devices belonging to that class are internally structured as insulated circulation contours – the fresh air needed for combustion is collected by the building facades where intake-diffusers are installed. Combustion products also unload there. The devices are appropriate for buildings where the quality of the ambient air is controlled by a central ventilation system, while rooms are impenetrable to the surrounding air.

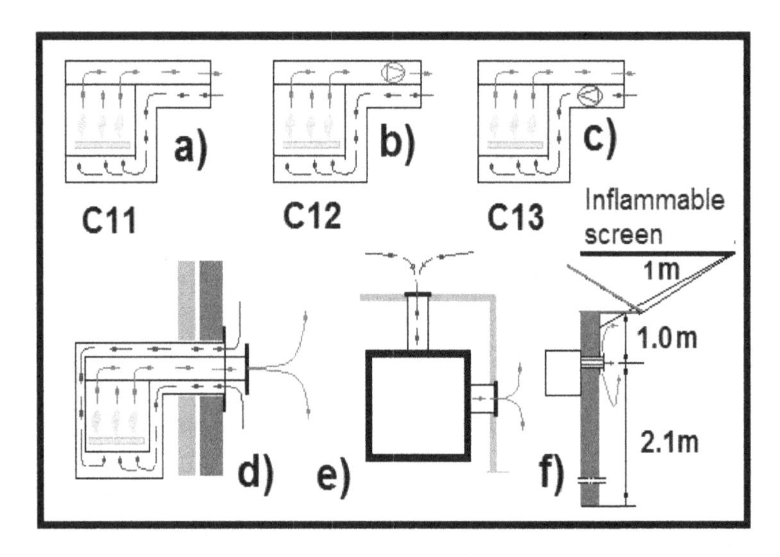

FIGURE 3.24 Facade ventilation of gas combustion chambers, class C: a) Class C11 with natural convection; b) Class C12 with forced convection and depressurized fuel device; c) Class C13 with forced convection and pressurized fuel device; d) Class C11 with natural convection and coaxial flues; e) Class C11 with natural convection and terminals on the neighboring facades; f) Assembly of a device close to a sloping roof.

Facade-ventilated gas devices classify pursuant to two criteria: the principle of formation of a smoke gas flux[41] and dynamic pressure balance at the intake/exhausting terminals. The second criterion is particularly applicable to and used by closed systems.

Pursuant to that criterion, a single gas device is considered as being balanced if the wind dynamic pressure[42] ($0.5 \times \rho \times w^2_{Wind}$) is one and the same at the intake and the exhausting diffusers of the terminal unit. Moreover, no flux of outdoor air would pass through the fuel chamber, if the intake-exhausting pressure difference is neglected. All gas appliances whose intake-exhaust diffusers are located on one and the same facade (for instance, the devices shown in Figure 3.24,a), b), c) and d)) can be considered as belonging to the class of "**dynamic-self balancing**" equipment, since their orientation implies identical dynamic pressure. Hence, gas devices, class C11, can operate steadily on the facades and under sole gravity.

Devices of the same type employing natural convection, but with diffusers located on **facades, whose orientation does not coincide with** the direction of the current wind, undergo different dynamic pressure at the intake and exhaust, as well as air imbalance. The imbalance results in the induction of internal compensating flow through the combustion chamber and unstable operation and turn-off of the device.

The improvement of the stability of imbalanced devices is attained by the incorporation of a pressurizing or suction fan, generating dynamic opposite pressure. On the other hand, the use of a fan yields an increase of the intake of outdoor air and a

[41] That criterion is traditional for gas devices and the basic types are illustrated in Figure 3.23.

[42] The value of the wind dynamic pressure varies in time depending on the wind direction and power, considering different building facades.

decrease of the pipe cross-section and dimensions of the intake/exhaust devices. As a result, the overall dimensions of the gas device with incorporated fans decrease.

In conclusion note that the choice of a system conveying combustion waste products from gas devices depends on the building type, device power and environmental health and safety requirements.

Example 3.4

Calculate the flues of two condense gas boilers with a power of 20 and 10 kW – Figure 3.25. The temperature of the smoke gases after the boilers is 75°C. The burner excess coefficient is k_{Air}=1.15. Boilers are installed in a pavilion, with denivelation from the roof ridge to the exhaust stack damper amounting to 6.6 m. The system of combustion products exhaustion is closed, unbalanced and of modular type. Select an appropriate smoke-suction fan.

Boilers are installed in a pavilion, with denivelation from the roof ridge to the exhaust stack damper amounting to 6.6 m. The system of combustion products exhaustion is closed, unbalanced and of modular type. Select an appropriate smoke-suction fan.

SOLUTION

The system exhausting combustion products is shown in Figure 3.25,a). It consists of a fresh air duct, a boiler structure, main flue and a smoke fan.

1. Determination of the prognostic flow rate:

The prognostic gas flow rate is found considering an efficiency coefficient of the condense gas boiler equals to 92%. The total prognostic fuel flow rate of the two boilers is calculated using Eq. (2.5):

$$\dot{V} = \frac{30 * 10^3}{0.92 * 50.68 * 10^6 \sqrt{0.58}} = 0.845 * 10^{-3}, m^3/s.$$

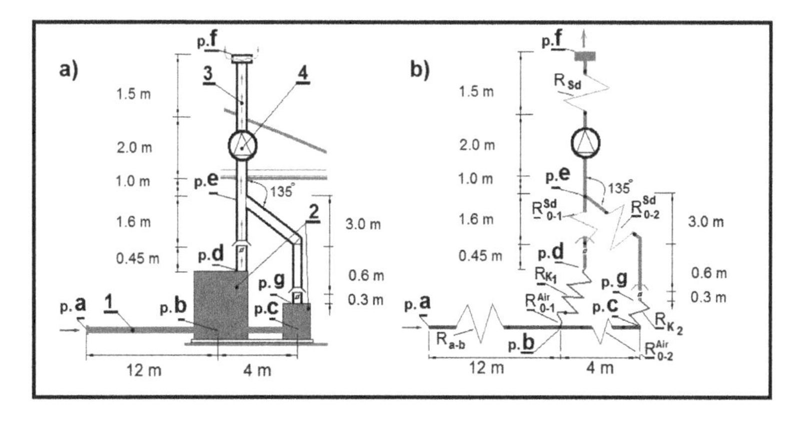

FIGURE 3.25 Closed unbalanced system exhausting gas combustion products: a) Vertical view; b) Equivalent scheme 1 – Fresh air duct; 2 – Boiler structure; 3 – Main flue; 4 – Fan.

The values of the air/gas ratio in Table 2.2. show **that the combustion of 1 volume of NG consumes 9.81 air volumes.**

The flow rate of air needed for the combustion of 0.845×10^{-3} m^3/s gas is:

$$\dot{V}_{Air}^{Th} = N_{AG} * \dot{V} = 9.81 * 0.845 * 10^{-3} = 8.29 * 10^{-3}, \quad nm^3/s.$$

Considering the excess coefficient α_{Air}=1.15, the **real air flow rate** will be:

$$\dot{V}_{Air}^{Real} = \alpha_{Air} * \dot{V}_{Air}^{Th} = 1.15 * 8.29 * 10^{-3} = 9.533 * 10^{-3}, \quad nm^3/s.$$

Smoke gas flow rate is 0.845×10^{-3} $m^3/s \times 10.854$ m^3=**9.17**$\times 10^{-3}$ nm^3/s[43].

Reduce the flow rate of the smoke gases to the operational temperature of gases in the flue system. Apply Charles's law (Sub-par. 1.4.2). The temperature after the condensing boilers is 70°C (see Example 2.11). Then:

$$\dot{V}_{SG_{70^0C}} = \dot{V}_{SG_{0^0C}} \frac{T_2}{T_1} = 9.17 * 10^{-3} \frac{(273+70)}{(273+0)} = 11.52 * 10^{-3}, \quad m^3/s.$$

2. Calculation of the initial cross-sectional area of the gas pipe.

We use the following empirical formula to calculate the initial cross-sectional area of the flues [14]:

$$A_0 = \frac{0.026 * Q_K}{\sqrt{h_{Ch}}} 10^{-4} = \frac{0.026 * 30000}{\sqrt{18}} 10^{-4} = 0.0184, m^2.$$

Assume a rectangular cross-section of the flue, with ratio 1:3 and dimensions 0.24×0.08 m. Then, the actual cross-seçtional area is equal to 0.0192 m^2.

Smoke gas velocity is:

$$w_{SG} = \frac{\dot{V}_{SG}}{A_0} = \frac{11.52 * 10^{-3}}{0.0192} = 0.6, m/s.$$

Assuming that the velocity of the fresh air is equal to that of the smoke gases ($w_{Air} = 0.6, m/s$), we find the flue cross-section using Eq. (2.8):

$$A_0 = \frac{\dot{V}_{Air}^{Real}}{w_{Air}} = \frac{9.533 * 10^{-3}}{0.6} = 0.0159, m^2.$$

Keeping the ratio 1:3 we find the dimensions of the flue rectangular cross-section – 0.22 × 0.07, m and the cross-sectional area – 0.0154 m^2. The averaged velocity of the air flowing in a flue with a cross-sectional area 0.0154 m^2 will be:

$$w_{Air} = \frac{\dot{V}_{Air}}{A_0} = \frac{9.533 * 10^{-3}}{0.0159} = 0.6, m/s$$

[43] The combustion of 1 m^3 natural gas generates 10.854 m^3 waste gas.

Channel	Dimensions m	Area m^2	Perimeter M	Equivalent diameter $D_h=4A_0/\Pi$, m
Flue	0.24×0.08	0.0192	0.64	$D_h^{\Pi\Gamma} = 0.12$
Air duct	0.22×0.07	0.0159	0.58	$D_h^{B-x} = 0.1096$

Air ducts and flues with a rectangular cross-section are calculated using equivalent diameters. They are specified in the above table.

3. Calculation of the gas-dynamic losses in the flue and the air duct, as well as the needed fan head.

The installation shown in Figure 3.25 formally consists of three sections connected in a series:

- Fresh air duct – **a** – **b**;
- Boiler structure – **b** – **e**;
- Main flue and fan – **e** – **f**.

Table 3.4 gives their basic geometric and kinematic characteristics, such as output, length, fluid operational velocity, Reynolds number and coefficient of linear resistance.

4. Calculation of the lift pressure and the gas-dynamic head of the fan.

Applying the law of gas-dynamic head in a pipe (Section 2.3.4) to the equivalent scheme in Figure 3.25,b), we find:

$$\Delta p_{Fan} + \Delta p_{EK} - \Delta p_{a-b} - \Delta p_{b-d} - \Delta p_{d-e} = 0,$$

and the needed gas-dynamic head of the fan (**Item 3** – Figure 3.25,b) Δp_{Fan} reads:

$$\Delta p_{Fan} = \Delta p_{a-b} + \Delta p_{b-d} + \Delta p_{d-e} - \Delta p_{EK}, \quad \text{Pa.} \tag{3.8}$$

TABLE 3.4
Geometrical and kinematic characteristics of a closed unbalanced EG system from Example 3.3.

№	Section output $\dot{V}\ m^3/s$	Length m	Velocity m/s	$Re = \frac{wD_h}{\nu}$	Coefficient of linear resistance[44] λ
a- b	$9.533 * 10^{-3}$	12.0	$w_{B-X} = 0.6$	4607	0.03195
b- e	$11.52 * 10^{-3}$	Cauldron structure	–	–	–
e- f	$11.52 * 10^{-3}$	4.5	$w_{\Pi\Gamma} = 0.6$	3494	0.042

[44] To calculate λ of channels with a rectangular cross-section 1:3, we use Konakov's formula: $\lambda = (1.8 * \lg Re - 1.5)^{-2}$ for Re >2320.

Here

- $\Delta p_{E\kappa}$, Pa – lift head of gravity;
- Δp_{a-b}, Pa – pressure losses in the fresh air duct **a–b** (**Item 1** – Figure 3.25,b);
- Δp_{b-d}, Pa – gas-dynamic pressure losses in the boiler structure **b–d** (Figure 3.25,b);
- Δp_{d-e}, Pa – gas-dynamic pressure losses in the main flue **d-e** (Figure 3.25,a).

Putting the respective expressions in Eq. (3.8) we obtain:

$$\Delta p_{Fan} = \left(\frac{L_{Air}}{D_h^{Air}} \lambda_{Air} + \xi_{Air} \right) \frac{w_{Air}^2}{2} \rho_{Air} +$$

$$+ \left(\frac{1}{\left(R_{01}^{Air} + R_{K1} + R_{01}^{GD} \right)} + \frac{1}{\left(R_{02}^{Air} + R_{K2} + R_{02}^{GD} \right)} \right)_{\kappa C}^{-1} * \frac{w_{SG}^2}{2} \rho_{SG} +$$

$$+ \left(\frac{L_{Ch}}{D_h^{SG}} \lambda_{SG} + \zeta_{SG} \right) \frac{w_{SG}^2}{2} \rho_{SG} - h_{Ch.} * g * \left(\rho_{Air} - \rho_{SG} \right), \text{ Pa}$$

(R_{K1} and R_{K2} are the internal resistances of the gas boilers; R_{01}^{Air}, R_{01}^{SG}, R_{02}^{Air} and R_{02}^{SG} are the resistances of the connecting pipes of the first and second boilers, conveying air and smoke gases[45]).

The different types of resistance along the route of the air and smoke gases are shown in Figure 3.25,b) while the discharged amounts are proportional to the power of the two boilers (2:1 in this case).

Resistance of the boiler structure	R_{0i}^{Air}	R_{Ki}	R_{0i}^{SG}	$R_{0i}^{Air} + R_{Ki} + R_{0i}^{SG}$
Boiler 20 kW	1.22	2.0	1.47	4.69
Boiler 10 kW	2.19	2.0	3.23	7.42
				$(1/4.68 + 1/7.42)^{-1} = 2.87$

Inserting the subsequent primary data, we find the necessary fan head:

$$\Delta p_{Fan} = \left(\frac{12.0}{0.1098} 0.03195 + 1.0 \right) \frac{0.6^2}{2} 1.293 + 2.87 * \frac{0.6^2}{2} 0.998 +$$

$$+ \left(\frac{4.5}{0.12} 0.042 + 1.4 \right) \frac{0.6^2}{2} 0.998 - 6.10 * 9.81 * \left(1.293 - 0.998 \right) =$$

$$= 1.045 + 0.515 + 0.534 - 17.65 = -15.55, \quad \text{Pa}.$$

Head Δp_{Fan} is negative. This means that the lifting head $\Delta p_{E\kappa}$ generated by gravity at ambient temperature of 0°C is sufficient to compensate pressure the gas-dynamic pressure losses in duct lines, and the system can operate without a fan.

[45] Smoke gas density at 70°C is found via density extrapolation at 300°C (0.63 kg/m^3) and 100°C (0.95 kg/m^3). The calculated value is 0.998 kg/m^3.

If the ambient temperature exceeds 50°C and air density drops below 1.093 kg/m^3, the gravity-generated head Δp_{EK} becomes negligible and one should switch-on the fan.

The reduction of flue[46] overall dimensions resulting in a respective increase of gas velocity yields an increase of the gas-dynamic losses and implies fan switch-on.

For instance, if gas velocity rises to 3 m/s, the needed fan head Δp_{Fan} is positive as proved by the following correction:

$$\begin{aligned}\Delta p_{Fan} &= \left(\frac{12.0}{0.06}0.0307+1.0\right)\frac{3.0^2}{2}1.293+2.87*\frac{3.0^2}{2}0.998+\\ &+\left(\frac{4.5}{0.07}0.0303+1.4\right)\frac{3.0^2}{2}0.998-6.10_*9.81_*\left(1.293-0.998\right)=\\ &=41.54+12.89+15.03-17.65=+51.8\,\text{Pa}.\end{aligned}$$

Then, a fan with head 55–60 Pa and flow rate of 12.0×10^{-3} m^3/s turns to be the appropriate device.

Note here that the calculation of the dimensions of the duct (pipe) cross-section and fluid velocity, respectively, is an optimization problem where the profit function is the cost of construction and equipment maintenance. Yet, the solution to this problem is beyond the task and scope of the present study.

3.5 DESIGN OF A PIPE NETWORK[47]

Pipe networks supply various gas devices with fuel gas from ECs. The interest in these systems has grown since the 60–70s of the 20th century, mainly in the USA, England and the EU. The change of the economic situation in our country during the period 2001–2010 stimulated the interest in the field. However, it subsided later, since biomass occupied the position of a preferable primary energy carrier.

3.5.1 Structure of Gas Pipe Networks in Residential and Public Buildings

Although current domestic gas consumption is minimal, i.e. under 5% of the total energy supply, the construction of urban and regional gas networks in district towns grows actual. This is so since gas prices are expected to drop in the next 3–8 years while larger gas amounts will be available on the market.

Regulation No 6 [3] of the Ministry of Regional Development and Public Works and Ministry of Energy sets forth the requirements for the design, construction and use of devices and equipment for gas transport and storage. It tackles also the specific

[46] Thus, the initial investments in a system exhausting gas combustion waste drop.

[47] The interior installations should be under strict state control to avoid incidents, such as that in East London in 1968, when gas explosion resulted in the collapse of a 22-storey residential building (CORGI).

features of gas installations undergoing a pressure of up to 0.01 MPa, including those in Paragraph 133 [3].

It is assumed that the intake point of a pipe network of a residential building is its exit from the station containing the measuring instruments (they are ownership of the gas supplying company). Pipe networks of administrative and public buildings start from a branch of the regional distributing gas pipe network.

A horizontal pipeline should run at 2.2 m above the floor. Vertical pipelines (risers) are installed in the following premises:

- Kitchens;
- Corridors;
- Stairwells.

Pipe networks can be exposed or hidden, installed in channels and wall gaps supplied with light dismountable covers. Floor channels should be covered by inflammable lids, and pipelines should rest on inflammable supports or consoles.

When passing through walls and beneath floors pipelines should be placed in a protective casing. Channels and gaps should allow pipeline inspection, repair, and ventilation. The same requirements hold to a pipe network running within the space above a suspended ceiling.

At their lowest points, channels should be supplied with a drainage system to prevent pipelines from the intake of acids or other corroding liquids. The installation of screwed or flange shut-off valves or plugs is prohibited. Low-pressure gas-distributing networks, operating under pressure lower than 0.01 MPa, should be installed right under the putty, and no cavities for pipe networks should be provided.

The installation of gas devices is not allowed in stairwells and sanitary-hygiene spaces (bathrooms, toilets etc.) without natural ventilation. Gas devices may be installed in dwellings with forced ventilation, where explosive or flammable materials with flammability limit higher than 61°C are stored. This is in conformity with Par. 140 of the Regulation. Yet, pipelines should not run inside ventilation and flues ducts and lift shafts.

Pipe networks of heating and water installations, as well as gas pipeline ducts, may pass through gas pipeline channels.

A transit passage of low- and medium-pressure pipelines is allowed through the building space if:

- Pipes are welded to each other;
- The use of pipe network fittings is avoided;
- Access to pipelines is provided.

Screwed or flange joints are used to install fittings or control-measuring instruments. Pipelines should not pass through electro-distributing stations and control rooms, storehouses or sub-terrain space used for the storage of explosive and flammable materials.

Dense concrete should be poured over a gas pipeline running in a concrete floor channel. Yet, this is allowed only in laboratories and public service enterprises. Pipe insulation with anti-corrosion and protective coating should precede this operation.

Passable sleeves (sealed pipe sleeves) should be mounted at pipe's either side, protruding 30 mm above the floor. No shut-off fittings are to be mounted there. If moisture percentage in the fuel gas is high, pipes should be thermally insulated. Their slope should be $3^0/_{00}$ with respect to the appliances.

The slope of pipeline sections equipped with gas-measuring instruments is as follows:

- At the gas flowmeter intake – the slope is oriented with respect to the intaking risers;
- At the gas flowmeter outlet – the slope is oriented with respect to the gas device;
- Gas risers should never be installed in storehouses, garbage shafts, lift and ventilation shafts, in flues, in mechanical or pump stations, in residential space (living rooms, sleeping rooms, dining rooms, bathrooms etc.);
- Pipelines may pass through massive foundations of buildings, shelters, stairwells or other channels. Then, pipes should be placed in a casing and the pipe-concrete looseness should be filled with bitumen/tarred paper or strong cement mortar. As for the sealed pipe sleeve, its end should be 50 mm below the floor or above the landing.

Pipe branches to gas devices in residential and public buildings are laid in floor channels. Yet, no fittings should be mounted on them.

Pipelines may pass through massive foundations of buildings, shelters, stairwells or other channels. Then, pipes should be placed in a casing and the pipe-concrete looseness should be filled with bitumen/tarred paper or strong cement mortar. As for the sealed pipe sleeve, its end should be 50 mm below the floor or above the landing.

- Pipe branches to gas devices in residential and public buildings are laid in floor channels. Yet, no fittings should be mounted on them.
- Gas pipes should run in a building accounting for the electrical installation and meeting the following requirements:

1. If a gas pipe crosses an electrical cable, the distance between them should be at least 0.1 m (the gap may be less than 0.1 m in residential and public buildings, if the gas pipe is insulated by a rubber or ebony sleeve);
2. If a gas pipe is parallel to an electrical cable, the distance between them should be at least 0.25 m;
3. The distance from the pipeline to the electric distribution board or sub-board should be at least 0.5 m. The pipeline should not touch other pipes, for instance water pipes, channels etc. After the installation, gas pipelines should be painted with light-brown oil and polyethylene paint.

Figure 3.26 shows an example of the installation of a gas device in residential buildings. An axonometric scheme of a pipe network of a three-storey building with six identical apartments is presented. Pursuant to the design, each apartment is equipped

[48] In this case, the urban distributing network is designed such as to undergo medium pressure.

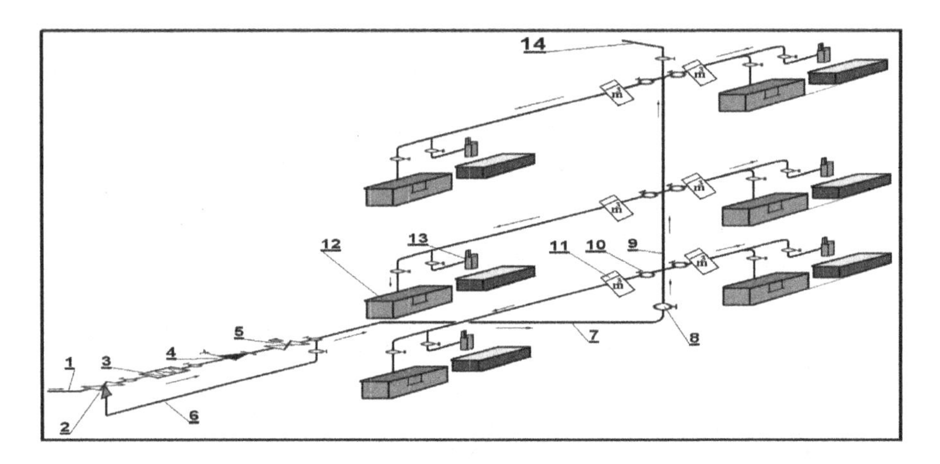

FIGURE 3.26 Axonometric scheme of the pipe network of a three-story building connected to a city gas distributing pipeline network – medium pressure: 1 – Intake pipe (medium pressure) of the city network; 2 – Triple-way valve; 3 – Filter (against moisture and mechanical impurities); 4 – Reverse-protecting valve; 5 – Pressure reducing with incorporated protective-cutting valve; 6 – Bypass joint; 7 – Horizontal distributing pipeline; 8 – Regulating valve; 9 – Gas riser; 10 – Connecting valve; 11 – Gas flowmeter in an apartment gas – measuring station (GMS); 12 – Gas cooker; 13 – Gas passable water heater; 14 – Blowpipe.

with gas appliances. The kitchen and the bathroom should have a cooker and a tankless water heater, respectively. A medium-pressure gas pipeline[48] supplies the building through a Gas Regulation & Measuring Station (GRMP), located in the building basement (see Figure 3.26). Filter (Item 3) cleans the mechanical impurities, while a regulator (Item 5) reduces pressure from medium to low pressure. Gas operational pressure after the regulator stabilizes at about 2.0 kPa. The pipe network is designed such as to convey the fuel gas to the remotest device losing 1.5–1.7 kPa, while the remaining pressure should compensate the internal gas-dynamic losses.

Fuel gas is flowed to the gas riser, having passed through the gas-regulating equipment via the horizontal distributing pipeline (the main internal gas-distributing pipeline – Item 7). Figure 3.26 shows a single gas riser (Item 9). Riser's end exits the building as a blow pipe (Item 14) through a shut-off valve used to block the system for eventual repair.

A different number of vertical risers, conforming to the device location, are installed in the common space of buildings, whose height is up to 15 m inclusive. Pursuant to Par. 141 of the Regulation, they cross storey- or inter-storey landings. Two vents are drilled on the facade at the stairwell lowest and highest level.

Gas-measuring panels (GMP) are mounted in stairwells, not less than 2 m above the staircase and/or the inter-storey landings. They should not interfere with the use of the landings. On the other hand, panels of NG flow meters are mounted on storey or inter-storey landings. They should not interfere with the use of the staircase too. The installation height should be such as to allow easy reading of the gas consumption.

Flow meter devices for NG should be ventilated through vents, providing a connection to the staircase well. No flow meters should be installed in rooms with increasing electrical transformers, whose nominal voltage exceeds 400 V.

Consider pipelines passing through stairwells and landings of buildings higher than 15 m. Then, vents should be drilled through the external wall at every 10 meters, in addition to the vents pursuant to Par. 141. Jalousie grids should cover them and their cross-section should be no less than 0.01 m^2.

In buildings equipped with separate emergency staircases and landings, pipe lines may pass through them pursuant to at least one of the following requirements of Par. 142:

- They should be installed in shafts or channels built from inflammable materials with fire resistance time amounting to 90 min;
- Shafts should be equipped with vents and isolated from the staircase air. Their ventilation should be individual or combined, and the vent cross-section should be 0.01 m^2;
- No ventilation is required if inflammable materials fill the shafts.

Pipelines are solidly mounted on inflammable supports and coated with putty no thinner than 0.015 m. Plastered pipes should be corrosion-protected by special varnish coatings or bands.

The use of a pipeline as an earthing system is prohibited. The antistatic bridges of gas devices should not be connected to the common earthing installation of the building.

Pipes of a network

Various pipes should be used in an NG pipework. Pursuant to Regulation No 6/2004, these are seamless pipes (Mannesmann pipes), electrically welded pipes, copper pipes, pipes from high-density polyethylene, pipes from plasticized polyvinyl chloride etc. However, above-ground pipelines should be built from metal pipes only.

Depending on the external diameter of copper pipes, their nominal thickness is found as follows: pipe diameter of up to 22 mm corresponds to pipe thickness of 1 mm; 22–42 mm pipe diameter corresponds to 1.5 mm pipe thickness; 43–89 mm pipe diameter corresponds to 2 mm pipe thickness; 90–108 mm pipe diameter corresponds to 2.5 mm pipe thickness; pipe diameter larger than 108 mm corresponds to 3 mm pipe thickness. Pipes are soldered or welded to each other, using high-temperature solder with melting temperature higher than 450°C.

The standard order of network pipes

	¼"	3/8"	½"	¾"	1"	1 ¼"	1 ½"	2"	2 ½"
d_p	Ø6	Ø10	Ø15	Ø20	Ø25	Ø32	Ø40	Ø50	Ø65
mm	6.35	9.52	12.7	19.05	25.4	31.87	38.25	51.00	63.5

Pipe networks undergoing a pressure of up to 10 kPa should be subjected to strength ($\mathbf{p}_{NWork}$=0.1 MPa) and density ($\mathbf{p}_{NWork}$=0.01 MPa) tests. Their exploitation start should meet the requirements of BDS EN 12327: 2013 and the Regulation pursuant to Article.200, Par.1 of the Energy Law.

3.5.2 Calculation of Pipe Networks

There are two steps of pipe network calculation:

- Assessment of NG consumption (by each apartment and by the entire building);
- Gas-dynamic calculation of pipe networks, consisting of two steps.

The **first step** specifies in detail the home appliances. Then, a list of the end users of NG is prepared considering the device "nominal power" – "kW"[49]. The pipe network is divided into individual sections stretching between spots **where gas consumption varies (assessed by the prognostic gas flow rate).**

To prepare a prognosis of gas consumption of a building, one needs to consider the probability of simultaneous operation of all devices. In buildings comprising up to 5 apartments, the effect of simultaneousness is accounted for in conformity with the respective coefficient in Table 3.5.

Larger buildings require a special calculation algorithm described below [16].

Assess NG flow rate $\dot{V}_j^{Ap}$ of the j-th apartment using the formula:

$$\dot{V}_j^{Ap} = \sum_{1}^{m} k_0 \dot{V}_i n_i, \ m^3/s \tag{3.9}$$

where:

- $\dot{V}_i$ m³/s – nominal flow rate of the i-th gas device;
- k_0 – coefficient of simultaneousness of identical gas devices used in the apartment (see Table P5);
- n_i – number of identical devices in the apartment;
- m – number of device types in the apartment.

TABLE 3.5
Coefficient of simultaneousness of small buildings comprising up to 5 apartments

Number of apartments	1~2	3~5
Cookers	1.0	0.8
Water heaters	1.0	0.89
Fridges	1.0	1.0

[49] Distinguish "Gross heat input" from "Net heat input" on the device label. Use a divisor 1.1 to transform the first quantity into the second one.

Calculate similarly the prognostic flow rate of NG $\dot{V}_j^R$ transmitted through the OT j-[th] section of a **vertical riser**, which feeds the apartment branches (see Figure 3.25):

$$\dot{V}_j^R = \sum_1^{mm} k_0 \dot{v}_i n_i, \quad m^3/s. \tag{3.10}$$

Here summation is performed with respect to the number of devices supplied by sections located above the j-[th] section of the riser. Perform calculations starting from the last storey, considering the remotest riser (the remotest end user) and proceeding downward to the main horizontal pipeline.

The prognostic gas flow rate $\dot{V}_j^M$ through the j-[th] main horizontal pipeline is found as a sum of the estimated NG flow rates of the feeding risers:

$$\dot{V}_j^M = \sum_j^m k_0 \dot{v}_j n_j \tag{3.11}$$

where ***m*** is the number of gas devices fed after the considered main pipeline.

The prognostic flow rates participate in the calculation scheme of the gas installation, which consists of:

- apartment branches and connections to the devices;
- risers;
- main pipeline branches.

Next step, one should perform two-stage gas-dynamic calculations:

a. **Preliminary (initial)** calculation of pipe diameters at a given **flow rate, pipe length and available pressure drop**;
b. **Final calculation of the pipework** considering the remotest end users, knowing pipe network geometry and topology and including the **estimated gas-dynamic losses**.

Gas-dynamic calculation of **low-pressure pipe networks** is performed using tables, nomograms or on a computer.

Gas pressure loss is due to friction in linear pipe sections and local resistance. We use a traditional formula [16]:

$$\Delta p_{GL} = \Delta p_{Line} + \Delta p_{Local} = 0.592 \frac{\dot{V}^2}{d_p^4} \lambda \frac{L_p}{d_p} + 0.592 \frac{\dot{V}^2}{d_p^4} \zeta = 0.592 \frac{\dot{V}^2}{d_p^4} \left(\lambda \frac{L_p}{d_p} + \zeta \right), Pa \tag{3.12}$$

where the friction factor is presented as:

$$\lambda = 0.11 \left(\frac{k_e}{d_p} + \frac{68}{Re} \right)^{0.25} = 0.11 \left(\frac{0.000015}{d_p} + \frac{763.33_* 10^{-6} d_p}{\dot{V}} \right)^{0.25} \tag{3.13}$$

Pressure loss in the pipe network due to local resistance Δp_{Local} is proportional to the value of the dynamic pressure $0.592\frac{\dot{V}^2}{d_p^4}$, expressed via the gas flow rate. It is adjusted using a coefficient of local resistance ζ. Its values considering the most popular pipe fittings are given in Table P6 of the Appendix.

3.5.2.1 Pipe Network Preliminary Calculation

The calculation algorithms employ laws discussed in Sub-par. **3.2.3,** namely:

- Law of balance between intake and discharge fluxes at knot points (Eq. (3.9) and Eq. (3.10));
- Law of the gas-dynamic head in a pipe contour (Eq. (2.15) and Eq. (2.16));
- Darcy's law (Eq. (2.18)).

The algorithms comprise two stages:

- Preliminary calculation of pipe diameters.

The designer situates the gas devices in the building scheme during this stage. He specifies the gas pipe route and measures the length of all pipe sections. Then, he makes a prognosis of the expected flow rates in the respective sections, calculating approximately pipe diameters (using Tables and nomograms: P3 and P4, P8 and P8A, P25 and P26, for instance);

- Final pipe calculation.

The designer uses an interactive technique to calculate pressure gas-dynamic losses in each section, putting the respected values in Eq. (3.12) and Eq. (3.1). Then, he checks whether the accuracy is satisfactory via Eq. (2.16) (±5% of the pressure drop in each gas contour, for instance).

We illustrate here three **basic techniques** of pipe diameter precalculation, i.e.:

- The use of **admissible gas-dynamic losses** in pipe networks-Δp_{GL}^{Acc};
- The use of **the available pressure drop in the pipe network –** Δh_{av};
- The use of **limit gas velocity in pipe network sections.**

3.5.2.1.1 Pipe Network Preliminary Calculation Considering Gas – Dynamic Pressure Losses

The **first technique (using admissible gas-dynamic losses** in pipes – Δp_{GL}^{Acc}) is based on the popular expression resulting from Darcy's law (Eq. (2.18)) and linking the basic design parameters of the pipe networks:

$$\dot{V} = \frac{\pi d_p^2}{4}\sqrt{\frac{2\Delta p_{GL}^{Acc}}{L_{Cal}\lambda\rho_{NG}}}, \quad m^3/s$$

(Here $\dot{V}$, m³/s is the prognostic gas flow rate, L_{Cal}, m – calculated length of the gas section, $\mathbf{d}_p$, m – its internal diameter; λ- coefficient of linear resistance, ρ_{NG}, kg/m³ – gas density; Δp_{GL}^{Acc}, Pa – admissible gas-dynamic losses in the pipe network).

The value of Δp_{GL}^{Acc} is specified conforming to the method of engineering design. The value of the typical admissible gas-dynamic losses in a pipe network is Δp_{GL}^{Acc} =100Pa (0.4" water gauge). The aim is to decrease the operational pressure in the pipe network and thus, the risk of internal gas leakage[50].

Rewrite it with respect to the diameter:

$$d_p = \left(0.81 * \dot{V}^2\right)^{0.2} \left(\frac{L_p \rho_{NG} \lambda}{\Delta p_{GL}}\right)^{0.2}, \text{ m} \tag{3.14}$$

Denote the second term by **K** [17] called in what follows the "pipe characteristic" of a certain pipe section. Then, we obtain the following formula used by designers [18] to calculate the pipework:

$$d_p = 0.959 * \left(\dot{V}\right)^{0.5} * \mathbf{K}, \text{ m} \tag{3.15}$$

Equation (3.15) is used if its two members are parametrically tabulated (see Tables P3 and P4). We use the following expression of the coefficient λ to calculate the pipe characteristic **K** of a certain pipework section:

$$\lambda = 1 + 3.6/d_p + 0.03 d_p. \tag{3.16}$$

The algorithm is as follows[51]:

1. Find the pipe characteristic **K** from Table P3 for each device, considering the prognostic flow rate and the calculation length of a certain pipeline section;
2. Having found the characteristics **K**, select from Table P4 the corresponding pipe diameter $\mathbf{d}_p$;
3. Use all determined values of **K** to find the respective diameters $\mathbf{d}_p$ of the common pipeline sections of the gas contours.

Then, select pipe nominal diameters using Table P4.

The following examples illustrate the approach – see Examples 3.5–3.9.

[50] The pipe network itself undergoes density test. A gaseous fluid (air or inert gas) is compressed applying pressure of up to 20.3 kPa (6" Hg) for 10 min. Avoid compression decrease. Do not use water in the test.

[51] Pipeworks of residential and public buildings should be precalculated.

Example 3.5

Precalculate the pipe network of a one-storey residential building whose horizontal scheme is shown in Figure 3.27, **knowing the power and the admissible gas-dynamic pressure losses** in a pipe network supplying:

- A gas cooker with power 10kW;
- A gas tankless water heater with power 6kW.

Gas kinematic viscosity and density are 14.3×10^{-6} m²/s and 0.73 kg/m³, respectively.

SOLUTION

1. Determination of the prognostic flow rate:

The specific gravity and Wobbe number are given in Table 2.1 – SG=0.58 and 50.68 MJ/m³, respectively. We find directly the prognostic flow rate using Eq. (2.5)[52] introducing the initial data:

$$\dot{V} = \frac{Q}{\eta_{GD} N_W \sqrt{SG}} = \frac{10 * 10^3}{0.96 * 50.68 * 10^6 \sqrt{0.58}} = 0.27 * 10^{-3}, m^3/s$$

The calculated flow rates of the cooker and the water heater amount to 0.27×10^{-3} m³/s and 0.16×10^{-3} m³/s. The coefficient of simultaneousness is 1.0 (Table 3.5).

2. Precalculation of pipeline diameters.

The selection of the preliminary diameter $\mathbf{d}_p$ of the gas pipelines is made using the systematized data for **K** in Table 3.6, assuming admissible pressure drop of 100 Pa and knowing the sectional flow rate $\dot{V}$, m³/s and the length of the gas pipelines in **m**. These quantities are specified in Table P3.

The diameter of the intake gas line **0-1** is found by summation of the values of **K** of the two supplied devices, since the entire gas amount passes through section **0-1**.

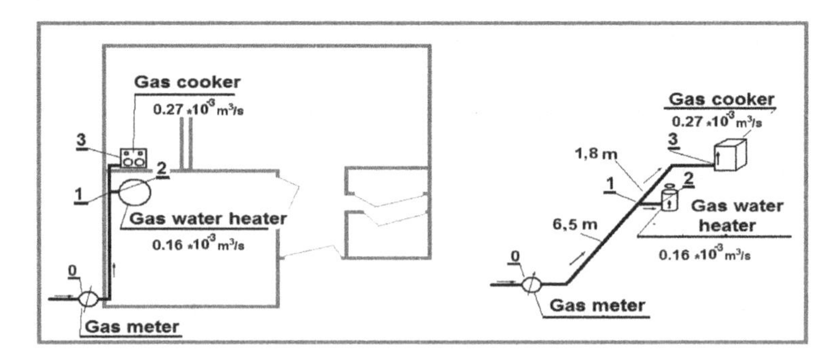

FIGURE 3.27 Gas installation in a one-story residential building: plan and axonometric scheme.

[52] The efficiency coefficient η_{GD} of the fuel device is assumed to be 96%.

TABLE 3.6
Selection of preliminary diameter of the pipeline in Example 3.4

№	Flow rate of a section, m^3/s	Length of gas pipe - L_{cal}, m	Pipe characteristic K, $m/(m^3/s)^{0.5}$	d_p
1-3	$1.0 \cdot 0.27 \cdot 10^{-3}$	1.8+6.5=8.3	714	½"
1-2	$1.0 \cdot 0.43 \cdot 10^{-3}$	0.5+6.5=7.0	1122	½"
0-1			714+1122=1836	¾"

Example 3.6

Calculate the pipe network of a one-storey residential building whose horizontal scheme is shown in Figure 3.28, knowing flow rates and **admissible gas-dynamic pressure losses** in a pipe network supplying:

- A gas cooker with power 20 kW;
- A gas tankless water heater with power 15 kW.

Gas kinematic viscosity and density are 14.3×10^{-6} m^2/s and 0.73 kg/m^3.

SOLUTION

1. Calculation of the prognostic flow rates.

Using Eq. (3.5) we calculate the device flow rates amounting to 0.54×10^{-3} m^3/s and 0.4 m^3/s, respectively. The coefficient of simultaneousness is 1.0 (Table 3.5).

2. Calculation of the initial diameters.

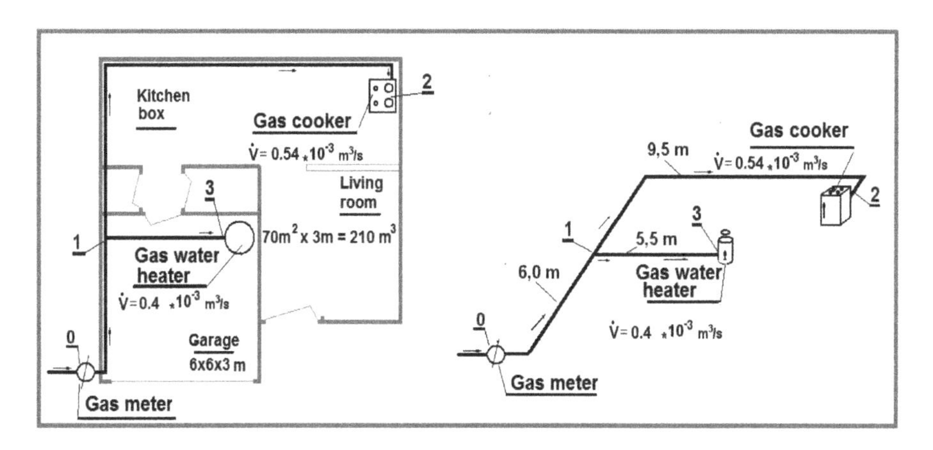

FIGURE 3.28 Plan and axonometric scheme of the pipe network of a one-family house.

TABLE 3.7
Preliminary values of pipe diameter in Example 3.5

Section No	Gas flow rate, m³/s	Gas pipe length - L_{Cal},m	Pipe characteristic K	d_p
0-1	$1.0{*}0.54{*}10^{-3}$	9.5+6.0=15.5	2030	¾"
0-3	$1.0{*}0.4{*}10^{-3}$	6.0+5.5=11.5	1210	¾"
0-1			2030+1210=3240	1"

The preliminary values of pipe diameter $\mathbf{d_p}$ are specified in Table 3.7 depending on the characteristic **K** and knowing the pipe section length L_{Cal},m and the gas flow rates in sections **0-1** and **0-3** – $\dot{V}$ m³/s, given in Figure 3.28.

The diameter of the intake gas section **0-1** is found by summation of the values of **K** of the supplied devices, since the entire gas amount passes through that section.

Example 3.7

Precalculate the gas installation of a one-storey residential building whose horizontal scheme is shown in Figure 3.29, knowing gas appliance powers and **admissible gas-dynamic pressure losses** in a pipe network that supplies:

- A gas cooker with power 45 kW;
- A gas refrigerator with power 18 kW;
- A gas tankless water heater with power 38 kW.

Pressure in the pipework is 0.5 kPa, while gas kinematic viscosity and density are 14.3×10^{-6} m²/s and 0.73 kg/m³, respectively.

SOLUTION

1. Prognostic flow rates:

Flow rates of the cooker, refrigerator and water heater amount to 1.21×10^{-3} m³/s, 0.47 m³/s and 1.03 m³/s, and the coefficient of simultaneousness is 1.0 (Table 3.5).

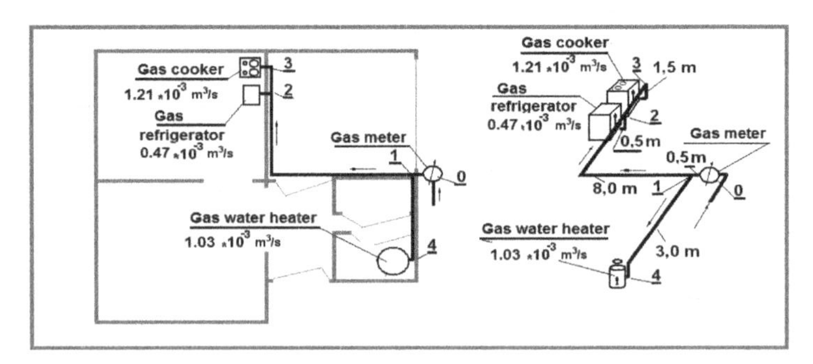

FIGURE 3.29 Pipe network of a one-family house – scheme of device location and axonometric scheme.

TABLE 3.8
Prognostic gas flow rates, K values and pipe diameter for Example 3.6

№	Flow rate of a section m^3/s	Length of the gas pipeline, - L_{Cal},m	Pipe characteristics K	d_p
3-0	$1.0*1.21*10^{-3}$	1.5+8+0.5=10.0	3630	1"
2-0	$1.0*0.47*10^{-3}$	0.5+8.0+0.5=9.0	1344	¾"
4-0	$1.0*1.03*10^{-3}$	3.0+0.5=3.5	1910	¾"
2-1			3630+1344=4974	1"
0-1			3630+1344+1910=6884	1 ¼"

2. Initial values of pipe diameters.

Table 3.8 specifies the values of **K** and d_{TP}, considering the prognostic flow rates and the length of lines **2-0**, **3-0** and **4-0.** The diameters of the common pipelines **2-1** and **0-1** are found by summation of the pipe characteristic **K** of the connected devices. For instance, we summate the values of **K** of the first and second devices, regarding pipeline **2-1.**

Yet, for pipeline **0-1**, we should summate the values of **K** of the three devices, since the entire gas amount passes through it despite its small length.

Example 3.8

Calculate the pipe network of a one-family house whose horizontal scheme is shown in Figure 3.30, knowing flow rates and **admissible gas-dynamic pressure losses** in the pipes supplying:

- A gas convective hearer with power 3 kW;
- A gas infrared radiator with power 1.5 kW;
- A gas fireplace with power 5 kW;
- A gas cooker with power 12 kW;
- A gas tankless water heater 8 kW.

The kinematic viscosity and gas density are 14.3×10^{-6} m^2/s and 0.73 kg/m^3, respectively.

SOLUTION

1. Prognostic flow rates:

They are calculated using Eq. (2.5), and the results are given in Table 3.9.

To precalculate pipe diameters, we use the coefficient of simultaneousness in Table 3.9. Its value is 1.0.

Calculation of the preliminary values of the pipeline diameter. The preliminary diameters of the gas pipes d_p are specified in the first 9 rows (column 5) of Table 3.10, depending on gas flow rates $\dot{V}$, m^3/s of the sections and considering the pipeline lengths.

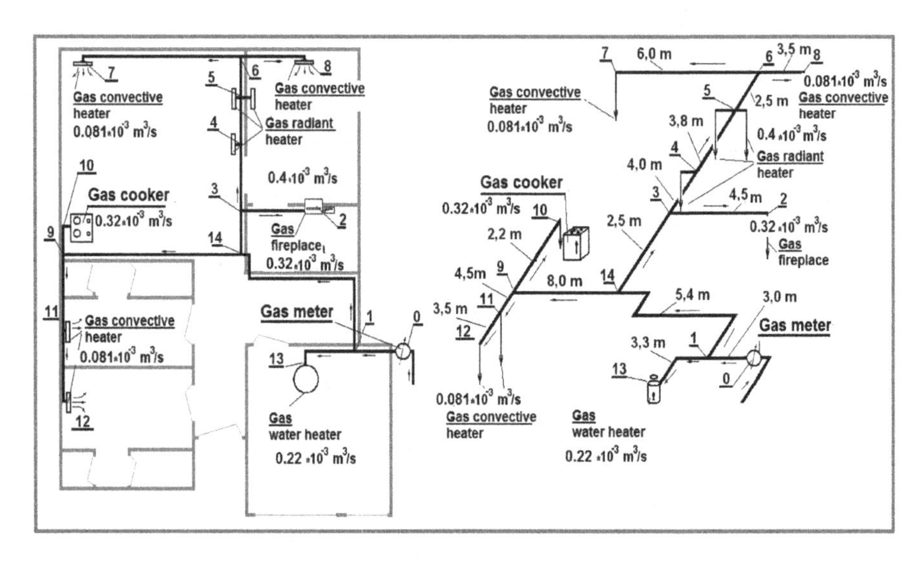

FIGURE 3.30 Sketch and axonometric scheme of the pipe network of a one-family house.

Table 3.10 also provides the pipe characteristics K (see Table P3) and pipe diameters $\mathbf{d}_p$ (see Table P4).

On the other hand, the diameter $\mathbf{d}_p$ of the common pipeline segments: 6-5, 5-4, 4-3, 3-14, 11-9, 9-14, 14-1 and 1-0 are found using data in Table P4 and summating the values of **K** of pipes supplying the devices.

Example 3.9

Choose a gas cogenerator (with a fuel cell) and calculate the pipe network of an EC supplying a residential building whose horizontal scheme is shown in Figure 3.31. The power of the fuel cell is 50 kW_{th} (the coefficient of "fuel gas – heat" transformation is 0.36). A gas convertor with a power of 3 kW also participates in the scheme. It provides thermal comfort to the controller's room. Pipe network pressure is 0.7 kPa, while Wobbe number and the specific gravity are 50.68×10^{-6} MJ/m^3 and 0.58 kg/m^3, respectively.

TABLE 3.9
Prognostic gas flow rate for Example 3.7

Gas device	Power, kW	Gas flow rate, m^3/s
Convective heater	3	0.081∗10^{-3}
Infrared radiator	1.5	0.4∗10^{-3}
Fireplace with power (η_{ry} = 40 %)	5	0.32∗10^{-3}
Cooker with power	12	0.32∗10^{-3}
Instantaneous water heater	8	0.22∗10^{-3}

TABLE 3.10
Preliminary values of pipe diameter in Example 3.7

№	Gas flow rate of a pipe section $\dot{V}$, m³/s	Gas pipe length - L_{cal}, m	Pipe characteristic K, $m/\left(m^3/s\right)^{0.5}$	d_p, m
8-0	$1.0 \cdot 0.081 \cdot 10^{-3}$	3.5+2.5+3.8+4+2.5+5.4+3=24.7	354	½"
7-0	$1.0 \cdot 0.081 \cdot 10^{-3}$	6.0+2.5+3.8+4+2.5+5.4+3=27.2	410	½"
5-0	$2.0 \cdot 1.0 \cdot 0.4 = 0.8 \cdot 10^{-3}$	0.5+3.8+4+2.5+5.4+3=19.2	3300	¾"
4-0	$1.0 \cdot 0.4 \cdot 10^{-3}$	0.5+4+2.5+5.4+3=15.4	1450	¾"
2-0	$1.0 \cdot 0.32 \cdot 10^{-3}$	4.5+2.5+5.4+3=15.4	1160	¾"
10-0	$1.0 \cdot 0.32 \cdot 10^{-3}$	2.2+8.0+5.4+3=18.6	1280	¾"
12-0	$1.0 \cdot 0.081 \cdot 10^{-3}$	3.5+4.5+8.0+5.4+3=24.4	375	½"
11-0	$1.0 \cdot 0.081 \cdot 10^{-3}$	4.5+8.0+5.4+3=20.4	320	½"
13-0	$1.0 \cdot 0.22 \cdot 10^{-3}$	3.3+3=6.3	552	½"
6-5			354+354=808	½"
5-4			808+3300=4108	1"
4-3			4108+1450=5558	1 ¼"
3-14			5558+1160=6718	1 ¼"
11-9			375+320=695	½"
9-14			695+1280=1975	¾"
14-1			6718+1975=8693	1 ¼"
1 – 0			8693+552=9245	1 ¼"

SOLUTION

1. Determination of the prognostic flow rate:

Using Eq. (3.5) we calculate the fuel gas flow rate considering the fuel cell and the convective heater in the controller's room:

$$\dot{V} = \frac{50 * 10^3}{0.36 * 50.68 * 10^6 \sqrt{0.58}} = 3.68 * 10^{-3},\ m^3/s$$

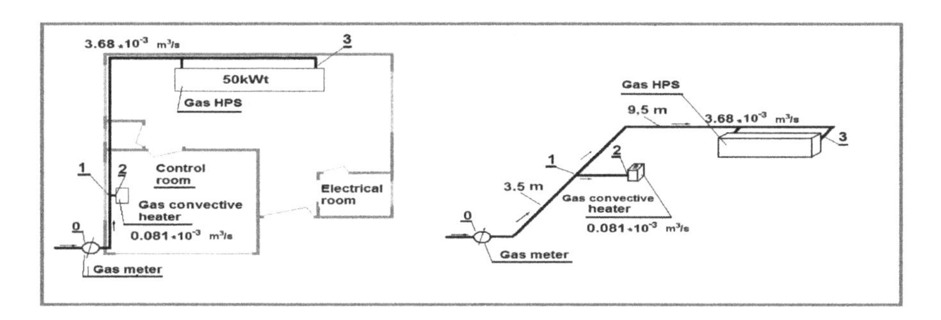

FIGURE 3.31 Scheme of the gas HPS and axonometric scheme of the gas supply.

TABLE 3.11
Preliminary calculation of pipeline diameters in Example 3.8

№	Flow rate of the section, m^3/s	Length of the gas line, - L_{cal}, m	Pipe characteristic K, $m/(m^3/s)^{0.5}$	d_p, m
0-3	1.0*3.68*10⁻³	3.5+9.5=13.0	12350	1 ½”
0-2	1.0*0.081*10⁻³	0.5+3.5=4.0	1344	¾”
0-1			12350+1344=13694	1 ½”

We find 3.68×10^{-3} and 0.081×10^{-3} m^3/s for the fuel cell and the gas convective heater, respectively.

2. Initial values of pipe diameters.

The calculated prognostic flow rates are given in the axonometric scheme (Figure 3.31) and Table 3.11.

Pipe preliminary diameters are also specified in Table 3.11 as dependent on **K**, considering the length of the gas pipeline and the prognostic flow rate $\dot{V}$, m^3/s.

3.5.2.1.2 Pipe Network Precalculation with Respect to the Admissible Pressure Drop in a Pipe Network

The second approach[53] of pipe network precalculation is employed when the investor's technical task specifies the exact values of the operational pressure at the gas intake and after the measuring instruments (when it exceeds 0.5 kPa). This method is a top priority in the calculation of gas installations in industrial buildings.

The admissible pressure drop in the pipework Δh_{av} is found applying the law of gas dynamic in the pipe contour described in Sub-par. 2.3.5 (an analog to Kirchhoff's second law concerning an open pipe contours).

In case of a unilateral open pipe contour (Figure 2.5,b) with two gas-dynamic resistances (Δp_{AF} – of the pipe network and $\Delta p_{GD.}$ – of the device), the algebraic expression of the law has a form similar to that of Eq. (2.17):

$$\Delta h_{av} = \left(p_{AF} - p_a \right) = \Delta p_{NetW} + \Delta p_{GD}.$$

Hence, the **admissible pressure drop in** a pipe network Δh_{av} is equal to or larger than the sum of the gas-dynamic losses there.

The admissible pressure drop is estimated via the expression:

$$\Delta h_{av} \geq p_{AF} - \Delta p_{GD} - 1.0_* 10^5$$

[53] The so-called differential method “DVGW-TRGI” is similar [25].

Here

- p_a, Pa – atmospheric pressure (p_a=1.0×10^5 Pa, because the pipe contours are open after the gas appliance);
- p_{AF}, Pa – absolute pressure after the measuring instruments;
- p_{NetW}, Pa – gas-dynamic pressure drop in the pipe network after the measuring instruments.

Then, the admissible pressure drop Δh_{av}, is equal to *the difference* between **the gas gauge pressure** after the measuring instruments – p_{AF} (2.0 kPa, for instance) and the **internal gas-dynamic losses** in the gas device – Δp_{GD}, specified by the manufacturer (gauge pressure of 5 kPa, for instance) and the atmospheric pressure.

$$\Delta p_{NetW} \leq \Delta h_{av} = p_{AF} - \Delta p_{GD} - p_a \quad (3.17)$$

Considering the specified conditions, **the admissible pressure drop in the pipe network** Δp_{NetW} should not exceed 1.5 kPa[54].

To precalculate the pipes considering the pressure drop, one should **initially** measure the length of the gas sections of pipelines L_M, stretching from the gas-regulating station to the remotest device. The results are inserted in column 3 of a table similar to Table 3.12 in Example 3.10.

The second step of design is a specification of the type and number of local resistances and the coefficients reducing the local resistances b_{eqv} to the pipe network linear length (see Table 2.4).

Then, one should calculate the length L_{Cal} of the pipeline sections using formula (2.22). The results are written in column 6 of a table similar to Table 3.12 of Example 3.10. Next, find the sum $\Xi\, L_{Cal}$ of the calculated lengths L_{Cal} of the gas pipeline sections that supply the remotest device (see the bottom of Table 3.12 of Example 3.10).

The fourth step is the calculation of the **specific** admissible gas-dynamic **pressure drop** $\Delta \bar{p}_{NetW}^{av} = \dfrac{\Delta h_{av}}{\Xi L_{Cal}}$ in the gas pipe supplying the remotest device. It is a ratio between the maximal admissible pressure dropΔh_{av} and the total length of the gas pipeline sectors supplying the remotest device – $\Xi\, L_{Cal}$.

Having calculated the **specific** admissible gas-dynamic **pressure drop** $\Delta \bar{p}_{NetW}^{av}$, we calculate the **admissible pressure drop** Δp_{NetW}^{j} in each pipeline section via the following product:

$$\Delta p_{NetW}^{j} \leq \Delta \bar{p}_{NetW}^{av} \times L_{Cal}^{j}, \text{ Pa} \quad (3.18)$$

The results are arranged in a table similar to Table 3.12.

---

[54] The admissible gas-dynamic pressure drop in a pipework amounts to about 350 Pa [14]. Its distribution between the respective sections is as follows:

- In the courtyard network - 100Pa;
- In the home flow meter -100Pa;
- After the flow meter -100Pa;
- In the reserve -50Pa.

TABLE 3.12
Prognostic flow rates and geometrics characteristics in Example 3.9

Section №	Flow rate, m^3/s	The measured length of the gas pipeline section L_{M}, m	Coefficient of local resistance b_{eqv}	Equivalent length of the gas pipeline section L_{eqv}, m	Calculated length of the gas pipeline section L_{Cal}, m
1	2	3	4	5	6
1 – 2	0.00083	3.5	1.2	4.2	7.7
2 – 3	0.00083	5.5	1.2	6.6	12.1
3 – 4	0.001058	3.0	0.2	0.6	3.6
4 – 5	0.001426	3.0	0.2	0.6	3.6
5 – 6	0.001795	3.0	0.2	0.7	3.7
6 – 7	0.001795	6.0	0.25	1.5	7.5
7 – 8	0.00286	6.0	0.25	1.5	7.5
8 – 9	0.00375	6.0	0.25	1.5	7.5
9 – 10	0.00491	6.0	0.25	1.5	7.5
10 -11	0.00607	6.0	0.25	1.5	7.5
11 -12	0.00720	12.0	0.25	3.0	15,0
				$\Xi\ L_{Cal} =$	**83.2 m**

The sixth step is the determination of the pipe internal diameter $\mathbf{d}_{p,j}$. This is done by solving the non-linear algebraic equation:

$$\Delta p^{j}_{NetW} - 0.065\left(\frac{0.0001}{d_{p,j}} + 0.0001238\frac{d_{p,j}}{\dot{V}_j}\right)^{0.25}\frac{L^{j}_{Cal*}\dot{V}_j^2}{d^5_{p,j}} = 0 \tag{3.19}$$

where:

- $d_{p,j}$, m – the sought internal diameter of the j^{-th} section;
- Δp^{j}_{NetW}, Pa – pressure drop in the j^{-th} section-see Table 3.13;
- $\dot{V}_j$, m^3/s – the prognostic gas flow rate in the j^{-th} section – see Table 3.13.

Various methods are used to solve the above equation. We shall specify in the next Sub-par. 3.5.2.1.2, a numerical method, known as Code AAKC “Derive5”. The results are given in a table similar to Table 3.13, while the graphical solutions are plotted in Figure 3.33.

Example 3.10

Precalculate pipe diameters of the pipe network shown in Figure 3.32, **knowing the gas flow rate and the specific available gas-dynamic head**. The pressure at the pipe network intake is 2.0 kPa, the nominal gas pressure before

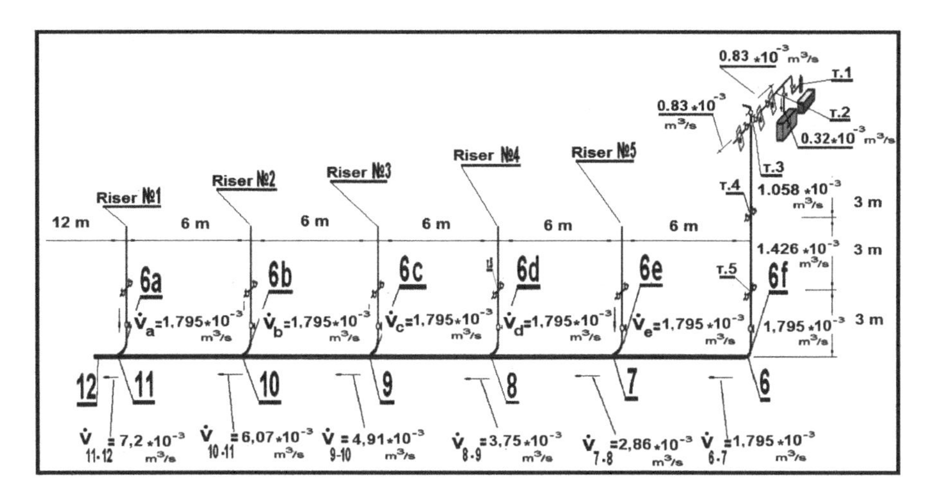

FIGURE 3.32 Determination of the real and equivalent length of the pipe supplying the remotest gas device.

the remotest gas device should not be lower than 0.5 kPa and gas density is 0.73 kg/m³.

SOLUTION

Draw a scheme of the section supplying the remotest gas device – see Figure 3.32. Insert in Table 3.12 the prognostic gas flow rate $\dot{V}$ of the sections supplying the remotest gas device, which is located at the end of section **1-2** (see Figure 3.32). Insert in the **3rd column** of the table the measured lengths of each individual section of the gas contour – L_M.

Column 6 of Table 3.12 is filled with the pipe lengths L_{Cal} calculated via Eq. (2.22) (the values of the intermediate local resistance reduced to the linear length of the pipeline sections are inserted in columns **4** and **5**).

TABLE 3.13
Precalculation of the diameters of pipes with admissible pressure drop

Section	**1 – 2**	**2 – 3**	**3 – 4**	**4 – 5**	**5 – 6**	**6 – 7**	**7 - 8**	**8 - 9**	**9 - 10**	**10 - 11**	**11 - 12**
Flow rate $*10^3$ m³/s	0.83	0.83	1.06	1.53	1.79	1.79	2.86	3.75	4.91	6.07	7.2
Admissible drop $-\Delta p^j_{NetW}$, Pa	139	218	65	65	67	135	135	135	135	135	270
Calculated diameter $*10^2$ m	**1.5**	**1.5**	**1.7**	**1.9**	**2.0**	**2.0**	**2.4**	**2.7**	**2.99**	**3.2**	**3.37**
Nominal diameter $*10^2$ m	1.6	1.6	2.0	2.0	2.0	2.0	2.5	3.2	3.2	3.2	3.9

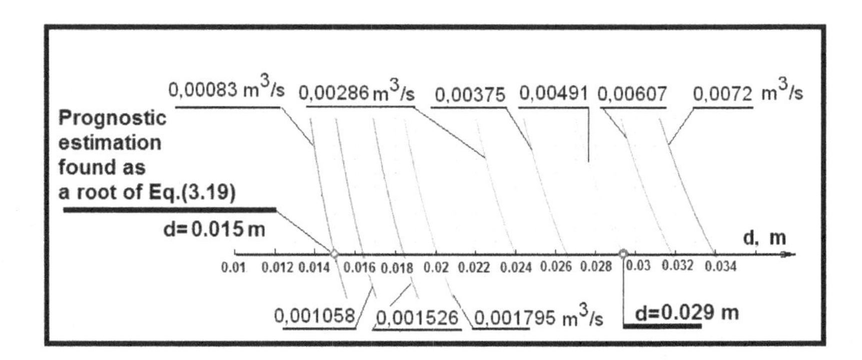

FIGURE 3.33 Graphical solution of Eq. (3.19) via AAKC DERIVE 5 code.

The calculated length of the **entire gas-distributing** branch, stretching from the building gas-regulating station to the remotest gas device, amounts to **83.2** m (see the bottom of Table 3.12). Hence, **the specific available pressure drop** amounts to 1500/83.2=**18.03** kPa/m, while the available pressure drop Δp^j_{NetW} of the respective section is the **product** 18.03 × $\boldsymbol{L_{Cal}}$.

Consider for instance section **11-12** with $\boldsymbol{L_{Cal}}$=15.0 m. Then, the available pressure drop is **18.03**×15.0=270.45 Pa. The remaining values are given in Table 3.13 (see the third row).

The next step is to precalculate the diameter $\mathbf{d}_{p,j}$ of the j-th pipeline section by finding the real roots of Eq. (3.19). We use the AAKC "DERIVE 5" Code for that purpose.

Figure 3.33 illustrates the graphical solution of Eq. (3.19), where the graph of functional (Eq. (3.19)) crosses the "zero" coordinate (axis d). The parameter used here is "pipe permeability" assessed by the gas flow rate $\dot{V}_j$, m³/s.

We assume a prognostic gas flow rate amounting to $\dot{V}_j$=0.0008 m³/s. Then, we find the value $\mathbf{d}_p$=0.015 m via extrapolation of the graphically found roots of Eq. (3.19).

Pipe diameter $\mathbf{d}_p$ should increase in order to increase the gas flow rate $\dot{V}_j$, keeping the specific gas-dynamic drop of pressure. For $\dot{V}_j\uparrow$ ($\dot{V}_j$=0.0049 m³/s, for instance), pipe diameter should increase to $\mathbf{d}_p$=0.029 m (see Figure 3.33)[55].

The real roots of Eq. (3.19) are given in row 4 of Table 3.13. The fifth row specifies the internal pipe diameters belonging to the standard series, whose values are the closest ones to the values of the precalculated diameters.

To perform expedient calculations, we recommend the use of Table P8. It is designed on the basis of the roots of Eq. (3.19). As seen, the preliminary diameter $\mathbf{d}_{p,j}$ is found knowing the **admissible pressure drop** of a pipeline section $\overline{\Delta p}^j_{NetW}$ (the values in the second row of the table) and the prognostic gas flow rate $\dot{V}_j$ of the j-th pipeline section. Similarly, we can use Table P8A employing an available specific gas-dynamic head within ranges 50–300 Pa/m as a parameter.

For instance, if the calculated available pressure drop Δp^j_{NetW} is 7.5 kPa, while the prognostic gas flow rate $\dot{V}_j$, m³/s is 0.0008 m³/s, we choose a diameter $\mathbf{d}_{p,j}$=0.016 m from Table P8.

[55] The closest pipe of the standard series is that with diameter d=1 ¼ "(d=0.032)".

3.5.2.1.3 Precalculation of Gas Pipes with Respect to Gas Limit Velocity in a Pipe Network

The third approach to the precalculation of gas pipes involves the value of the **limit velocity** $\mathbf{w}_{NG}$, m/s[56], based on the law of gas flow continuity discussed in Par. 2. (Sub-par. 2.2.2) and presented by Eq. (2.8). An elementary algebraic transformation yields the expression:

$$A = \frac{\dot{V}}{w_{NG}}, \tag{3.20}$$

where:

- A, m^2 – area of the cross-section of the gas pipe ($A=\pi d_p^2/4$);
- $\mathbf{W}_{NG}$, m/s – limit velocity averaged throughout the cross-section[57].

Redoing Eq. (3.20) with respect to the pipe diameter[58], we find a practically applicable formula, if we have precalculated the diameter of the gas pipe. It reads:

$$d_p = 1.128\sqrt{\frac{\dot{V}}{w_{NG}}} \tag{3.21}$$

We adopt a limit velocity $\mathbf{w}_{NG}$ averaged throughout the cross-section, avoiding the emergence of noise in the gas pipes.

Example 3.11

Precalculate the diameters of pipes shown in Figure 3.32, knowing flow rates and limit velocity of gas flow $\mathbf{w}_{NG}$=3.0 m/s in the pipe network.

SOLUTION

The precalculation of section **11-12,** whose prognostic flow rate is 7.2×10³ m³/s, will illustrate the respective algorithm.

Using Equation (3.21) we find the diameter – $d_{11-12} = 1.128\sqrt{\frac{7.2_* 10^{-3}}{3}} = 0.055$ m.

Considering the standard series of nominal diameters, we choose d_{11-12}=60 mm.

The calculated preliminary diameters are given in Table 3.14, considering all remaining sections of the scheme shown in Figure 3.31. However, data in Table 3.13 show differences owing to the adoption of different estimation criteria. The selection of an appropriate limit velocity of the gas flow $\mathbf{w}_{NG}$ would yield a total agreement between the two criteria.

---

[56] The velocity should not exceed 7 m/s [14].

[57] Gas flowing velocity should not exceed ≈6 m/s.

[58] Internal pipe diameter.

TABLE 3.14
Preliminary calculation of pipe diameters using a limit velocity

Section	1 – 2	2 - 3	3 – 4	4 – 5	5 – 6	6 – 7	7 – 8	8 - 9	9 - 10	10 -11	11 -12
Flow rate $*10^3 m^3/s$	0.83	0.83	1.058	1.526	1.795	1.795	2.86	3.75	4.91	6.07	7.2
Diameter $*10^2$ m	1.9	1.9	2.1	2.2	2.7	2.7	3.48	0.4	4.56	0.51	0.55
Internal diameter, m	0.020	0.020	0.025	0.025	0.025	0.025	0.039	0.04	0.05	0.05	0.06

3.5.2.2 Final Pipe Network Calculation by the Estimation of the Gas-Dynamic Pressure Losses in Pipes Supplying the Remotest End Users and Knowing Pipe Network Geometry and Topology

We finalize the calculations by finding the total gas-dynamic pressure loss in all contours, including those supplying the remotest devices. Consider Eq. (3.17)). Then, the following condition should be obeyed:

$$\Delta p^{av}_{NetW} - \Delta p_{GD} \geq \Delta p_{GL}, \quad (3.22)$$

The formula given below is more convenient for use:

$$\sum_{1}^{n} 0.81\rho \frac{\dot{V}_j^2}{d_{p,j}^4}\left(\frac{\lambda L_M^j}{d_{p,j}} + \Xi\varsigma_j\right) \leq \left(\Delta p^{av}_{NetW} - \Delta p_{GD}\right). \quad (3.23)$$

It is derived introducing Eq. (2.18) and Eq. (2.8) into inequation (Eq. (3.22)). Summation takes place along all pipeline segments following the route of fuel gas transfer – from the measuring instruments to the remotest end user.

Calculate all preliminary pipe diameters and design the topology and geometry of the pipe network. Then, perform a fine adjustment of pipe diameters.

The model algorithm consists in a cyclic check of the validity of inequality (Eq. (3.23)). The check should proceed until attaining a good engineering accuracy within the range /hav-Δ**p**NetW/≤ 5%.

Examples 3.12 and 3.13 illustrate the final gas-dynamic calculation of pipeworks. The objects discussed are pipe network components. The first example deals with the calculation of a vertical riser – see Figure 3.32. The second example illustrates the calculation of a horizontal main line supplying six risers identical with the riser in Example 3.10 – see Figure 3.32.

Example 3.12

Perform the final calculation of a pipe network assessing the gas-dynamic losses in a riser of a three-storey residential building (Figure 3.34). Pipe diameters are known in advance and specified in Table 3.15. All apartments have:

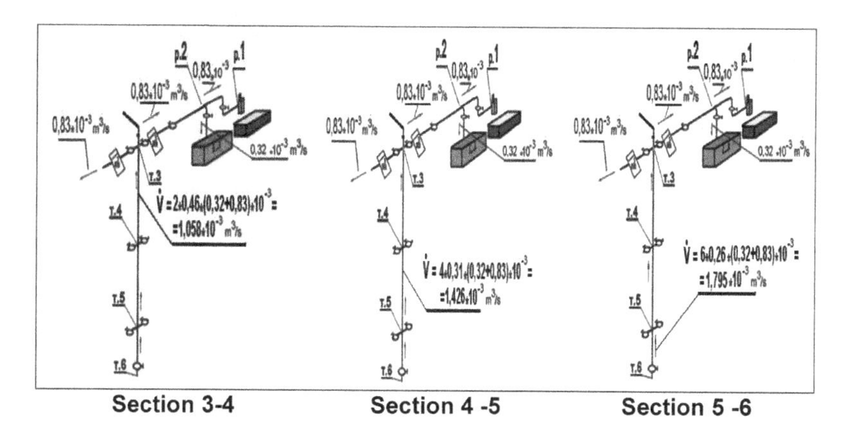

FIGURE 3.34 Assessment of the prognostic gas flow rate of natural gas in the three riser sections.

- A cooker with four burners;
- A gas tankless water heater.

The available pressure head is the difference between gas pressure at point 6 ($\mathbf{p}_{p.6}$=0.5 kPa) and the pressure drop in the device ($\Delta \mathbf{p}_{GD}$=0.3 kPa). The kinematic viscosity and gas density are 14.3×10^{-6} m^2/s and 0.73 kg/m^3, respectively.

SOLUTION

1. Assessment of the prognostic consumption of NG (by each apartment, by each storey and by the entire building).

To assess the prognostic gas flow rate through each apartment branch we need to account for the probability of a joint operation of all gas appliances.

1.1 An individual apartment

Table P7 specifies the following appliances:

- A gas cooker with four burners: 0.32×10^{-3} m^3/s;
- A gas tankless water heater in a bathroom: 0.83×10^{-3} m^3/s

Sum: 1.15×10^{-3} m^3/s.

TABLE 3.15
Prognostic flow rates through riser sections

Section No	Number of supplied identical devices	Coefficient of simultaneousness (Table P4)	Prognostic flow rates through the riser $-\dot{V}$, $m^3/s \cdot 10^3$ (see Fig. 2.24)
3 – 4	2	**0,46**	0,46.(2·0,32+2·0,83) = 1,058
4 – 5	4	**0,31**	0,31·(4.0,32+4·0,83) = 1,426
5 – 6	6	**0,26**	0,26·(6.0,32+6·0,83) = 1,795

Then, we assess the prognostic gas flow rate through the gas devices using the coefficient of simultaneousness k_0 specified in Table P7 (here k_0=0.72):

$$\dot{V}_{ap} = 0.72 \times (0.32 + 0.83) \times 10^{-3} = 0.72 \times 1.15 \times 10^{-3} = 0.83 \times 10^{-3} m^3/s. \quad (3.24)$$

1.2 Riser sections

The prognostic gas flow rate through the respective riser sections is found using formula (2.10), modified in the form:

$$\dot{V}_{i-j} = k_0 \times \sum (n_1 \times V_{GD1} + \ldots + n_m \times V_{GDm}),\ m^3/s. \quad (3.25)$$

Section 3-4: $n_{3\text{-}4}$=2, a k_0=0.46 (see Figure 3.34):

$$\dot{V}_{3-4} = 0.46 \times (2 \times 0.32 + 2 \times 0.83) \times 10^{-3} = 1.058 \times 10^{-3},\ m^3/s.$$

The results of the remaining riser sections are specified in Table 3.15 and illustrated in Figure 3.35.

2. Assessment of pressure losses in pipes connecting the remotest end users.

2.1 Available gas-dynamic pressure drop.

The available gas-dynamic pressure drop Δh_{av} is equal to the difference Δh_{av}=$p_{p.6} - \Delta p_{GD}$=0.5-0.3=0.2 kPa.

(It is seen in Example 3.10 that gas pressure at point 6 is $\mathbf{p}_{p.6}$=0.5 kPa, while $\Delta \mathbf{p}_{GD}$=0.3 kPa).

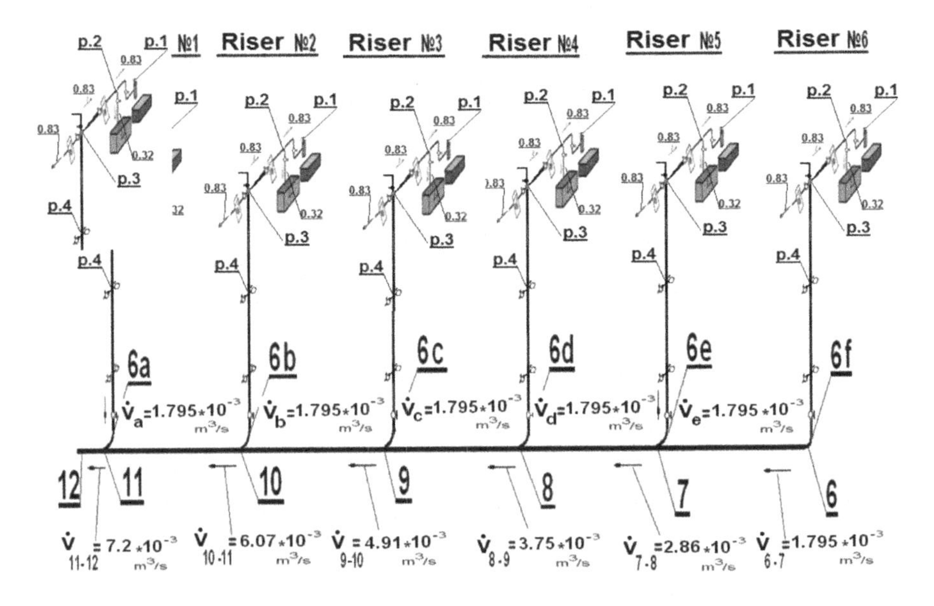

FIGURE 3.35 A gas distributing main, supplying six vertical risers of a pipe network.

TABLE 3.16
Final riser calculation

№	$\dot{V}\cdot 10^3$ m³/s	d_pm	L_Mm	Ξ ζ-	Re-	λ-	Δp Pa	List
1	2	3	4	5	6	7	8	9
1 – 2	0. 83	0.015	3.5	9.8	4926.9	0.0416	157.1	**2.2.2+4+1**
2 – 3	0.83	0.025	5.5	10.5	2956.1	0.04459	21.2	**2+2+5+1.5**
3 – 4	1.058	0.025	3.0	2	5078.9	0.0423	12.05	**2**
4 – 5	1.426	0.025	3.0	2	5078.9	0.0395	21.0	**2**
5 – 6	1.795	0.025	3.0	2	6393.1	0.03826	19.4	**2**
							Ξ=245.5 Pa	

2.2 Final pipe network calculation assessing the gas-dynamic losses.
Pressure losses are calculated using Eq. (3.12). They concern the individual risers systematized in Table 3.16.

Consider, for instance, section **3-4** with flow rate 0.001058 m³/s, pipe diameter 0.025 m and Re=5078.9. Then, we find the following values of the friction factor and pressure drop: λ=0.0423 and 12.05 Pa.

The total pressure losses in the riser are a sum of gas-dynamic pressure losses of the respective sections given in the 8th column. They amount to 245.55 Pa. The assessed gas-dynamic losses exceed by 22.8% the available gas-dynamic head of 0.2 kPa.

Hence, we should **correct the pipe diameter** of sections **4-5** and **5-6** as well as that at the beginning of the riser line. The total loss in those sections amounts to 53.2 Pa (21.0+32.2) Pa. Then, we repeat the gas-dynamic calculations involving the new diameters of the two riser segments – they are equal to ø **1 ¼** (or 0.03187 m).

The results are given in Table 3.17.

Gas-dynamic losses fluctuate around a value of 14.4 Pa, while their total sum in the riser is 231 Pa. Losses exceed the available pressure by 31 Pa (15.5%), which is an unsatisfactory result.

Final riser calculation

№	$\dot{V}$.10³ m³/s	d_pm	L_M m	Ξ ζ -	Re -	λ -	Δp Pa	List
1	2	3	4	5	6	7	8	9
1 – 2	0.83	0.015	3.5	9.8	4926.9	0.0416	157.1	**2.2.2+4+1**
2 – 3	0.83	0.025	5.5	10.5	2956.1	0.04459	21.2	**2+2+5+1.5**
3 – 4	1.058	0.032	3.0	2	2942.6	0.0561	9.3	**2**
4 – 5	1.426	0.032	3.0	2	3966.0	0.0505	15.4	**2**
5 – 6	1.795	0.04	3.0	2	3993.9	0.0435	19.4	**2**
							Ξ=222.4Pa	

Hence, new correction of the riser diameter should be done. Additional corrections, for instance in section **1-2**, could reduce 3.93 times the gas-dynamic losses in the section, but the emerging reserve of pressure is inexpedient, since it will be consumed in device adjustment.

TABLE 3.17
Correction the pipe diameters

№	$\dot{V}\cdot10^3$ m³/s	D_pm	L_Mm	$\Xi\zeta$ -	Re -	λ -	Δp Pa	List
4 – 5	1.426	0.032	3.0	2	3966.0	0.0505	15.4	**2**
5 – 6	1.795	0.032	3.0	2	4992.3	0.0472	23.4	**2**

Example 3.13

Perform final calculation of a pipe network by assessing the gas-dynamic losses in a **horizontal main gas line** supplying six risers of the pipe network of a three-storey residential building (Figure 3.35). Pipe diameters are known in advance and specified in Table 3.16. As in Example 3.12, all apartments have:

- A gas cooker with four burners;
- A gas tankless water heater.

The available gas-dynamic pressure drop is formed by gas pressure at point 12 (see Figure 3.35), which is $\mathbf{p}_{p.12}$=1.6 kPa, and $\Delta\mathbf{p}_{GD}$=0.3 kPa. Gas kinematic viscosity and density are 14.3×10⁻⁶ m²/s and 0.73 kg/m³, respectively.

SOLUTION

The 2D scheme of the pipe network is shown in Figure 3.35.

As in Example 3.12, we solve the problem by: (i) finding the prognostic flow rates in the gas line sections and (ii) performing gas-dynamic calculations.

1. Assessment of the necessary NG passing through the pipeline and supplying the risers.

The prognostic flow rate through the "i-j"-th section (i-j varies from 6-7 to 11-12) of the main pipe is calculated via the formula:

$$\dot{V}_{i-j} = k_0 \times \sum (n_1 V_{GD1} + \ldots + n_m V_{GDm})\ m^3/s. \tag{3.26}$$

where:

- $\dot{V}_{i-j}$, m³/s – prognostic flow rate through the i-j-th section;
- k_0 – coefficient of simultaneousness of the operation of the supplied devices, specified in Table P5;
- n_{i-j} – number of appliances in the apartments, supplied by the i-j-th pipe section;
- m – number of the types of supplied devices.

Consider, for instance, section **7-8** of the horizontal main pipe (see Figure 3.35), supplying two risers that connect 12 ps. gas cookers and 12ps. tankless water heaters. Assume that their coefficient of simultaneousness is k_0=0.207. Then, the

TABLE 3.18
Prognostic low rate through the horizontal pipeline

Section	Number of supplied risers	Coefficient of simultaneousness (Table P4)	Prognostic flow rate through the section, $m^3/s.10^{-3}$ (see Fig. 3.34)
6 – 7	1	0.26	0.26*(6*0.32+6*0.83) = 1.795
7 – 8	2	0.207	0.207*(12*0.32+12*0.83)=2.86
8 – 9	3	0.181	0.181*(18*0.32+18*0.83)=3.75
9 – 10	4	0.178	0.178*(24*0.32+24*0.83)=4.91
10 -11	5	0.176	0.176*(30*0.32+30*0.83)=6.07
11 -12	6	0.174	0.174*(36*0.32+36*0.83)= 7.2

prognostic flow rate will be: $\dot{V}_{7\text{-}8}$=0.207×(12×0.32+12×0.83)=2.86×10^{-3} m^3/s. The results are given in a table form (Table 3.18) and inserted in the pipeline scheme (Figure 3.35).

Having found the prognostic flow rate $\dot{V}$, m^3/s, knowing the diameters[59] and geometry of the pipe network (length and local resistance), we calculate the drop of the gas-dynamic pressure.

2. Calculation of pressure losses in risers supplying the remotest end users

2.1 Available gas-dynamic pressure drop.

The available pressure drop compensating the gas-dynamic head in the pipe network h_{av} is equal to the difference

$$p_{p12} - \Delta p_{GD} \quad \text{(Eq. (3.17)):}$$

(According to Example 3.11, gas pressure at point 12 is $p_{p.12}$=1.6 kPa, and $\Delta \mathbf{p}_{GD}$=0.3 kPa).

$$h_{av} = 1.6\text{-}0.3 = 1.3 \text{ kPa}$$

2.2 Final calculation of the pipe network via the estimation of the gas-dynamic losses.

The calculated total gas-dynamic losses in the six segments of the horizontal pipeline are found using formula (Eq. (3.12)) and inserted in Table 3.19. We add here the value of the total gas-dynamic losses in the vertical riser of Example 3.10. Note that riser (Item 6) in the present example is its replica. The sum of the total gas-dynamic losses of the segments of the horizontal pipeline (8th column of Table 3.19) amounts to **536.3** Pa.

The total gas-dynamic losses throughout the entire gas contour, stretching from the measuring instruments of the gas-regulating station to the remotest user supplied by riser – Item 6 (the water heating device at point 1), are:

$$245.55 + 536.3 = 781.85 \text{ Pa.}$$

[59] See Examples 2.2–2.9.

TABLE 3.19
Gas-dynamic calculation of the pipeline

№	$\dot{V} \cdot 10^3$ m³/s	dm	lm	Ξ ζ -	R -	Λ -	Δp, Pa	List
1	2	3	4	5	6	7	8	9
6 - 7	1.795	0.025	6	7.0	6393.1	0.038261	79.0	**1+2+4**
7 - 8	2.86	0.025	6	1.0	10186.22	0.03535832	117.59	**1.0**
8 - 9	3.75	0.04	6	1.0	8347.533	0.0353338	20.49	**1.0**
9 - 10	4.91	0.04	6	1.0	10929.7	0.03361564	33.69	**1.0**
10 -11	6.07	0.05	6	1.0	10809.5	0.0331926	17.39	**1.0**
11 -12	7.2	0.05	12	2.1	12821.8	0.032157	268.17	**1.1+1**

TABLE 3.20
Correction of the gas-dynamic calculations of a main gas line using reduced pipe diameters

№	$\dot{V} \cdot 10^3$, m³/s	d_pm	L_Mm	Ξ ζ -	Re -	λ -	Δp, Pa	List
1	2	3	4	5	6	7	8	9
1 – 2	0.83	0.015	3.5	9.8	4926.9	0.0416068	157.1	2.2.2+4+1
2 – 3	0.83	0.015	5.5	10.5	4928.9	0.0416068	207.5	2+2+5+1.5
3 – 4	1.058	0.025	3.0	2	5078.9	0.0423	12.9	2
4 – 5	1.426	0.025	3.0	2	5078.9	0.0395	21.0	2
5 – 6	1.795	0.025	3.0	2	6393.1	0.038261	32.2	2
6 – 7	1.795	0.025	6	7.0	6393.1	0.038261	79.0	1+2+4
7 – 8	2.86	0.025	6	1.0	10186.2	0.035358	117.6	1.0
8 – 9	3.75	0.025	6	1.0	13356.1	0.033966	195.0	1.0
9 - 10	4.91	0.025	6	1.0	17487.5	0.03278	324.0	1.0
10 -11	6.07	0.04	6	1.0	13511.9	0.032406	49.9	1.0
11 -12	7.2	0.05	12	2.1	12821.8	0.032157	48.2	1.1+1

Since the available gas-dynamic pressure drop is 1.3 kPa, there is a reserve in the pipe network amounting to 518.15 Pa. Hence, one can **reduce pipe diameters**, thus decreasing the initial investments in the project.

The corrected diameters are given in Table 3.20. The total hydraulic losses amount to **1243.5 Pa**, while the diameters of the pipe sections decrease in accordance with the values in Table 3.21.

The difference between the available gas-dynamic pressure drop (1.3 kPa) and the total gas pressure losses (1243 Pa) amounts to 56.5 Pa (4.3%).

It is admissible from a practical point of view, since it will be consumed to overcome the internal gas-dynamic losses and adjust the devices.

4 Main Building Systems Operating with Fuel Gas

As discussed in Sub-par. 1.1, a gas pipe network is one of the components of the logistic infrastructure of the building energy system (the second logistic component is the electrical circuit). Pipe networks are designed to distribute and supply fuel gas to end users (gas devices), as part of the main or auxiliary energy systems of a building. Fuel gas may feed systems for thermal and visual comfort, food preparation and storage, hygiene maintenance etc.

We discuss in what follows all systems in a building, which can be "charged" via energy carriers. We consider the theory of engineering installations, arrangement of the installation components within the frame of the building topology and the available equipment. Our considerations follow a particular succession starting with a system providing thermal comfort.

4.1 SYSTEM FOR THERMAL COMFORT CONTROL

Considering the energy carriers and their energy convergence, **the systems controlling the thermal comfort** (heating or air conditioning systems) present installations transferring energy from:

- Air to air;
- Water to air;
- Steam to air;
- Electricity to air;
- Gas to air.

The issue of comfortable environment in a particular building space yields the necessity of providing thermal comfort to the entire building via the establishment of thermal homogeneity. For now, the concept of providing thermal comfort throughout the building is dominating[1], bringing particular technical advantages. Yet, the solutions prove to be unprofitable from economical and energy point of view. Note that the realization of a thermal homogeneity within the entire building space is physically impossible. This is due to a number of effects such as:

- Effects of the inhomogeneous boundaries of the thermal area owing to the building envelope (mainly to vitreous areas and thermal bridges);

[1] Building spaces with approximately identical thermal characteristics and thermally controlled by a single sensor.

- Effects of 3D bodies (components of the technological equipment);
- Effects due to the occupation schedule, building exploitation and thermal lag of structural components (floor, massive external walls).

The most important effects are those due to the inhomogeneous boundaries of the thermal zones. However, they result from the incomplete insulation of components such as walls, thermal bridges and various vitreous areas. Together with structural and surface defects, the incomplete insulation yields also asymmetric radiative heat exchange between the building components and people working nearby, thus violating the thermal homogeneity. The availability of air flows – natural and forced, yields vertical temperature inhomogeneity (stratification) in the areas and a feeling of asymmetric thermal discomfort. Local sources of low radiative temperature (with surface temperature not higher than 500°C) such as furnaces, heat generators and thermal devices generate the same effect, i.e. radiation asymmetry. Office equipment, including computers, copying and fax machines, provide additional thermal loads and discomfort in the occupied space. These "boundary conditions" complicate the work of designers, forcing them to arrange intuitively the type, power and location of thermal appliances. Note that the choice of gas devices yields also additional issues, such as the choice of a system conveying the products of combustion and guaranteeing thermal as well as air comfort to the occupants.

4.1.1 Need for Thermal Comfort and Accessible Levels of Thermal Loads of Occupied Areas

There is a specific issue facing the designers, i.e. what is thermal comfort and how can one control it in an occupied area via the building engineering installations. As known, the stable vegetative activity and capacity for work of humans are due to appropriate dissimilating chemical processes and metabolism running in the human body. However, metabolism is the process of releasing heat per unit time – metabolic heat flux Q_M[W]. Human skin and lungs release the heat excess. Figure 4.1,a) illustrates the heat fluxes released by people at rest or work. Their intensity depends on a number of factors including the state of the environment and the individuals present. Conditionally, fluxes are[2]:

- Sensitive (Q_S);
- Latent (Q_L).

The temperature of the human body t_{hb} should be within limits $36.0 < t_{hb} \leq 37.2$°C to guarantee the normal running of the metabolism. There is a number of receptors

[2] Sensitive heat fluxes (via thermal conductivity, convection and radiation) emerge at a difference between the ambient temperature and that of the human body. Latent heat fluxes (via release of moisture by the skin and lungs) emerge as a result of the difference between the humidity of the ambient air and the body water percentage.

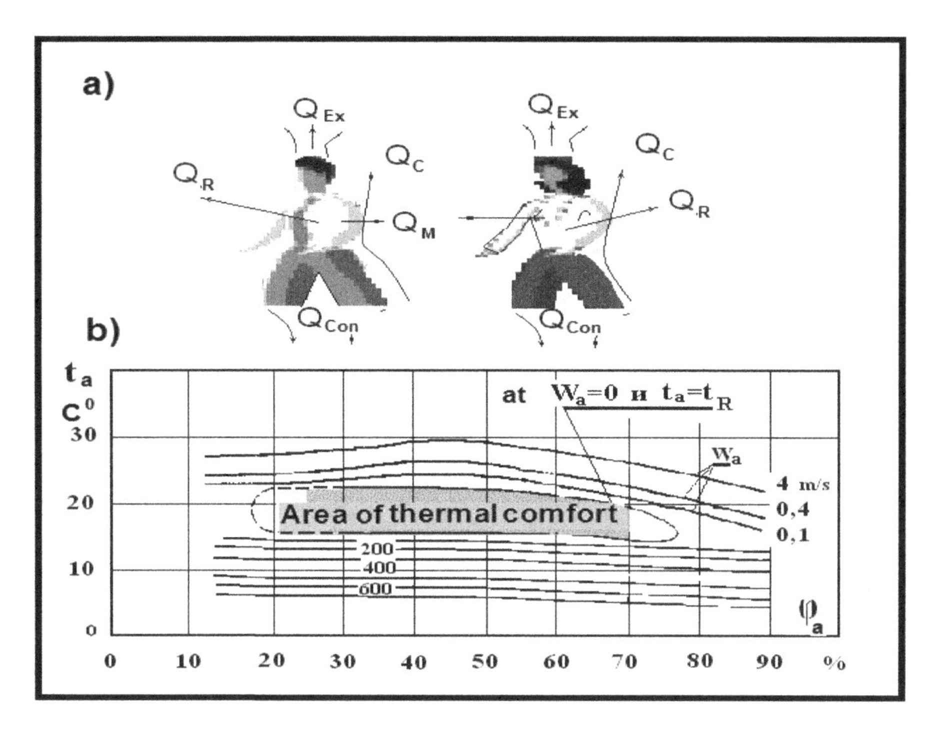

FIGURE 4.1 Scatter of the metabolic heat: a) Structure of the metabolic heat exchange Q_M: – via thermal conductivity Q_T; – via convection Q_C; – via radiation Q_R; – via exudation Q_{Ex}; b) Bioclimatic chart.

located on the human skin and reacting to the change of the ambient temperature. Hence, one feels hot or cold due to the skin receptors, and he feels good when the skin temperature t_{Sc} varies within limits $31.0 < t_{Sc} \leq 34.0°C$.

The energy analysis of the processes using the "Law of energy conversions (the first principle of thermodynamics)" applied to the conversion of chemical energy into work (vegetative and labor) proves that human body releases thermal "waste". Part of the heat maintains the human body temperature. The rest of it scatters in the environment through skin and lungs as energy waste (see Table 4.1).

TABLE 4.1
Structure of the heat exchanging fluxes running between the human body and the surroundings

Through the skin			Through lungs	
Radiation	Convection	Exudation	Heating	Exudation
42%	26%	18%	7%	7%

Thermal comfort exists when there is a balance between the generated and released thermal fluxes from and to the human body described by the equation:

$$Q_M \mp Q_R \mp Q_R \mp Q_T - Q_{Ex} = Q_S + Q_L = \pm\Delta Q = 0$$

Here

- Q_M – heat flux due to metabolism;
- $Q_{C,}$ $Q_{R,}$ Q_T and Q_{Ex} – heat fluxes due to convection, radiation, conductance and exudation, respectively;
- Q_S and Q_L – sensitive (due to the temperature gradient) and latent (thanks to moisture evaporation) heat fluxes.

Several mechanisms of biophysical regulation run in a human body with temperature $t_ч$, but within certain limits:

- At $\Delta Q > 0$ – increase of skin temperature and sweating;
- At $\Delta Q < 0$ – increase of skin surface resistance via pricking sensation on the skin (paresthesia) and muscle metabolism (twitch) yielding thermal balance.

If the system of biophysiological regulation fails or does not restore the thermal balance, the human body overheats or undercools, and the result in both cases is lethal.

Note that the metabolism and the released heat depend on a number of factors including the physical parameters of the environment, human activity, human weight, gender, age, emotions etc. Table 4.2 gives as an illustration the mean values of the metabolic heat flux, considering four types of activities and different values of ambient temperature. The metabolic heat flux depends also on people's age, health, feeding habits etc.

Regardless of the large number of factors affecting the human body's thermal balance (i.e. the conditions of thermal comfort), the engineering practice uses only four of them to create and regulate indoor comfort:

- Air temperature t_a;
- Air velocity w_a;

TABLE 4.2
Dependence between the total heat flow released by human bodies, the ambient temperature and the inhabitants' activity

	15⁰	**20⁰**	**25⁰**	**30⁰**	**35⁰**
At rest	145 W	116	93	93	93
Easy work	157 W	151	145	145	145
Medium work	208 W	203	197	197	197
Hard work	290 W	290	290	290	290

- Radiative temperature T_R[3];
- Relative humidity ϕ_a[4].

The favorable combination of those factors yields a "friendly" and comfortable indoor environment[5]. Figure 4.1,b) shows the bioclimatic chart of A. Olgyay [10] found by empirical testing of sitting individuals (reading, writing, working on the computer, wearing 1 Klo clothes and acclimatized to the conditions of Central Europe).

When air temperature and relative humidity are such as to fall into the hatched area (Figure 4.1,b), people in the area experience thermal comfort. If the specified parameters are outside the area, people experience cold or overheating. Hence, there will be a thermal comfort in the area for all combinations of t_a and ϕ_a, belonging to the intervals $18 < t_a < 24°C$, $25 < \phi_a < 70\%$, and the comfort probability will be up to 90%.

Except for the specification of the operative comfort in an occupied area, the bioclimatic chart also discloses **the mechanisms of attaining comfort** outside the hatched areas, outlining a so-called "standard approach" at $w_a=0$ и $t_a=T_R$, namely:

i. One can provide comfort to the areas below the "hatched" area at a temperature lower than 18°C, if the requirement $t_a=T_R$ is violated and the area is irradiated by an infrared energy flux with intensity higher than 70–75 W/m^2.
ii. Thermal comfort is attainable in areas above the "hatched" area if condition $w_a=0$ is violated, i.e. if occupants' bodies release heat via forced convection. Note that air velocity w_a depends on t_a, but it should not exceed 0.2–0.4 m/s. At the same time, many people experience discomfort due to the airflow arising at a high-convective velocity. The operation of that mechanism proceeds even after the reversion of the temperature gradient (i.e. at $t_a > t_{hb}$) from the body to the surroundings. This is so, since forced convection of air stimulates the diffusion of water vapors from the human skin thus maintaining occupants' thermal balance.

Specialized literature has published a large amount of data on the relation between the air temperature $\boldsymbol{t_a}$ and the mean radiative temperature of the surrounding walls $\boldsymbol{T_R}$ at thermal comfort. Note here the occupants' activity. Following A. Colmar [38], thermal comfort is attained if the following conditions hold:

- Calmly staying individuals – $\boldsymbol{t_a+T_R=38°C}$;
- Warmly dressed individuals – $\boldsymbol{t_a+T_R=32°C}$;
- Individuals performing easy work – if $\boldsymbol{t_a+T_R=27.75°C}$.

---

[3] The radiative temperature is the mean value of the radiative temperature of the surrounding surfaces (walls, windows, floor, ceiling).

[4] These factors can be united in a complex containing information about the thermal conditions in the room. It is called "sensing temperature" t_{Sen}, depending on the active area of a dressed individual, the convection and radiation of his body, and the heat exchange during exudation.

[5] The plot of the field of the possible combinations of those factors providing thermal comfort is called a bioclimatic chart proposed by the Italian architect Olgyay V. in 1963.

The described mechanisms of attaining thermal comfort operate under varying environment, for instance varying intensity of solar radiation, velocity and direction of wind or outdoor air temperature. Pursuant to the main architectural project, these variations should be filtrated by the building envelope, which modulates them conforming to the demand for indoor comfort [41]. The arrangement of the building envelope as a filter of environmental energy penetration is an individual trend of building design and is known as a "passive regulation" of the effects of climate changes. The alternative approach of design is known as "active regulation". It assumes that the leading factors affecting the thermal comfort are the building installations – heating or air conditioning.

The active maintenance of thermal comfort in an occupied area within acceptable limits[6] is based on the control of the surrounding **radiative temperature T_R** and temperature (t_a), humidity (ϕ_a) and air velocity (w_a). This is done by the introduction of energy (heat or cold) "generated" by the building installations. Consider the **mechanisms of "correction" of the parameters of** the indoor environment within the limits defined by the bioclimatic chart. Note that it is rooted in the control of the above four factors, and the control is performed by a system for building thermal control (a heating or air conditioning installation).

4.1.2 Classification of Systems Providing Thermal Comfort

There are two most powerful factors providing thermal comfort in a building – radiative temperature T_R and air temperature t_a. They are set forth considering the systems providing thermal comfort, namely heating systems that use natural gas as an energy carrier – see Figure 4.2.

As seen in Figure 4.2, gas heating systems are classified into three basic groups depending on the mechanism regulating the thermal comfort in an occupied area:

- convective (Figure 4.2,a);
- convective-radiant (Figure 4.2,b) and
- radiant (Figure 4.2,c).

These three systems differ from each other with respect to the mechanism creating thermal comfort and maintaining the thermal balance of occupants' bodies. Regarding the first group, thermal comfort arises thanks to the regulation of the air temperature t_a and utilization of the heat released by the heating device after the end of the fuel gas combustion. Consider the structurally-aggregated heat exchanger (Item 1) in Figure 4.2. There, inflowing cold air accumulates the released heat. Gas convective heaters are screened and supplied with heating but not with radiating surfaces.

[6] Acceptable values of temperature, relative humidity and air velocity maintaining thermal comfort depend on a number of factors, including occupants' activity, age and gender. They are normatively related to the function of the respective premises.

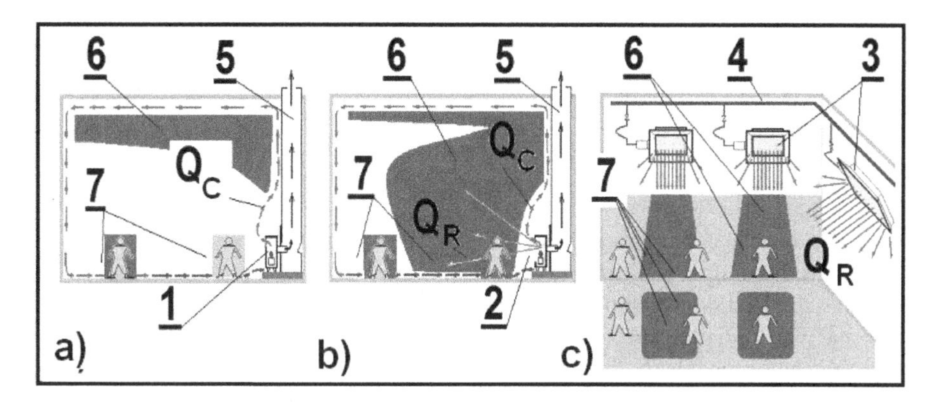

FIGURE 4.2 Mechanisms of heat transfer from gas devices and the creation of thermal comfort in an occupied area: a) Plot of hot air flow (convection), induced by a gas convective heaters; b) Combination convection-radiation flux generated by a gas radiator-convector; c) Radiation flux from a gas ceramic radiators 1 – Gas convective heater; 2 – Gas radiator-convector; 3 – Gas radiator; 4 – Gas-distributing main line; 5 – A system exhausting gas combustion products; 6 – Infrared radiation field; 7 – Occupied area.

They effectively heat the circulating room air, while flues convey the combustion waste products (Item 5, Figure 4.2,a). Heating systems exclusively using convective heaters form a layer of hot air in the higher zones and increase the ceiling temperature, thus affecting the total radiative temperature in the occupied area. The heating of the indoor air takes place slowly.

Gas convective heaters should not be mounted on facade walls and under windows, since their location should agree with that of the flues. Hence, purely convective heating systems are seldom used. Small power gas convectors (with power below 7.5 kW) are used in a so-called "baseboard system", whose function is to compensate part of the thermal fluxes running along the building perimeter and cut the cold infiltrating flows.

The second type is a heating system using radiator-convectors – see Figure 4.2,b). As illustrated by its name, the heating structure is designed such as to radiate a significant amount of heat, affecting the total radiative temperature T_R. Part of the "generated" heat (15–20%) is transferred to the circulating air, while the rest of it leaves the building through the flue, being carried by the gas combustion products.

Radiator-convectors have a wider impact on an occupied area, since the gas combustion chamber efficiently releases heat via radiation. It is equipped with infrared radiators. Its heating effect is similar to that of classical stoves and fireplaces on firewood and coal. Moreover, some heaters create decorative flames, i.e. visual effects similar to those created by casual fireplaces.

The third type of a gas heating system is that providing heat via radiation – see Figure 4.2,c). It attains thermal comfort at low air temperature by changing the radiative temperature T_R in the occupied area. If the air temperature in the occupied area is outside the comfort zone, belonging to the interval 11°C $< t_a <$13°C, the radiative temperature (T_R) should be controlled within the following limits to maintain thermal comfort:

- **27 < T_R < 25°C** – calmly staying individuals;
- **21 < T_R < 19°C** – warmly dressed individuals;
- **20.4 < T_R < 18°C** – individuals performing easy work.

Consider devices called gas radiators. They generate an intensive release of photon energy from ceramic red-hot, yellow-hot or white-hot cores. The heat released by fuel gas combustion accumulates and produces shining "cores" with high temperature and wide radiation spectrum. The energy flux mostly consisting of infrared photons falls on occupants' bodies making occupants feel comfortable. The energy flux "irradiates" the occupied area and distributes throughout it following Wien's cosine. Its largest part runs along the normal to the irradiating surface. Several radiators cover the entire occupied area. Then, heating of a certain control surface is assessed by the summation of the heat released by neighboring devices.

There are two types of gas radiant heaters:

- Flat radiant heaters;
- Tube radiant heaters.

Well-positioned[7] and calculated radiators should generate about 150–160 W/m^2 at the level of the control surface and no more than 240 W/m^2 at the level of occupant's head. The installation of radiators provides options to solve problems of the thermal comfort in high rooms or outdoor, without heating the air in the occupied area.

According to another criterion, the type of gas heating systems depends on the energy carrier flow, i.e. we have:

- Heating systems with natural convection;
- Heating systems with forced convection.

All heating devices and pipe components are calculated to provide power meeting the current needs in the heated area. Heaters are installed considering so-called "calculation conditions" found on the basis of the climate zone, the regime of occupation and the building class.

We shall consider in what follows schemes of gas heating installations and devices, grouped as follows:

- Systems "water-air";
- Systems "gas -air".

4.1.2.1 "Water-Air" Fuel Gas Systems

That class belongs to the set of central water installations, which are widely used in mass construction in Europe. To understand the principle of their operation, we show in Figures 4.3 and 4.4 two schemes of installations, type "water-air" with a

[7] About 4–5 m above level 0° (keep the directions of the manufacturer).

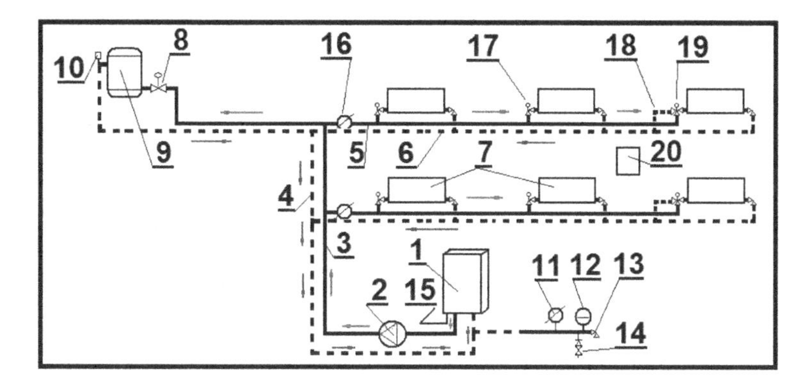

FIGURE 4.3 Heating system "water-air": 1 – Tankless gas water heater installed on a wall; 2 – Circulation pump; 3 – Supply riser; 4 – Return riser; 5 – Distributing horizontal pipe; 6 – Collecting horizontal pipe; 7 – Radiator; 8 – Valve controlled by a thermostat; 9 – Heat accumulator; 10 – Automatic venting valve; 11 – Manometer; 12 – Membrane expanding tank; 13 – Balance valve; 14 – Link to the drain system; 15 – Pipe draining the condensate; 16 – Heat energy meter of the individual circular contour; 17 – Two-way thermostat valve; 18 – Bypass; 19 – Three-way thermostat valve; 20 – Controller.

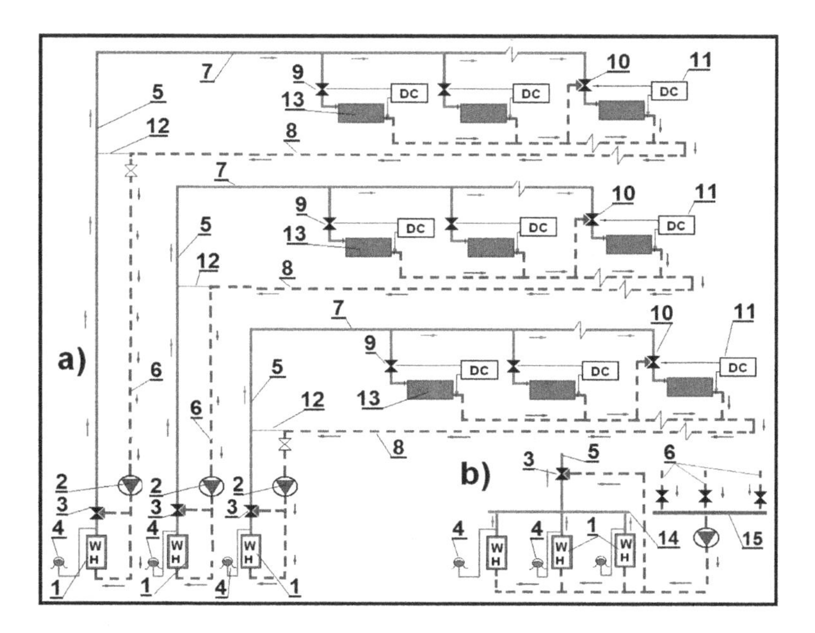

FIGURE 4.4 Heating system "water-air" with a block aggregate consisting of passable gas heat generators: a) Independent circular contours; b) Central distributing system: 1 – Tankless gas heat generator mounted on a wall; 2 – Circulation pump; 3 – Three-way thermostat valve; 4 – Membrane expanding vessel; 5 – Supply riser; 6 – Return riser; 7 – Distributing horizontal pipe; 8 – Collecting horizontal pipe; 9 – Valve controlled by the console; 10 – Three-way valve; 11 – Console of SAC; 12 – Bypass; 13 – Radiator; 14 – Distributing tube collector; 15 – Gathering tube collector.

horizontal distribution of the heat transferring medium. Power consumption is measured at each end user (apartment, office or another building segment).

The difference between that class of fuel gas systems and the central water heating installations lies in the type of the central heat generator only. We use here gas boilers or water heaters.

One can apply the scheme of a gas heating installation shown in Figure 4.3 to a multistory building, performing a serial multiplication of the hot water circles of a pipe network consisting of horizontal distributors. The scheme shows only two circles of users. It is a combined "heating-hot water" installation for household use.

Except for classical heaters such as radiators, convectors etc., the scheme includes a gas tankless water heater (Item 1) and a hot water accumulator (Item 9). Another important component is the heat energy meter (Item 16) included in each circle of users and used to account for the consumed power. The offered heating system is equipped with a device regulating the power of the individual appliances and that of the entire system. A bypass (Item 18) of the three-way valve (Item 19) keeps the hydraulic stability of the individual circular contour when the heating devices turn off. Assume that the gas consumption of the heating bodies terminates.

Assume also that the difference between the temperature of the hot water in the supply and return risers drops. Then, one can reduce hot water circulation in the risers by reducing the number of revolutions of pump 2.

The second scheme of a gas heating system "water-air" is the one shown in Figure 4.4. It illustrates the use of a block aggregate of tankless gas heat generators for wall installation in two variants:

- Independent circular contours;
- Common distributing network.

The first variant is a modification and improvement of the scheme in Figure 4.3. It is based on the introduction of groups controlling the block aggregate of tankless water heaters.

Executive devices such as a three-way valve (Item 3) and a circulation pump (Item 2), as well as sensors and digital computer consoles, participate in the groups.

When the difference between the temperature of the supply riser water and that of the return riser water becomes negligible (for instance, 5°C), the system for automatic control (SAC) [consisting of sensors and digital console 11 plus the executive devices] activates the three-way valve – Item 3, which opens the bypass channel. This deflects the water flow running through the tankless heater (Item 1) and it dies down. Then, the automatic control turns off the fuel system, thus saving fuel.

Consider burner shut-off. Then, the heat carrier flux in any user's heating circle proceeds through the return riser (Item 6) and through the bypass to the supply riser (Item 5). It reaches the heaters and the horizontal collecting pipe (Item 8) through the distributing horizontal pipe (Item 7). The circulation of the heat carrier proceeds until it cools down so as to open the channel of the three-way valve. Thus, the fluid can flow through the water heater.

The scheme of the second variant in Figure 4.4,b) shows tankless gas heat generators of the block aggregate. They are parallel to each other, and the heat carrier distributes and enters a common pipe network through the tube collectors – Items 14 and 15. A three-way valve (Item 3) controls the passability of the bypass measuring the difference between the temperature of the hot water in the supply riser and its temperature in the return riser. The number of operating tankless heaters is controlled, shutting off their fuel systems. This is done depending on the hot water intake through the group. The three-way valve (Item 3) controls the flow rate.

4.1.2.2 Convective Systems "Gas-Air"

Considering gas convective systems, type "gas-air", the thermal comfort is regulated by changing the temperature of the air circulating in the occupied area. The heat generated by the combustion of fuel gas heats the air in devices of two types:

- Gas air-preparing handlers (called gas-air handlers in what follows);
- Local devices.

Gas-air handlers are the main components of systems providing thermal comfort throughout the building. They are known as central air heating systems. They are in fact complex engineering installations, comprising also hot air ducts equipped with distributing diffusers and SAC. Modern gas-air handlers allow quantitative and qualitative control of the parameters of the circulating air, varying its temperature and output.

The second type of gas devices such as convectors, radiator-convectors, fireplaces etc. participate in the regulation of local thermal comfort – in a particular room or building space. Yet, they can heat the entire building by means of an air duct network.

4.1.2.2.1 Central Gas-Air Heating Systems

The main component of a heating system, type "gas-air", is a gas-air handler comprising:

- Gas-air heating section including a burner (Item 2) and a combustion chamber (Item 3) – see Figure 4.5;
- A fan section (Item 7);
- Heat-exchangers (Items 5 and 6);
- Gathering (Item 8) and distributing (Item 4) sections (plenums).

Gas-air handlers are installed in a mechanical room in a basement, on technical floors or on the roof. They should be situated at the center of the thermal load.

The heat generated by fuel gas combustion in the combustion chamber (Item 3, Figure 4.5) is transferred by the smoke gases to the air flowing around the primary (Item 5) and the secondary (Item 6) heat exchangers under the head of a fan (Item 7). The primary heat exchanger is designed such as to use the main thermal potential of the smoke gases until reaching the dew point of water vapors present in the gas combustion products.

The secondary heat exchanger is incorporated in air-preparing handlers of the new generation. It extracts the hidden (additional) energy of water vapors via

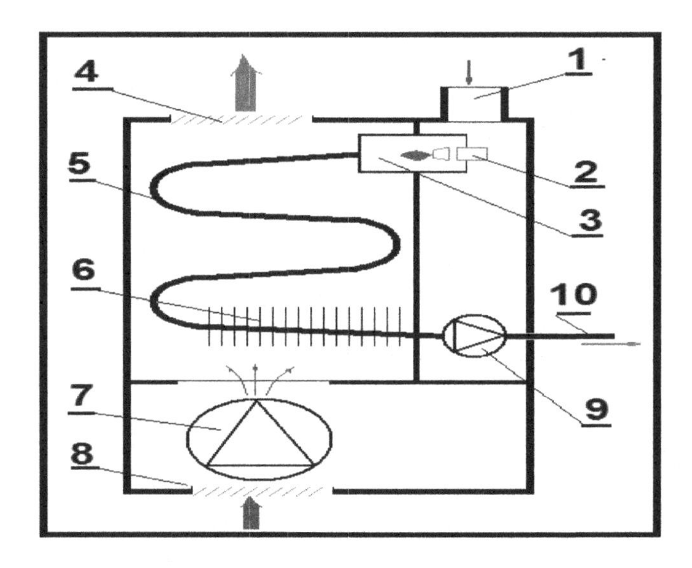

FIGURE 4.5 Air-preparing handler 1 – Intake of combustion air; 2 – Burner; 3 – Combustion chamber; 4 – Air-distributing plenum and louvered outlet of hot air; 5 – Primary heat exchanger; 6 – Secondary heat exchanger; 7 – Intake fan; 8 – Intake of heated air; 9 – Smoke gas fan; 10 – Exhaustion of gas combustion products.

condensation (phase transition). Smoke gases cool down to 70°C, proceed their way through the flue and leave the building sucked by the smoke fan (Item 9). The air heated by the smoke gases heads through the distributing plenum (Item 4) and hot air ducts (the distributing system) to the heated premises.

Figures 4.6 and 4.7 show various types of gas air-preparing handlers operating under the air dynamic head in air ducts. Similar to Figure 4.5, a fan (Item 7) operates creating pressurization in the devices along the route of the heated air. Thus, the mixing of air with combustion products, released at eventual leakage in the heat exchangers, is avoided. The three handlers in Figure 4.6 allow a convenient connection of the air-distributing plenum (Item 3) to the air ducts of the distributing system.

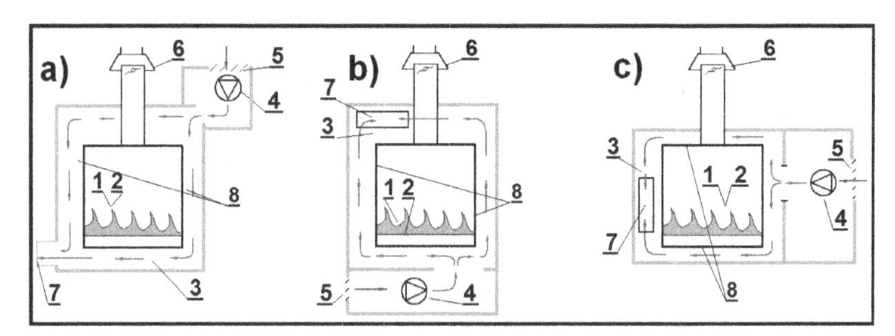

FIGURE 4.6 Gas-air handlers, operating under pressure head: a) Opposite-streaming with side release; b) Unidirectional with upper release; c) Crossed current with lateral release: 1 – Combustion chamber; 2 – Burner; 3 – Distributing plenum; 4 – Fan; 5 – Intake filter; 6 – Draught diverter; 7 – Hot air outlet; 8 – Heat exchanger.

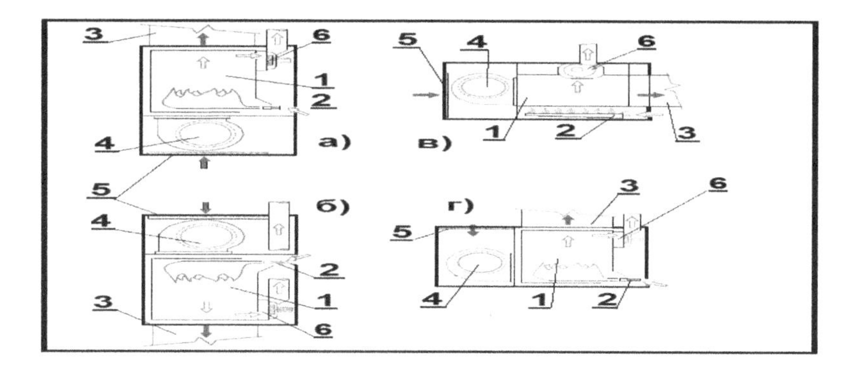

FIGURE 4.7 Gas-air handler with a fuel system subjected to depressurization generated by a smoke fan: a) Parallel with upper conveyance; b) Unidirectional with lower conveyance; c) Crossed current with lateral conveyance; d) Crossed current with upper conveyance. 1 – Combustion chamber; 2 – Burner; 3 – Plenum; 4 – Fan; 5 – Inlet filter; 6 – Smoke fan.

The connection is realized by means of flanges or louver opening (Item 7) of lower, upper or lateral position.

In contrast to the devices in Figure 4.6, the fuel systems of the devices in Figure 4.7 operate under depressurization performed by a smoke fan (Item 6). The installation of a separate fan decreases the probability of mixing the fresh air flux with the combustion products. When the heating installations need higher power, one should use monoblock gas burners and contact heat exchangers of chamber type.

The distributing hot air ducts of the thermal comfort system depend on the horizontal plan of the building and on the type and configuration of the air-distributing plenum (see Figure 4.8). The variants in Figure 4.8,a) and b) fit longer buildings, while the variant in Figure 4.8,c) fits square buildings.

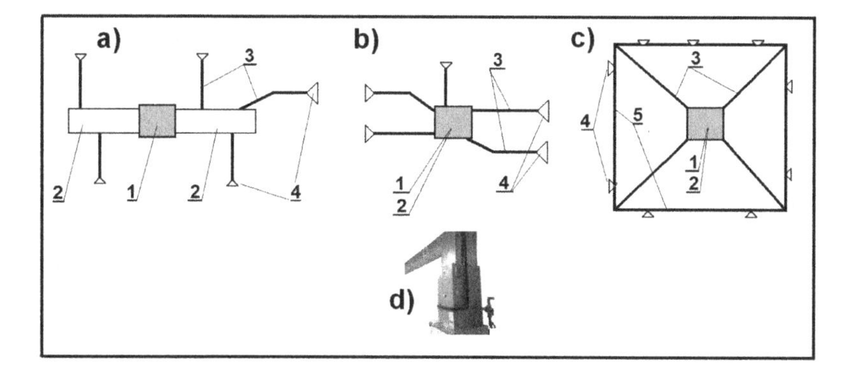

FIGURE 4.8 Types of hot air ducts of "gas-air" heating systems: a) Elongated plenum with radial air ducts; b) Shortened plenum with radial air ducts; c) Shortened plenum with ringwise distributing air duct; d) Axonometric view of a gas-air heating handler: 1 – Gas air heater; 2 – Air-distributing plenum and outlet of hot air; 3 – Radial air ducts; 4 – Diffuser with register grid; 5 – Air-distributing ring.

The installation of vertical connecting air ducts depends on a number of architectural-structural characteristics of the building – number of floors, the distance between floors, construction system, class and overall dimensions of the air ducts. Yet, parameters prescribed by the investor's technical-economical project are of crucial importance.

The design and construction of an air-distributing system is unnecessary in other types of gas heating installations, which use local gas convective heaters, radiator-convectors, stoves, fireplaces or infrared radiators in heating occupied building areas.

4.1.2.2.2 Local Gas-Air Heating Systems

They are classified as central systems with respect to gas supply. At the same time, they are considered to be local systems with respect to heat generation and distribution in the occupied space. Thus, they have essential advantages over other heating systems where heat is centrally generated (by a central heating plant – CHP, thermal electric power plant or building boiler stations) and regionally distributed, but with excessive thermal and gas dynamic pressure losses. This proves that local heating systems "gas-air" are significantly more economical, effective and low-inertial as compared to central heat distributing systems, type "water-water", "water-air", "steam-water" or "steam-air".

As stated, there is a direct analogy between the operation of local systems compounded of heating units such as stoves, fireplaces and other devices using traditional fuel, and the operation of heating systems, type "gas-air". The heating devices are situated in heated rooms such as to provide the needed thermal comfort within the entire occupied area.

The difference consists in that the fuel needed by the systems of the first type is supplied periodically and in portions (accompanied by cleaning of the fuel system). On the other hand, devices, type "gas-air", are continuously supplied with fuel by the central urban gas network, as described in Par. 3. This makes them more attractive as compared to the traditional heating stoves and fireplaces.

Quite often, the installation of heating units in the building plan follows a strategy of: (i) attaining a homogeneous thermal comfort, (ii) "compensating" the "radiation" reducing it to cold 3D objects and (iii) blocking cold air fluxes directed to an occupied area[8].

In premises where the equipment is often rearranged, the use of movable gas devices, type convector-radiators, supplied with a gas tank, is an adequate designer's solution. Those gas devices are smokeless, Class A (see Figure 3.17) and have the power of up to 7.5 kW. Hence, they do not need a flue system to release combustion products. Moreover, they stimulate the natural ventilation and create an area of thermal comfort around them, 3–4 m wide.

A typical movable device consists of two sections: a section where the combustion and heat-exchanging components are arranged and a section where the gas tank is installed. These sections are separated from each other by an inflammable compact screen – see Figure 4.9. The first section comprises a fireplace equipped with a

[8] The strategy of creating a homogeneous thermal comfort yields waste of energy and material resources under deficit of primary energy sources (see [9]).

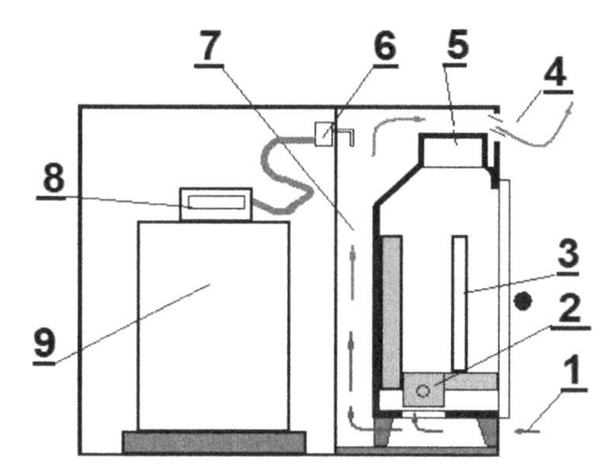

FIGURE 4.9 Movable gas heater, type radiator-convector: 1 – Cold air intake; 2 – Gas burner; 3 – Radiating ceramic plate; 4 – Louvers for warm air outlet; 5 – Catalytic converter of smoke gases; 6 – Control valve/igniter; 7 – Heat exchanger; 8 – Gas regulator and hose; 9 – Gas tank under pressure.

burner (Item 2, Figure 4.9), a radiator (Item 3), a heat exchanger (Item 7), a smoke gas catalytic converter (Item 5) and louvres directing the warm air (Item 4).

Most gas devices are equipped with duplex burners (Figure 2.13,e), while the fireplace is installed in a box and built of fire-resistant bricks[9] radiating heat to the occupied area. Special ceramic radiators are installed in some gas devices, in front of the burner. They undergo high temperature (1000–1100°C), increasing device efficiency by 15–20% and its functionality to operate under non-traditional conditions (see Sub-par. 4.1.2.3). The heat exchanger (Item 7) is a flat metal casing installed above the fireplace, and the heated air flows around it. Since it operates under significant temperature differences, its heating surfaces undergo the destructive effect of material fatigue. A special control valve (Item 6) stands between the two sections avoiding eventual fuel leakage from the tank, which would cause an explosion.

There is nothing special in the principle of operation and structure of the gas heater, distinguishing it from typical gas devices that operate under natural convection. Figure 4.9 shows that intake slots of cold air (Item 1) supporting gas combustion in the burner (Item 2) are perforated through the device bottom. Part of the air circulates within the heat exchanger (Item 7) too. The dimensions of the suction slots are such as to provide a gas flow rate of about 60–70 m^3/h or realize a double air exchange per hour.

After combustion, subsequent products leave the chamber through the smoke gas converter (Item 5), where they additionally oxidize in a catalytic medium. Thus neutralized, combustion products mix with the air, heated in the heat exchanger (Item 7), and enter the inhabited volume. The basic thermal flux (about 85%) is transferred to the indoor air via natural convective recirculation, while the ceramic plate (Item 3) radiates only 15% of it. The closing plate should be sealed along its edges by means of a heat resistance rubber adhesive, withstanding a temperature of about 100°C.

[9] Fire resistant bricks increase the thermal inertia of the fireplace and prevent the gas device from overheating and cracking.

TABLE 4.3
Gas flow rates needed by a movable gas radiator-convector

Power	Natural gas N_w=50.68; SG=0.58			Propane N_w=80.14; SG=1.78		
kW	$\dot{V}_{NG}$ m³/s	$\dot{V}_{B\text{-}X}$ m³/s	$\dot{V}_{д.г.}$ m³/s	$\dot{V}_{NG}$ m³/s	$\dot{V}_{B\text{-}X}$ m³/s	$\dot{V}_{д.г.}$ m³/s
2.5	$67.6*10^{-9}$	$862.1*10^{-9}$	$733.7*10^{-9}$	$30.3*10^{-9}$	$936.8*10^{-9}$	$21996*10^{-9}$
7.5	$202.8*10^{-9}$	$2586.3*10^{-9}$	$2201.2*10^{-9}$	$90.8*10^{-9}$	$2810.6*10^{-9}$	$65993*10^{-9}$

In support of the assertion that the needed flow rates of fuel gas and combustion air, as well as the release of combustion products, are negligible, Table 4.3 gives the theoretical estimations of the needed flow rates of two fuel gases: natural gas and propane[10]. Data corresponding to two representative power values (2.5 and 7.5 kW) show that there exists a direct linear proportion between the sought power and the amount of generated combustion products. Another fact as seen in the table is that under identical power, the fuel consumption of a gas device using propane is twice lower. Yet, the amount of released smoke gases is significantly larger. Hence, the device is not appropriate for use in premises where the demand for clean air is higher. To improve the air in the occupied area, catalytic gas transformers (Item 5) are installed in some movable gas devices along the route of smoke gases.

The discussed movable heaters have the following advantages:

- Use of replaceable gas bottles;
- Easy mobility; no need of special installation;
- Appropriate for large ventilated premises only.

The disadvantages are due to the combustion products freely entering the occupied area. They yield an increase of moisture in the air. Yet, moisture is in a gaseous state when and until the air temperature is higher than the dew point. When however moist air approaches and contacts thermal bridges, building angles or cold equipment, moisture condenses and is absorbed by the wall components. This phenomenon yields the risk of freezing, wall destruction and growth of mold and fungus.

The use of gas devices is prohibited in bedrooms and nurseries. When their use is inevitable, effective control of the air quality should be performed and rooms should undergo forced ventilation.

Most projects do not foresee the rearrangement of the internal equipment and the gas devices. In that case, one may use stationary or free devices (Figure 4.10) or devices equipped with flues (Figure 4.11).

[10] The estimations are made with a coefficient of air excess α =1.3.

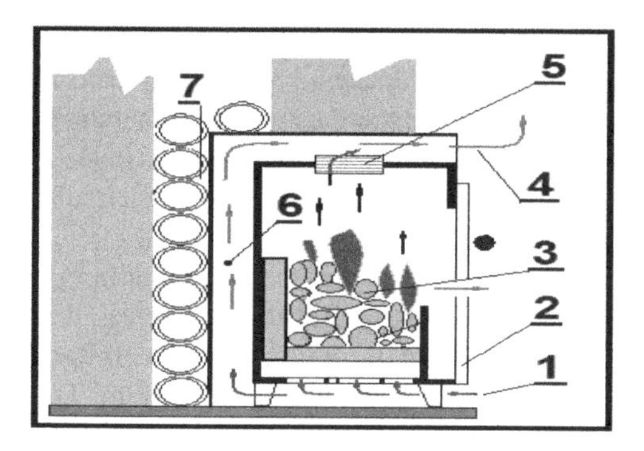

FIGURE 4.10 Convective heater without a flue: 1 – Cold air intake; 2 – Sealed glass panel; 3 – Decorative coal; 4 – Warm air outlet; 5 – Catalytic converter of smoke gases; 6 – Heat exchanger; 7 – Insulation.

Pursuant to the mechanisms of heat transfer to an occupied area, the used gas devices are:

- Convective heaters (Figures 4.10 and 4.11,a);
- Radiator-convectors (generating combined heat exchange (Figure 4.11,b).

The efficiency of convective heaters amounts to 45%, at average while that of combined devices is by 10–15% higher. The increase of efficiency is due to the incorporation of a heat exchanger (Item 13) and a radiating surface (Item 2, Figure 4.11). This contributes not only to the more efficient use of energy, released during combustion, but also to a significant improvement of the thermal comfort in the area of occupation (note the "long distance" penetration and distribution of the radiative heat flux in the occupied area).

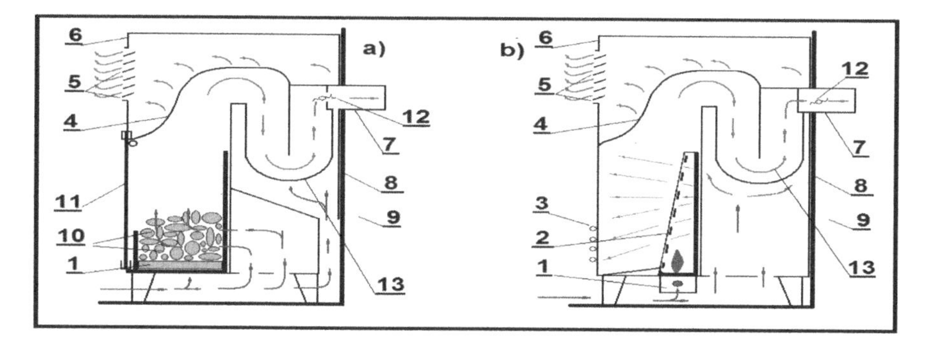

FIGURE 4.11 Gas heaters: a) Type "decorative convector"; b) Type "radiator-convector": 1 – Burner; 2 – Radiator; 3 – Leader; 4 – Gas flux converter; 5 – Warm air louvers; 6 – Box; 7 – Smoke gas outlet; 8 – Closing plate; 9 – Wall; 10 – Decorative burning coal; 11 – Sealed glass panel; 12 – Flow converter; 13 – Heat exchanger.

The combustion and heating air enters the space under the gas device in both cases. The air flux splits into two branches. The first one ejects into the burner, mixes with the fuel and participates in the combustion and generation of smoke gases. The combustion products are directed to the heat exchanger (Item 13), where combustion heat is transmitted to the second air flux, intended to recirculate back and heat the premises.

Considering the radiator-convector (the second heater in Figure 4.11,b), a special ceramic body (Item 2) accumulates part of the heat released in the burner, and its temperature reaches 850–900°C. After heating, it emits the accumulated energy in the form of photons of the visible and the infrared spectra. The rest of the released thermal energy is transferred to the air and to the combustion products. The energy flux through the heat exchanger (Item 5) heats the recirculating indoor air, which passes through the aperture covered by louvers (Item 5) and returns back to the heated space. Smoke gases, cooled down to 100–120°C in the heat exchanger, leave the device through the outlet (Item 7) under the action of the flue draught.

The available flue draught of a typical family house is about 10 Pa (0.1 mBar). It is sufficient to convey the combustion products out of the device. The available flue draught of multistory buildings increases significantly. Hence, they should be equipped with an additional resistance valve (sliding rule), installed at the device exit. Practice shows that the choice of a gas device depends on a number of factors, including space function, occupation schedule, location of premises within the building, the structure of the envelope, as well as the location, area and number of vitreous components offered by the designer.

Device thermal capacity (power) should be such as to meet the "calculated" need for thermal energy of the respective settlement. One should also account for the structure of the building envelope. Various heaters, such as open local heating fireplaces, stoves etc., are used to heat living and guest rooms, providing visual comfort. Their structure is shown in Figure 4.12. It does not differ essentially from the structure of heaters discussed so far. Except for the components creating luminescent

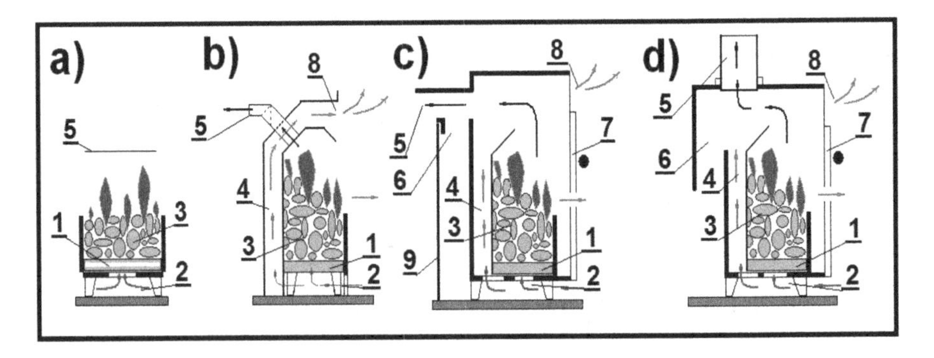

FIGURE 4.12 Gas heating fireplaces/stoves with decorative flames: a) Gas fireplace under a canopy; b) Open gas fireplace; c) Gas fireplace with a closing plate; d) Gas fireplace with a direct connection to the flue: 1 – Burner; 2 – Cold air intake; 3 – Decorative coal; 4 – Heat exchanger; 5 – Smoke gas outlet; 6 – Draught diverter; 7 – Sealed glass panel; 8 – Warm air outlet; 9 – Closing plate.

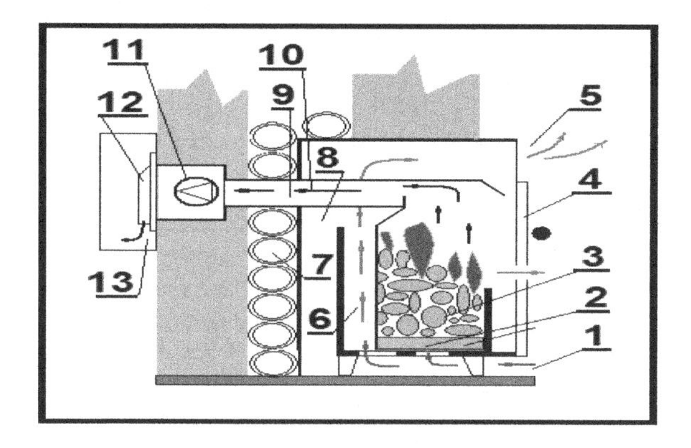

FIGURE 4.13 Forced gas radiator-convector: 1 – Cold air; 2 – Burner; 3 – Decorative coal; 4 – Sealed glass panel; 5 – Warm air outlet; 6 – Heat exchanger; 7 – Insulation; 8 – Flow reverser; 9 – Exhausting smoke pipe;10 – Flue; 11 – Fan; 12 – Terminal; 13 – Smoke gas outlet; 14 – Fire resistant shield.

effects (radiating segments (Item 3) and transparent coatings (Item 7)) gas heaters are remarkable by their specific connection to the system conveying combustion products (see Figure 4.13).

For instance, an open gas fireplace (Figure 4.12,b) is installed directly in a structural niche, with specific wall smoothness and density, designed to decrease the eventual loss of pressure head and avoid suction of additional "needless"air.

Radiator-convectors are installed using a closing plate, which is put in front of the assembly aperture (Figure 4.13,c) or in a box. A hollow volume emerges behind the plate and it collects sediments, soot or other impurities. Profile segments (Item 4, Figure 4.13,c) are installed in appropriate flue channels to create an ejection effect at the device-flue connection.

Gas radiator-convectors can be installed directly in the flue channel (Item 5, Figure 4.13,e) or can be connected to the flue via a pipe (Item 5, Figure 4.13,e). Minimal flue dimensions correspond to the structural apertures of the device. Yet, they should not be under 0.125 m.

To increase the efficiency of gas devices, one must impregnate the installation niches. All technological or casual apertures should be closed. Hence, during the construction of gas fireplaces and the installation of face panels, one should check whether there are holes in the niche and through the fireplace cast iron block. Unnecessary flue pipes installed in structural niches should be closed and the closing plates should be sealed.

Another approach to flue impregnation is the use of dynamically balanced gas devices for facade installation – see Figure 4.14. They operate successfully under the power of 35–60 kW.

Note that the conveyance of combustion products and combustion air is normal with respect to the facade. It takes place through industrially manufactured flues and fittings operating under natural or forced convection (Figure 4.14,a and b). Yet, the forced conveyance is more reliable. A fan sucking smoke through the closest facade

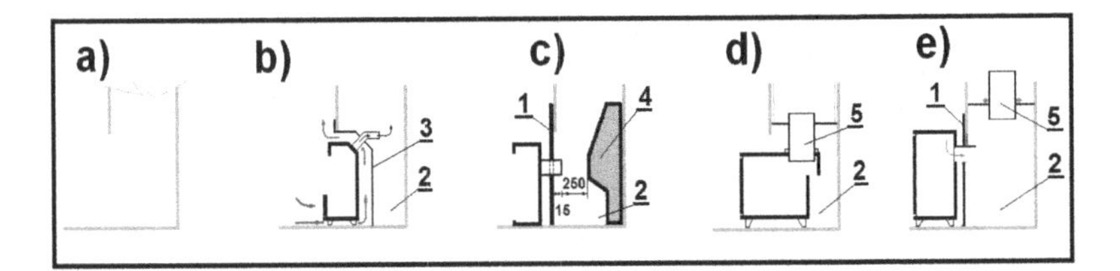

FIGURE 4.14 Assembly schemes of gas heaters in a flue niche: a) A constructional flue niche – vertical profile; b) An open gas fireplace; c) A radiator-convector with a closing plate; d) A gas fireplace with a direct connection to a flue; e) A gas fireplace with a closing plate: 1 – Closing plate; 2 – Drainage space; 3 – Back wall of the device; 4 – Profile segment; 5 – Vertical flue.

is used for that purpose – see Item 11 (Figure 4.14). Moreover, highly efficient heat insulation (Item 7, Figure 4.14) is mounted in the clearance between the fan metal box and the external wall. The resistance against heat scatter should be R100 m^2K/W.

The dimensions of smoke flues and fresh air ducts should be specified depending on the installed power (see Table 4.4). Table 4.4 recommends some expressions to calculate the dimensions of the air ducts of balanced gas devices – see the Example 4.1.

To avoid the "shortcut" between the fresh air suction nozzle and the smoke gas terminal, one should leave at least 0.3 m between them.

Example 4.1

Find the dimensions of a horizontal air duct of a balanced device with power 48 kW.

SOLUTION

The cross-section of the air duct is calculated via the expression:

$$A_0 = 1.41\sqrt{0.45Q_{GD}} = 1.41\sqrt{0.45.48000} = 20722\ mm^2 = 0.0207\ m^2 \quad (4.1)$$

If the cross-section is a square, its dimensions are 0.15 × 0.15 m.

Balanced devices do not initiate infiltration or exfiltration in the rooms. They only stimulate the recirculation of the indoor air by the heat exchanging surfaces.

TABLE 4.4
Cross-section of air intake ducts of balanced gas devices

Lower level mm^2	Horizontal mm^2	Upper level mm^2
$1.22\sqrt{0.45Q}$	$1.41\sqrt{0.45Q}$	$1.58\sqrt{0.45Q}$

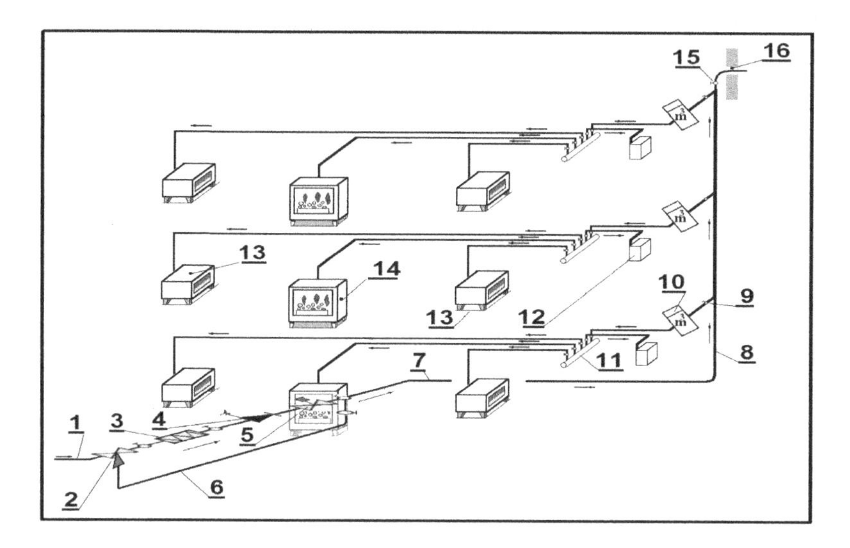

FIGURE 4.15 Gas supply of end users via a horizontal pipe network: 1 – Intake pipe; 2 – Three-way valve; 3 – Filter; 4 – Reversive valve; 5 – Pressure regulator; 6 – Bypass; 7 – Horizontal main pipe; 8 – Supply riser; 9 – Apartment branch and slider; 10 – Gas-measuring instrument in GMP; 11 – Gas distributing tube; 12 – Tankless water heater; 13 – Gas fireplace;14 – Radiator-convector; 15 – Emergency control valve; 16 – Vent in staircase cell.

Premises heated by balanced gas devices undergo density control (mainly during opening and closing of doors and windows). Technically, this is done by installing microcontact switches blocking the device operation during the occurrence of an event. Except for turn-off switches of combustion systems, premises with balanced devices are equipped with double building envelopes (twin walls) and significant thermal insulation – see Figure 4.15.

As stated, central gas networks feed local convective heating installations. As in water installations, designers employ two schemes of gas supply to end users:

- Vertical supply (see Figure 3.35);
- Horizontal supply (Figure 4.15).

According to the first scheme, vertical gas distribution starts at the level of the horizontal main pipe (Item 7, Figure 3.35), to which vertical risers (Item 8) are connected. Fuel gas consumption is accounted for at the intake of each apartment branch.

According to the second scheme, fuel is supplied to each floor through a supply riser (Item 8, Figure 4.15). There is a horizontal branch to each apartment (Item 9), with a gas flowmeter mounted at its inlet (Item 10).

Indoor fuel distribution to the devices takes place through the gas distributing tube (Item 11).

Heating installations with gas convectors or radiator-convectors combine the properties of the systems controlling thermal and air comfort. This is so, since they remove unhealthy emissions in the inhabited area on one hand, and improve the thermal comfort on the other hand. In fact, they induce depersonalization in the area around the devices, yielding the inflow of fresh air. However, Sub-par. 4.2 will clarify this issue.

4.1.2.3 Radiant Heating Systems "Gas-Air"

Gas radiant heating systems are applied in high premises or open space and also in premises with lower demand for indoor thermal comfort (for instance, poultry farms, pig-breeding farms, dairy farms etc.).

As already noted, there are two types of gas radiators: overhead radiant heaters and radiant tube heaters[11]. There are various options for their incorporation in a system for thermal comfort control. We consider in what follows some of them.

4.1.2.3.1 Overhead Radiant Heaters – Flat Radiators

There are two groups of these devices:

- Ceramic plaque heaters – Flameless;
- Catalytic heaters.

Visually, they are similar to each other but they operate employing different heat-generation methods.

Radiant heaters of the first group use a heat generator, which is a modification of a traditional ejector burner. The heat generator operates with a small coefficient of air excess (α=1.05) and the **combustion is flameless**. At the same time, it generates high temperature (above 1000°C), and the radiated energy has a small Lagrange multiplier and belongs to the short-wave section of the spectrum (about 2.5–5.0 μm).

Figure 4.16,a) shows the structure of ceramic plaque heaters. In contrast to classical ejector burners, there is no heat exchanger and a flat ceramic head is the main heat-exchanging component (Item 1, Figure 16,a) functioning as a radiator[12].

The initial section of the fuel system is identical to that of the ejector burners (see Figure 2.14). It consists of: ejector (Item 5), mixing pipe (Item 4), distributing chamber (Item 2) and radiator's head (Item 1). Fuel gas flows through the nozzle of the ejector (Item 5) with high velocity, ejects the combustion air, mixing with it in the pipe (Item 4) and discharges through the distributor (Item 2) and the nozzles of the head of the radiator (Item 1) as a bundle of coaxial jets. The gas-air mix burns in a thin layer (1–1.5 mm thick). Burning temperature is in the interval 1123–1473 K.

The head of the radiator accumulates a larger portion of the released heat increasing its temperature to a value close to the burning limit. When burning is flameless, the burner head becomes white hot and starts emitting secondary radiation to the occupied area. The intensity of the secondary radiation amounts to about 40–60% of the nominal thermal power of the gas device. A screening net installed in front of the head intensifies the emission by 50–60% – the net is fabricated from heat-resistant steel.

[11] Pursuant to some classifications, they are classified high- and low-intensive ones with respect to the intensity. Other divide them into light, dark and super-dark with respect to the radiation.

[12] The radiant heaters are fabricated from fire-resistant clay or volcanic lava.

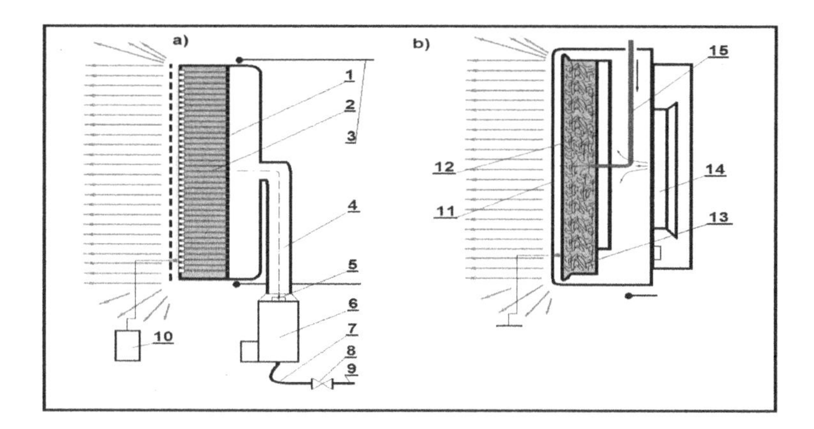

FIGURE 4.16 Overhead radiant heaters – flat gas radiators: a) Ceramic plaque flameless heater; b) Catalytic heater:1 – Radiant ceramic head – burner; 2 – Distributing chamber; 3 – Carrier; 4 – Mixing pipe; 5 – Ejector; 6 – Multifunctional gas valve; 7 – Flexible connector; 8 – Isolation valve; 9 – Gas network; 10 – Control box; 11 – Radiant surface; 12 – Reactor chamber; 13 – Fibers with catalyst; 14 – Fan; 15 – Gas pipe.

The heads of the radiant heaters (monolithic, porous, brush-like or channel-like) are made of ceramics, metal or metal-ceramic oxides. The apertures[13] have different diameters and shapes (Figure 4.17). Various catalytic extensions are used to decrease harmful emissions. The efficiency coefficient of a one-row radiant heater is 50–52%, while that of a two-row radiant heater – 54–57%.

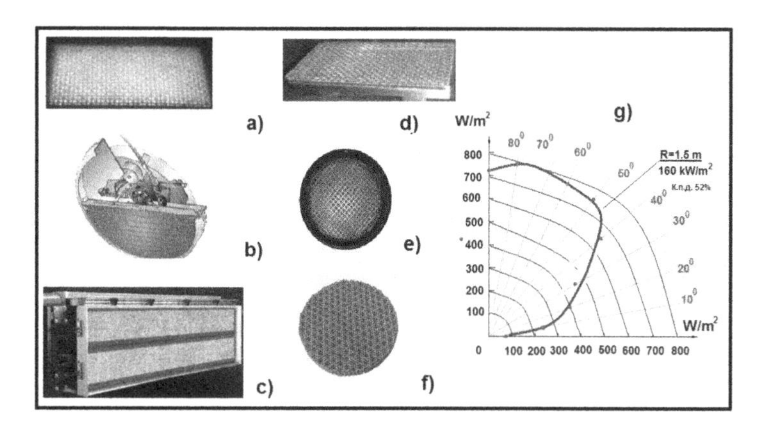

FIGURE 4.17 Overhead radiant heaters – flat gas radiators: a) Ceramic plaque heater; b) Axonometric section of a catalytic radiator; c) Porous gas radiator; d) Screening net; e) Catalytic overhead heater; f) Disk-shaped head; g) Indicatrix of the density of radiation emitted by the surface of the radiant head GI-01.

[13] The number of apertures with diameter 0.8–1.55 mm on a standard plate is 625 – 1625. The distance between the apertures is about 0.5 mm.

The structure of a catalytic overhead radiant heater is shown in Figure 4.16,b). Fuel enters the chemical reactor (Item 13) through the pipeline (Item 15). The oxygen necessary for the reaction comes from the ambient air blown by the fan (Item 14).

Visually, these heaters are similar to the overhead ceramic radiant heater but they operate employing different technology. As already stated, combustion takes place in a catalytic medium without a flame. Hence, overhead radiant heaters are often called catalytic. The combustion temperature is moderate – about 600–800°C, and combustion intensity is moderate too. The generators are suitable for premises with smaller height, about 4.0–4.5 m, and their thermal power is in the range 7–10 kW.

The thermal flux exchange between the overhead radiant heaters and various bodies (including inhabitants' bodies) is traditionally assessed as:

$$q_{GR,i} = C_0 \varepsilon_i A_{GR} \phi_{GR,i} \left[\left(\frac{T_{GR}}{100} \right)^4 - \left(\frac{T_i}{100} \right)^4 \right], \text{ W/m}^2 \tag{4.2}$$

Here

- C_0 – Stephen-Boltzmann constant (C_0=5.67×10^{-8} W/(m^{2}K^4));
- ε_1 – coefficient of grayness of the radiant body ($C_0\varepsilon_i$=4.8-5 kW/m^{2}K^4 of a ceramic overhead heater; $C_0\varepsilon_1$=4.0 kW/m^{2}K^4 of a nichrome grid);
- A_{GR}, m^2 – area of the gas radiator;
- φ_{GR} – irradiation coefficient (see Figure 4.18);
- T_{GR} – mean temperature of the gas radiator surface;
- T_i – mean surface temperature of the ith body.

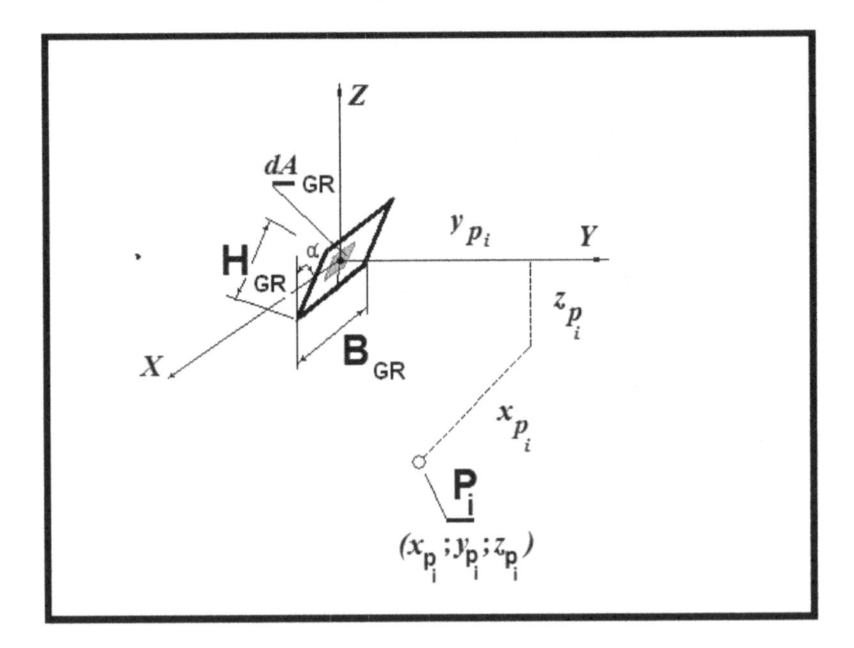

FIGURE 4.18 Scheme of the thermal irradiation of point P_i by the control surface of an overhead radiant heater.

To provide thermal comfort using a single overhead radiant heater, one needs to vary the intensity of radiated energy (for instance, variation of the density of the radiation flux - $\boldsymbol{q}_{GR}$, the distance to the body or the angular coefficient of radiation – $\boldsymbol{\varepsilon}_{GR\text{-}T.P}$) until the attainment of acceptable temperature of sensation t_{Sen} from a subjective or hygienic point of view.

The following inequality expresses this condition[14]:

$$t_{Com}^{D} > t_{Sen} = t_a + \frac{aq_{P_1}}{\alpha} > t_{Com}^{Up},\ ^\circ C \tag{4.3}$$

where

- t_{Sen}, °C – actual temperature of sensation;
- t_a, °C – air temperature;
- t_{Com}^{D}, °C – lower temperature limit of thermal comfort;
- t_{Com}^{Up}, °C – upper temperature limit of thermal comfort;
- a – coefficient of solar energy absorptivity;
- α, W/m^{2}K – coefficient of heat exchange;
- q_{P_1}, W/m^2 – thermal flux radiated by the gas device to point P located on the control surface (see Figure 4.18).

Considering an empirical relation between the coefficient of solar energy absorptivity and the heat exchange found in the laboratory of Shrank Company, the inequality takes the form:

$$t_{Com}^{D} > t_{Sen} = t_a + 0.0716{*}q_{P_1} > t_{Com}^{Up},\ ^\circ C \tag{4.4}$$

Performing some algebraic transformations we obtain an expression, which is convenient in the selection of overhead radiant heater parameters

$$q_{P_1} = 13.97\left(\bar{t}_{Sen} - t_a\right),\ W/m^2 \tag{4.5}$$

keeping the restriction:

$$q_{GR}^{D} \geq q_{GD}^{Nom} \geq q_{GR}^{Up},\ W/m^2$$

Here

- $\bar{t}_{Sen} = 0.5\left(t_{Com}^{D} + t_{Com}^{Up}\right)$,°C – mean temperature of sensation with respect to a specific activity, clothing and a representative excerpt of eventual occupants;
- t_a,°C – air temperature specified by the technical-economical project;
- q_{P_1}, W/m^2 – thermal flux to point P located on the control surface (see Figure 4.18), analytically found via expression (4.9);
- q_{GR}^{Nom}, W/m^2 – nominal heat flux of radiation of a standard gas device (q_{GR}^{D} and q_{GR}^{Up} heat flux values that should not decrease or be exceeded).

[14] The proposed expression is an analogue of the so-called "air-solar" temperature t_{Sol}.

The analytical determination of the heat flux q_{P_i} to point $P(x_P, y_P, z_P)$, located on the control surface is performed by geometrical projection of the radiant area $\boldsymbol{A}_{GR}$ of the gas device on a semispherical surface with radius r = 1 (Figure 4.18).

Consider an elementary area dA_{GR} of the radiant surface with center M(0, 0, 0) – Figure 4.18. Assume that the change of the radiation density at point **P** obeys the cosine law of Lambert, reading:

$$dq_P = q_S dA_{GR} \cos(\theta), \tag{4.6}$$

where

- q_S, W/m^2 – nominal radiated flux with active area A_{GR} (A_{GR}=H.B);
- dq_P, W/m^2 – intensity of the radiation flux at point **P** emitted by the elementary surface dA_{GR};
- θ – angle of observation defined by the normal $\rightarrow$ and the straight line MP (Figure 4.18).

The analytical expression (4.6) can be simplified if the radiation flux q_S is presented as a superposition of two components:

- $dq_P^{A_{GR}^h}$, W/m^2 – due to the horizontal projection of the radiant surface;
- $dq_P^{A_{GR}^v}$, W/m^2 – due to the vertical projection of the radiant surface.

Expression (4.6) reads:

$$dq_P = dq_P^{A_{GR}^v} + dq_P^{A_{GR}^v} = q_S \cos\theta^h . dA_{GR}^h + q_S \cos\theta^v . dA_{GR}^v =$$
$$= q_S \cos\theta^h . dA_{GR} \cos\alpha + q_S \cos\theta^v . dA_{GR} \sin\alpha, \text{W/m}^2.$$

Cosines of the observation angles θ^H and θ^V are expressed by means of the coordinates of the control point **P**:
and

$$\cos\theta^V = \frac{z_P}{\sqrt{x_P^2 + y_P^2 + z_P^2}}.$$

Angle α is between the radiant surface/plane and plane XOZ (see Figure 4.18).

Hence, after integration over the entire surface of the radiant heater with area A_{GR} (A_{GR}=H.B), we obtain an expression of the intensity of the radiation flux at point **P**:

$$q_P = q_S \frac{H.B}{\sqrt{x_P^2 + y_P^2 + z_P^2}} \left(\sqrt{x_P^2 + y_P^2} . \cos\alpha + z_P . \sin\alpha\right), \text{W/m}^2 \tag{4.7}$$

as well as an expression for the angular coefficient of irradiation:

$$\varepsilon_{GR\text{-}p.P} = \frac{q_P}{q_S} = \frac{H.B}{\sqrt{x_P^2 + y_P^2 + z_P^2}} \left(\sqrt{x_P^2 + y_P^2} . \cos\alpha + z_P . \sin\alpha\right), \text{W/m}^2 \tag{4.8}$$

or

$$q_S = \frac{q_P}{\varepsilon_{GR\text{-}p.P}}, \mathrm{W/m^2}, \tag{4.9}$$

where

- B_{GR}, m – width of the gas device;
- H_{GR}, m – height of the gas device.

When the gas radiant heater is installed vertically and near the wall, expression (4.8) takes the following form assuming that α=0°:

$$\varepsilon_{GR,p.P} = \frac{q_P}{q_S} = \frac{H_{GR}.B_{GR}\sqrt{x_P^2 + y_P^2}}{\sqrt{x_P^2 + y_P^2 + z_P^2}}, \mathrm{W/m^2} \tag{4.10}$$

Regardless of the simpler expressions, used to calculate the angular coefficients of irradiation, the employment of computer codes is necessary.

Figure 4.18,g) illustrates the physical radiation in quadrant 0–90° of a gas device, type "GI-01", with thermal intensity 160 kW/m² [38]. The experimental evidence shows that the frontal intensity of radiation departs from Lambert's cosine law (maximal intensity occurs at 70°), which may be due to the screening grid. The integral efficiency coefficient is 52%. In reality, the angle of installation of a flat ceramic overhead heater with respect to the horizon is up to β=60°, if necessary (see Figure 4.19).

Considering a particular installation consisting of a gas radiant heater, one faces **two actual tasks:**

- Synthesis of the installation;
- Analysis of the radiation needed to attain thermal comfort in an occupied area.

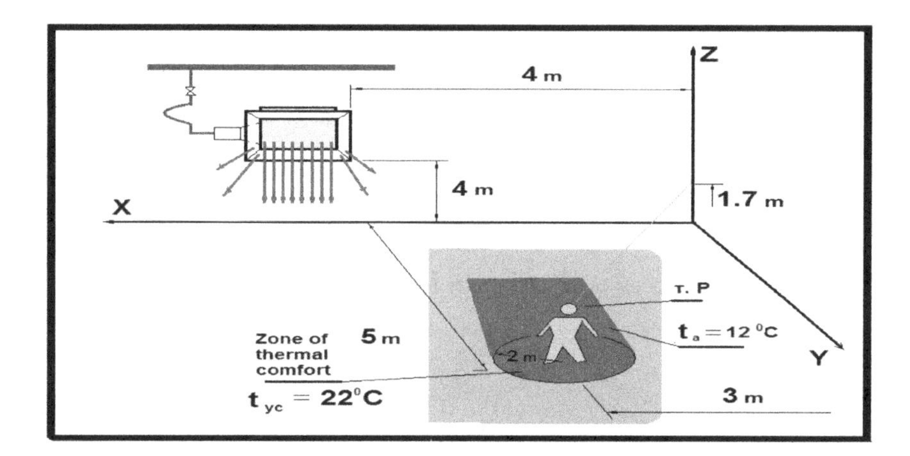

FIGURE 4.19 Calculation of an overhead radiant heater in a micro-zone of thermal comfort.

The first task requires finding the **acceptable parameters of the indoor environment** (air temperature t_a and temperature of sensation t_{Sen}, air velocity w_a and humidity (φ_a), considering the activities in the occupied areas. One should also find the nominal power of the **radiant bodies, their number and positioning** in the inhabited area. Moreover, the degree of irradiation of occupants' heads should also be considered, and an axonometric scheme of the installation should be prepared. Note that in the case of radiant heaters the application of the principle of thermal balance to rooms and the thermal area seems to be inappropriate. Hence, one cannot use effectively Eq. (4.2), while expressions (4.4) and (4.5) would assess the admissible temperature of sensation t_{Sen}.

Calculation of the device consists in the estimation of its nominal radiation power q_{GR}^{Nom} and hence, the determination of its thermal power Q_{GR} and dimensions. Substitute for that purpose the numerator of Eq. (4.10) for Eq. (4.5). Then, we find the modified Eq. (4.11) needed to calculate the nominal radiation flux q_{GR}^{Nom}:

$$q_{GR}^{Nom} = \frac{q_P}{\varepsilon_{GR,p.P}} = \frac{13.97\left(\overline{t}_{Sen} - t_a\right)}{\varepsilon_{GR,p.P}}, W/m^2 \tag{4.11}$$

The nominal thermal power of the gas device is:

$$Q_{GR} = \frac{q_{GR}^{Nom}}{\eta_R} A_{GR}, W \tag{4.12}$$

where

- η_R – device coefficient of the transformation of the released thermal energy into radiation energy (specified by the technical passport of the device). Typical values are: $0.5 < \varepsilon_{GR} < 0.6$;
- $A_{GR,}$ m^2 – active radiation surface of the gas device (see the commercial catalog of the manufacturer).

Control of the degree of irradiation of occupants' heads is performed at the end of the calculation procedure, since occupants experience indisposition and pain under excessive irradiation. The maximal radiation flux exceeding the acceptable norm and absorbed by occupants' head is assessed as:

$$q_{Head}^{max} = 10.2\left(t_{Head}^{Com} - T_R\right), W/m^2,$$

while the following condition should check the irradiation sufficiency:

$$q_{Head} = 233 - 12.9t_a, W/m^2. \tag{4.13}$$

We illustrate the discussed algorithm by the following cxample.

Example 4.2

Calculate the nominal radiation power of an overhead radiant heater positioned in a room-pavilion where the temperature of sensation should be maintained as t_{Sen}=+22°C. The air temperature should be t_a=+12°C in a micro-zone of comfort with radius 2 m and around point P with coordinates 3; 5; 1.7 m. The overhead radiant heater is 4 m away from the wall and 4 m above the floor (see Figure 4.19).

SOLUTION

The installation of the radiant heater should be such as to prevent the radiated flux from meeting the cold surface of the outer walls.

The main calculations concern the center of the comfort micro-zone – Point P (3; 5; 1.7). Control of the irradiation of the occupants' heads is performed with respect to the nearest point to the radiator.

- Calculation of the irradiation coefficient $\varepsilon_{GR,p.P}$ using Eq. (yp.4.8). $\varepsilon_{GR,p.P} = 0.37$;
- Calculation of the nominal radiation power q_{GR}^{Nom} (yp. 4.11):

$$q_{GR}^{Nom} = \frac{13.97(22-12)}{0.37} = 377.6, W/m^2$$

(Select the type and dimensions of the radiator. The active radiant surface A_{GD} and the transformation coefficient η_{GR} are taken from the manufacturer's catalog).

- Calculation of the nominal thermal power of the gas device – Eq. (4.12):

$$Q_{GR} = \frac{377.6}{0.55} 0.4 = 274, W;$$

- Check of the irradiation of occupants' heads – Eq. (4.13):

$$q_{Head} = 233 - 12.9.12 = 78.2, W/m^2;$$

- The radiation flux at point P is – Eq. (4.9):

$$q_{GR,p.P} = 377.6 \times 0.37 = 139.7, W/m^2.$$

Since $q_{GR,p.P} > q_{Head}$, the overhead radiant heater should be shifted from the center of the comfort micro-area (if for instance $\varepsilon_{GR,p.P} = 0.2$), so as to decrease the intensity of the radiated flux by 61.5 W/m^2.

If the condition of comfort in the micro-area (expression 4.4) is violated and the radiation flux is insufficient, additional radiators may join the comfort micro-area – see the illustration in Figure 4.20. To assess the total radiation effect, one should superpose the fluxes radiated by all heaters installed in the rooms.

The **second set of tasks** related to radiant heating installations are so-called checks of the degree of thermal comfort in micro-areas or premises, also known as tasks of the analysis. The subsequent problems are solved performing automated numerical simulations in order to assess the distribution of the temperature of sensation t_{Sen} at different points P_i of the comfort micro-area or throughout the global inhabited area. The results are plotted as isotherms. Expression (4.14) seems to be convenient in the calculation of $t_{Sen}(P_i)$. It is derived from expressions (4.4) and (4.9):

$$t_{Sen}(P_i) = t_a + 0.0716 * q_{P_i} = t_a + 0.0716 * q_{GR}^{Nom} * \varepsilon_{GR,p.P}(x_{P_i}, y_{P_i}, z_{P_i}) \quad (4.14)$$

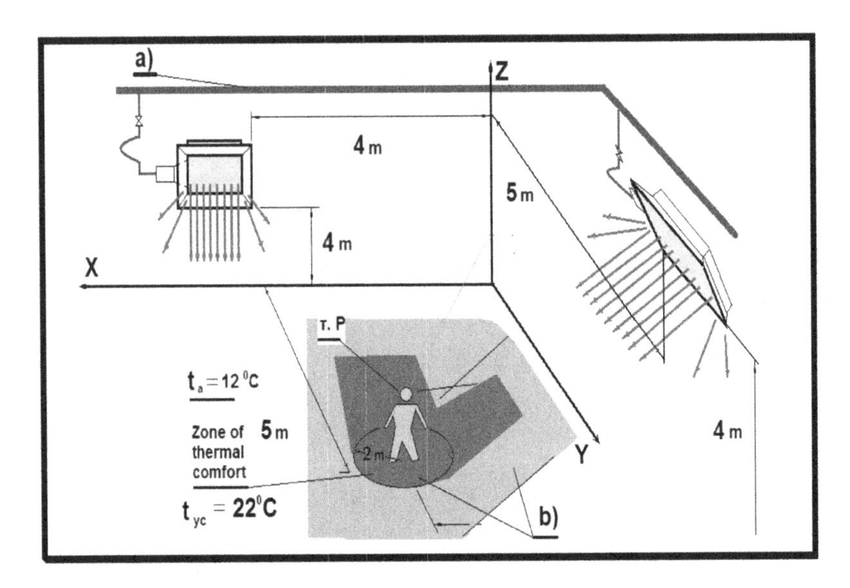

FIGURE 4.20 Comfort micro-area – comfort is maintained by a set of overhead radiant heaters: a) Pipework with connected radiant heaters; b) Isolines of the temperature of sensation in the inhabited area.

At the same time, we calculate the deviation of $\boldsymbol{t}_{Sen}$ at each point P_i from the limit values t^{D}_{Com} and t^{Up}_{Com}. Thus, we locate sites of thermal discomfort by outlining the sub-areas where the deviation of $\boldsymbol{t}_{Sen}$ exceeds 3%.

The results presented as 2D or 3D maps of the discomfort areas are enclosed in order to arrange the workplaces of the occupied areas. If necessary, the information gained is used to install additional radiant heaters or replace the existing heaters with more powerful ones.

Except that overhead radiant heaters do not have heat exchangers, they also lack flues, and combustion products freely diffuse into the surrounding medium. This specificity introduces certain restrictions imposed on the heated volume. Design recommends that the ratio "heated volume/installed power" be 10:1 m³/kW in flueless radiators.

Example 4.3

Check whether one can install an overhead radiant heater with total power of 32 kW in a space with dimensions 20 × 15 × 5 m.

SOLUTION

Perform the check by solving the empirical inequality:

$$V_{Heated} \geq 40Q_{Setting},\ m^3. \tag{4.15}$$

Substituting the respective data, we find:

$$20 \times 15 \times 5 \geq 40 \times 32$$

TABLE 4.5
The acceptable concentration of air pollution

Polluter	%	Ppm
Carbon dioxide	**0.28**	**2800**
Carbon monoxide	**0.001**	**10**

or

$$1500 \geq 1280\,\text{m}^3,$$

i.e. we can install the device.

One should control whether the concentration of carbon dioxide is admissible – 0.28% or 2800 **ppm,** to maintain air comfort – see Table 4.5 and Sub-par. 4.1.2 for details. Carbon dioxide is treated as an original indicator of air comfort. Except for carbon dioxide CO_2, combustion products also contain carbon monoxide CO and nitric oxides NO_x.

4.1.2.3.2 Radiant Tube Heater

Except for ceramic overhead heaters, radiant heat installations include also radiant tube heaters belonging to the group of so-called "dark and sub-dark radiators". They operate at a temperature within the range $200 < t_w < 500°C$, while the generated energy flux is with high Lagrange multiplier consisting mainly of infrared photons.

The structure of such a radiator is shown in Figure 4.21. The typical radiant tube heaters are 7.0 m long and U-shaped. A reflector (Item 6) is mounted on the combustion pipe (Item 8), whose diameter varies from 0.06 to 0.25 m. It is fabricated from polished aluminum or stainless steel.

A gas burner of ejector type is mounted at one end of the fuel pipe. A vacuum fan with a nominal flow rate of about 4000 m^3/h is installed at the other end. Frequency varying invertors control the flow rate in some devices.

A gas-distributing pipe network supplies the fuel, and it is installed in the heated area considering the location of the gas radiant tube heaters. The devices are connected to the local branch of the pipe network by means of an isolation valve (Item 4) and a flexible pipe connector (Item 5).

The supplied fuel runs into the ejector (Item 7) with high velocity, mixing with the air. The mix ignites and burns in the radiant tube (Item 8), while tube 9 conveys the combustion products. Smoke gases carry the released heat heating the tubes. On the other hand, tubes radiate the accumulated energy in the surrounding space increasing the radiation temperature $\boldsymbol{T}_R$ in the occupied area.

Since the wall surface temperature $\boldsymbol{t}_w$ of the radiant tube heaters is significantly lower than that of the ceramic plaque heaters, their assembly height is smaller than

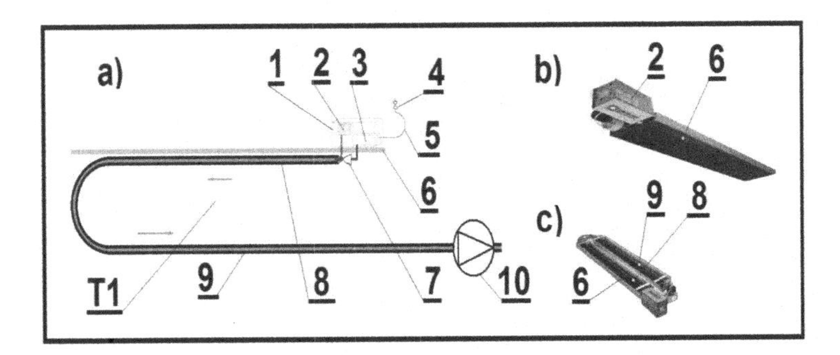

FIGURE 4.21 Radiant tube heaters: a) Scheme of the structure of a radiant tube heater; b) and c) Axonometric view of radiant tube heaters 1 – Intake for combustion air; 2 – Block of combustion control; 3 – Multifunctional gas valve; 4 – Isolation valve; 5 – Flexible connector; 6 – Reflector canopy; 7 – Gas ejector and burner's head; 8 – Combustion radiant tube; 9 – Fuel tube; 10 – Vacuum fan.

that of the plaque heaters. This makes the radiant tube heaters suitable to buildings with a smaller distance between the stories[15].

Gas overhead radiant heaters can be installed individually or in groups (successively or in parallel)[16]. They create separate comfort micro-areas or provide comfortable homogeneity of the entire area – the parameters of the latter are selected pursuant to the function of the inhabited area. Exemplary connections of radiant tube heaters are shown in Figures 4.21 to 4.24.

The first one is a scheme of gas overhead radiant heaters with serially connected flues and a common vacuum fan. The radiant heating bodies are arranged in a straight

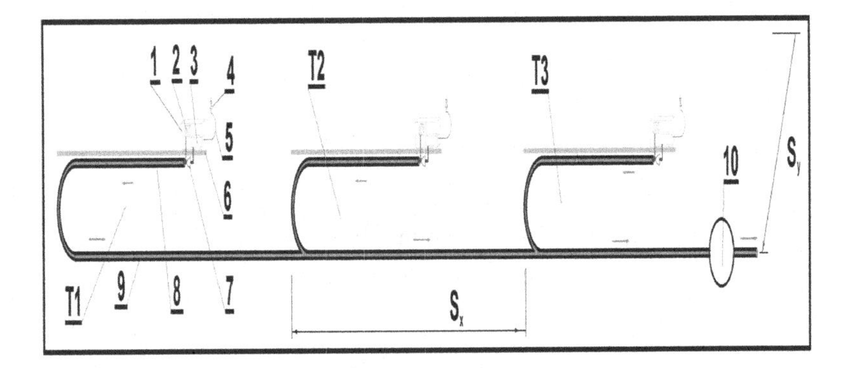

FIGURE 4.22 A radiant heating system compounded of tube heaters serially connected to a central flue T1-T3: 1 – Air Intake; 2 – Block controlling the ignition; 3 – Multifunctional gas valve; 4 – Isolation valve; 5 – Flexible gas connection; 6 – Reflector canopy; 7 – Gas ejector and burner head; 8 – Radiant tube; 9 – Flue tube; 10 – Vacuum fan.

[15] The survey of literature on radiant heating systems shows that they are widely applied in various animal farms breeding cows, pigs, poultry etc.

[16] Proposed power values are 50, 100, 115, 150, 180, 200, 250, 300 and 400 kW.

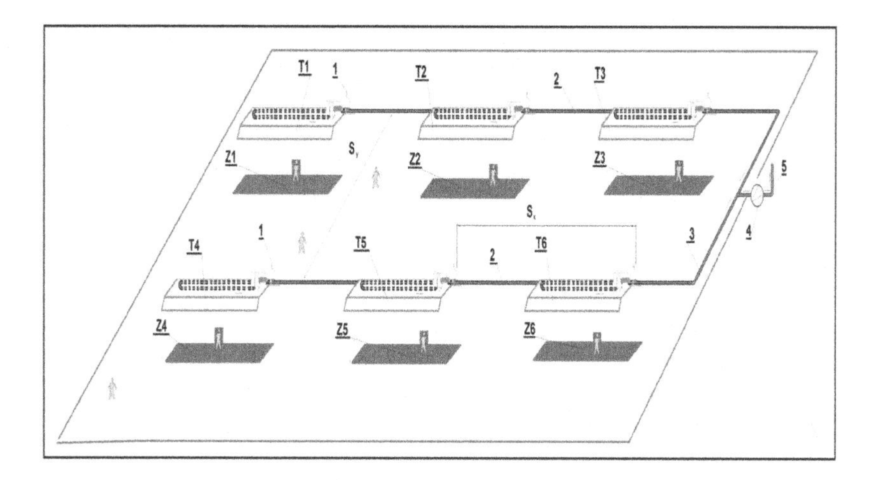

FIGURE 4.23 Scheme of a gas radiant heating system compounded of two parallel branches of tube heaters T1-T3 and T4-T6 serially connected to a central flue system. Z1-Z6 – areas of thermal comfort: 1 – Flexible gas connectors; 2 – Serially connected flue tubes; 3 – Collector; 4 – Vacuum smoke fan; 5 – Chimney.

line with step S_X. There may be several lines of radiant heaters in an occupied space, where the installation step is S_Y.

The overhead radiant heaters are installed in a chess-board order or at the knots of a rectangular grid. The situation of the gas radiant bodies should be optimized by varying steps S_X and S_Y till the attainment of a minimal mean quadratic deviation of the thermal sensing ($\boldsymbol{t}_{\mathrm{Sen}}$).

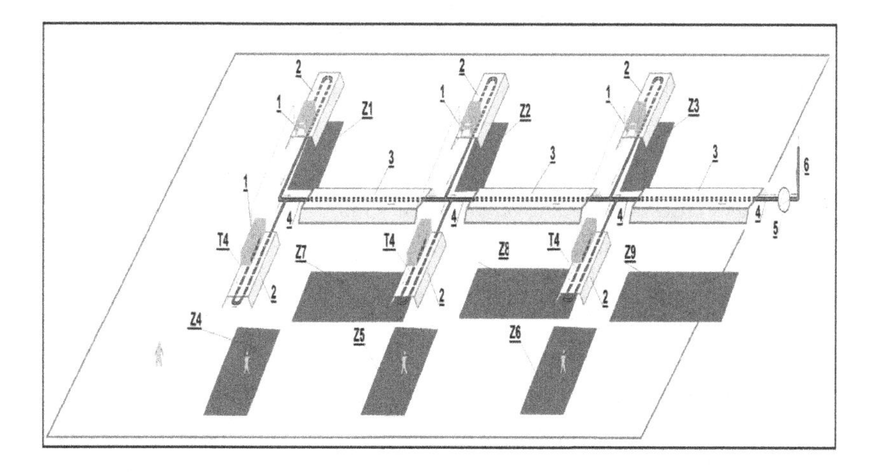

FIGURE 4.24 Scheme of a gas heating system compounded of pipe radiators T1-T6 connected in parallel to a central flue. Z1-Z6 – areas of thermal comfort: 1 – Flexible gas connection; 2 – Pipe radiators; 3 – Flue reflector; 4 – Central flue; 5 – Suction smoke ventilator; 6 – Chimney.

Equation (4.4) is used to assess $t_{Sen.}$ To calculate the angular coefficient of irradiation, we recommend the use of the following expression. It is derived by simplifying Eq. (4.8):

$$\varepsilon_{GR,p.P} = \frac{q_P}{q_S} = \frac{z_P H.B}{\sqrt{x_P^2 + y_P^2 + z_P^2}}, \tag{4.16}$$

and it is applicable to horizontally oriented radiant heaters.

Since several fans, corresponding to the number of the serial groups, exhaust the combustion products as shown in Figure 4.22, the next Figure 4.24 shows the installation of a single suction valve. This is a definite advantage of the installation, but the latter requires precise regulation of the gas-dynamic resistances of the flues of individual sections and branches. This is done by means of orifice plates or regulating sliding valves.

Note that the diameters of the flue pipes should increase, due to the total increase of the flow rate of smoke gases along the route of their conveyance (Second law of Kirchhoff) and the occurrence of high gas-dynamic pressure losses. The mean gas velocity through the pipe cross-section should not exceed 4–6 m/s, since a higher velocity decreases the efficiency of the gas devices.

Figure 4.24 shows an alternative connection scheme of gas radiators, where a single central flue (Item 4) and a fan (Item 5) convey the combustion products. The new component here is the flue reflector (Item 3), whose function is to facilitate the operation of the main gas radiators in the central area of the premises and to increase the energy efficiency of the installation.

The qualities of the pipe radiators discussed so far prove the advantages of band radiators used in heating installations. These are:

- Flexibility with respect to the building structure;
- Control of the radiation power by means of a microprocessor system – account for the temperature of the smoke gases and the radiation temperature within the occupied area.

4.1.3 Advantages and Disadvantages of Gas Radiant Heating Installations

Gas-air radiant heating installations occupy a market niche, which continuously expands. This is due to certain characteristics of the devices, such as:

- Relatively small investments (low consumption of metals and labor during installation) – it is 2–7 times lower than the capital investments in a traditional installation;
- Economic exploitation – exploitation costs are 6 times lower than those of the heating systems "water-air";
- Energy-effective impacts (saving up to 40–50% of primary energy carriers);
- "Weak-inertia" devices. They immediately start operation.

Gas overhead radiant heating installations are used in sites, which undergo unfavorable meteorological impacts – wind, rain or snow. Advantages: (i) capable of creating outdoor areas of thermal comfort; (ii) do not generate convective fluxes and flows; (iii) in contrast to convective systems, no air stratification is observed; (iv) significant operational durability (over 10000 h).

Our considerations prove that the current technological level of gas systems for thermal comfort allows their wider use in residential, industrial and agricultural buildings.

4.2 GAS EQUIPMENT AND CONTROL OF INDOOR AIR QUALITY

4.2.1 Indoor Air Quality and Methods of Its Attainment

The necessity of high indoor air quality (IAQ) in the occupied areas of a building is one of the vital needs of humans. Fresh air is the second mandatory component of the building comfort. Oxygen is a vitally important element not only for power devices and stations, but also for the functioning of the human body. At the same time, oxidation wastes (for instance, carbon dioxide (CO_2) and water) should be "expelled" out of the human cells and the entire human body[17].

The already discussed dissimilation processes running in the human body are accompanied by oxidation at cell level, where oxygen is needed for the regular running of the processes. Under normal indoor conditions, a man needs about 7–7.5 l/s fresh air.

The IAQ in a building space is high if the following three conditions are obeyed:

a. Lack of unknown contaminants or presence of polluters with non-hazardous concentration;
b. More than 80% of individuals do not experience discomfort due to the ambient air;
c. No smell and release of hazardous amounts of contaminants.

In occupied areas (residential buildings, offices or nearby technological stations) the IAQ is of essential importance for people's activity and health. Lack of fresh air yields fatigue, weakness, headache or dizziness of occupants [20]. An acceptable IAQ is achieved by restricting the concentration of all known contaminants to some specified acceptable levels.

The reasons for air contamination fall into three groups:

- Release of physiological waste by humans and mammals – carbon dioxide and water vapors;
- Indoor desorption of contaminants from building materials, "wet" technologies or technological equipment (including gas devices);
- Building pollution by gaseous or aerosol contaminants introduced by the outdoor air (due to environmental pollution, radon leakage etc.).

[17] From an ecological point of view, a human being is a specimen of the species Homo sapiens, occupying the top of the food chain, consuming oxygen and food and releasing CO_2, H_2O and excrements.

The ambient air is a mixture of oxygen (23%), nitrogen (76%), carbon and nitro oxides, water vapors, aerosols etc. Some of the air components, such as CO, Cl, SO_2, HS, CS_2, NO_x, are aggressive to the human tissues. They are poisons that may interact with the eye's mucosa, respiratory organs or open skin areas. Under low concentrations and short time of contact they cause inflammation while under high concentration – wounds, various allergies and other permanent injuries.

Being part of the atmosphere, gases CO_2, NO_2, water vapor, methane (NH_3) are neutral compounds with respect to the human physiology. Although nontoxic, their presence in an occupied area is unwanted, since they decrease the oxygen concentration.

Carbon dioxide (CO_2) is a natural product of the dissociative processes running in the human body. It has no color or smell, but is heavier than the air and concentrates near the earth's surface or slightly above the floor. Its concentration in buildings depends on the number of occupants. It can penetrate the occupied areas through voids or through the diffusers of an air conditioning system.

The indoor air contaminants are due to products released by operating gas devices or to the infiltration of smoke gases, water vapors, carbon dioxide or monoxide. Products due to human metabolism, as well as cigarette smoke, also pollute the air. The acceptable concentration of carbon dioxide in an occupied space should not exceed 5.04 g/m^3 (or 2800 ppm) – see Table 4.5.

Methane and other fuel gases may also behave as contaminants under a state of emergency. Normally, they are non-toxic, but like carbon dioxide, they decrease the amount of oxygen in the building[18] and have explosive potential. Their physical properties are discussed in Sub-par. 2.1.

Quite often, the incomplete oxidation of fuel releases a significant amount of carbon monoxide. It has no color and smell but is strongly toxic. A concentration exceeding 10 ppm is considered to be hazardous. Being strongly aggressive, carbon monoxide is easily absorbed by the human body. People, having inhaled CO, experience headache, nausea and weakness[19].

Sites potentially endangered by CO release are closed parking lots and garages, loading docks, crossroads regulated by traffic lights, spots of traffic jam where the engine velocity is idling or low.

Another toxic gas is the cigarette smoke, which can pose serious problems to allergic occupants, since it causes spasms and suffocation. Cigarette smoke stimulates also the occurrence of various throat diseases, including laryngitis and tumors.

Dust in the air in the form of aerosols causes diseases of the respiratory system known as "cones", of which the silicosis is the most spread one among people working or living in dusty areas.

Finally, formaldehyde gases can be present in the indoor air of a new building. The time of their release is long (12–13 months) and they cause respiratory problems to hypersensitive occupants expressed in asthmatic and allergic reactions, eyes

[18] When methane concentration in an occupied space is over 20%, people experience suffocation. Long-term stay in rooms gassed with methane yield collapse of the human nervous system.

[19] They should be immediately hospitalized.

and skin inflammation etc. Their acceptable concentration is 1 ppm per 8 hours of peoples' presence or 0.1 ppm per longer time.

The formation of indoor air is spontaneous or controlled by environmental impacts (solar light, wind, pollution). If gas devices are present, air needs oxygen for humans, as well as for gas combustion.

Note also that the IAQ is an important issue in industrial buildings, where technological devices release toxic gases.

In public buildings, where population density may reach 3–10 individuals/m^2, the neutralization of the products of human metabolism (CO_2, H_2O, methane or unpleasant smell) is mandatory, not only due to the decrease of oxygen concentration, but also due to the hazards of mass infections and pandemics.

In short, air contamination is a result of:

- Human activity;
- Presence of animals;
- Contaminants emitted by building materials (paints, plasters), as well as volatile organic compounds (VOCs);
- Emissions owing to indoor processes (VOCs);
- Growth of fungal material (mold);
- Smells and large particles.

Hence, indoor air may contain soothe, smoke, sand, metal powder, organic or vegetable waste in the form of fibers, mold spores, bacteria, viruses or plant pollens[20]. When bacteria or viruses are present, occupants may contract a flue, pneumonia and other respiratory diseases[21].

The basic cleaning methods (air ventilation) of a building are four (applied individually or in a combination):

1. Removal or modification of the sources of contamination;
2. Use of outdoor air to ventilate the indoor space (space ventilation);
3. Local ventilation;
4. Air cleaning.

The first method is the most effective one with respect to some contaminants that can be visually identified, such as cigarette smoke, combustion products, paints, mold spores or smoke gases. It consists in the elimination of the contamination source (for instance, no smoking in public buildings and restaurants). Considering other contaminants (including CO_2 due to human and mammals' activity), with no fixed source or with sources spread throughout the building, this method is inapplicable.

[20] There are several agents in occupied building areas, which affect human health: gaseous pollutants, suspended material particles, viruses and bacteria, allergenic agents, mold, fungi and radioactive materials.

[21] Hardening and calcification of the alveoli.

TABLE 4.6
Prescriptive flow rates of the outdoor ventilation air

Room type	Flow rate of the outdoor air- $\dot{V}$	
	l/s – individual	l/s –m^2
Kitchen – up to 20 individuals	**8**	
Dining room – up to 70 individuals	**10**	
Bedrooms		**15**
Living rooms		**15**
Baths		**18**
Offices – up to 7 individuals	**10**	
Conference halls–up to 50 individuals	**10**	
Corridors		**0.25**
Public toilets		**2.5**
Swimming pools		**2.5**

The second method is applicable to contaminants released by humans or mammals. There are two procedures providing an acceptable IAQ via the outdoor air:

- Application of a specific ventilation rate (prescription of a specific rate at which outdoor air should be delivered to various sites – for example, see Table 4.6);
- Improvement of the IAQ.

The improvement of the IAQ consists in the restriction of the concentration of all known contaminants by means of:

- Quantitative evaluation;
- Subjective evaluation.

The quantitative evaluation of IAQ involves acceptable levels of indoor contaminants generated by various sources. The use of outdoor "fresh air" to "refresh" the air in an occupied space is illustrated in Figure 4.25, which shows a general scheme of space ventilation.

An intake **confuser** sucks the outdoor air, which undergoes preliminary cleaning in the filter (Item 2), having passed through the regeneration section (Item 1)[22]. Consider the case when the temperature of the air in the ventilated rooms is lower than that of the supplied air (for instance, by more than 10°C). Then, the latter warms up (under the head of the supply fan (Item 3)) in the heat exchanger (Item 4) that thermally processes the fresh air – see also Figure 4.7.

Thus prepared, the air is supplied to the building to remove the contaminants[23]. Having mixed up with the indoor air, the "fresh" air is sucked by the return fan

[22] It preliminarily heats the fresh air at the expense of the "conditioned" polluted air. Economizers, thermal pumps or contact heat exchangers are used in his case.

[23] To avoid the formation of standstill and recirculation areas, the air is supplied through the diffusers (Item 5), "sticking" it to the ceiling or walls (effect of Coanda).

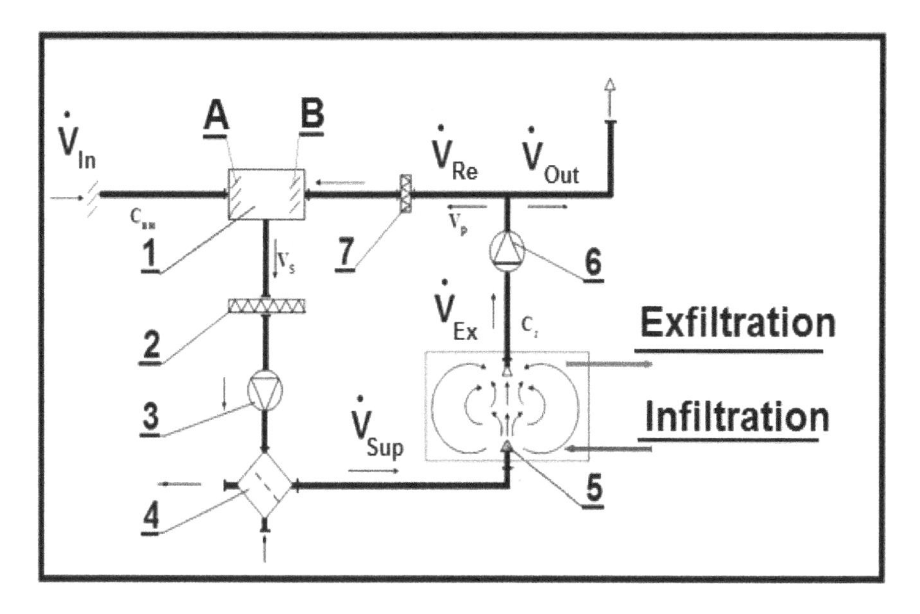

FIGURE 4.25 Space ventilation: 1 – Energy regenerator; 2 – Filter; 3 – Supply fan; 4 – Heat exchanger; 5 – Object (contaminated volume); 6 – Return fan; 7 – Filter.

(Item 6) and is exhausted. Note that it contains contaminants, but their concentration is lower than the admissible limit[24].

The flow rate of outdoor fresh air necessary for ventilation is calculated using two methods[25]:

- Method of determining the prescribed flow rate of ventilation;
- Method of attaining acceptable IAQ.

The first one recommends the intake of outdoor air in different rooms, in the entire building and in the area of occupation. The air flow rate is given in Table 4.6 (see also Table P9). The following example illustrates the method[26].

Example 4.4

Estimate the flow rate of the outdoor air necessary for the ventilation of the household shown in Figure 4.26.

[24] If the contaminant is not life-threatening (there are no bacteria or viruses or toxic gases or radon) part of the air is used in mixing economizers. In the opposite case, it is used in heat exchangers or thermal pumps while the polluted air is discharged from the building after pollutant elimination.

[25] The use of the first one is more convenient in the design although that it yields higher prescribed flow rate.

[26] We use the following formulas (A_0 – area; N – number of inhabitants; V_{St} – recommended value specified in Table 4.6): $\dot{V}_{BH}$ = A0.V˙St; $\dot{V}_{BH} = N.\dot{V}_{St}$

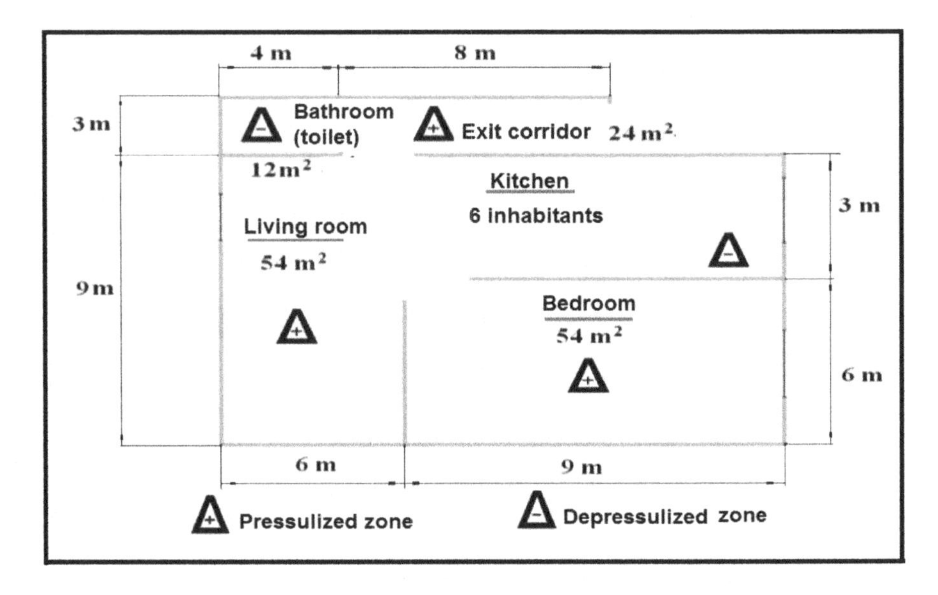

FIGURE 4.26 Plan of a one-room apartment.

SOLUTION

The flow rate of the needed ventilation air (called "nominal" flow rate) is calculated using formulas (4.17) and (4.18), while the results are given in Table 4.7.

The second method of finding the flow rate of the outdoor air is applicable just when the building is erected and brought into exploitation. We apply two variants:

- **Variant A:** calculation;
- **Variant B:** performance of experiments and subjective evaluation of the IAQ by odor observers.

TABLE 4.7
Estimation on the necessary fresh air flow rates in Example 4.4

Premises	Characteristics A_0 / N-m²/number of inhabitants	Prescriptive flow rates $\dot{V}$ l/s-m² / l/s-N	Nominal flow rate $\dot{V}$ m³/s
Living room	54m²	15	0.810
Bedroom	54m²	15	0.810
Kitchen	6 inhabitants	8	0.048
Bathroom/ toilet	12m²	18	0.216
Exit corridor	24m²	0.25	0.006

Pursuant to **variant A** we use the following formula:

$$\dot{V}_{Intake} = \frac{N_t}{60\left(C_i - C_e\right)},\ m^3/s, \tag{4.19}$$

where

- $N_{t,}$ l/min – flow rate of the generated pollutant;
- C_i,%, ppm – acceptable concentration of contaminants;
- C_e,%, ppm – concentration of contaminants in the supplied outdoor air.

Example 4.5

Find the necessary flow rate of outdoor air needed to ventilate a boiler room whose plan is shown in Figure 3.13, if the flow rate of the generated CO_2 is N_{CO2}=0.2025 l/min; the acceptable contaminant concentration is C_i=0.28% (2800 ppm); concentration of pollutants in the supplied outdoor air is Ce=0.03% (300 ppm).

SOLUTION

We introduce the data in expression (4.19):

$$\dot{V}_{Intake} = \frac{0.2025}{60\left(0.0028 - 0.0003\right)} = 1.35,\ l/s.$$

This is the minimal flow rate, needed by the ventilation of the boiler room. If "ventilating – indoor air" mixing is efficient (75%), the above value should be multiplied by 1.25.

The nominal flow rate of the supply fan should be 6075 m^3/h. Its head is selected in the course of the flue calculation.

To realize **variant B** of the method "Procedure of attaining an acceptable flow rate of the ambient air", observers specify the flow rate of outdoor ventilation air pursuant to their tolerance to certain air contamination. They select a flow rate, at which 80% of them do not sense the presence of a pollutant. However, the selection is highly subjective, since observers are appointed at random.

Most designers follow the first procedure, involving a **Ventilation Volume Rate**. However, one should operate with larger volumes of outdoor air in that case, facing also higher operative costs. Yet, the procedure is convenient for use.

The third method of contaminant removal – **local ventilation**, is quite similar to the second one in character and capabilities. The main requirement is the presence of a contaminant, whose location and type (generation) are specified (for instance, owing to specific technological equipment or appliance).

The method runs in two steps:

- Keeping or "encapsulating" the contaminant in the source vicinity;
- Local ventilation and neutralization.

The first step is realized by the installation of fencing walls or by the creation of a vacuum in the vicinity of the contamination source. Most often, local depressurization emerges around the device that generates a toxic substance. Source insulation is done by means of cabins, anemostats, umbrellas, canopy, capsules etc. The released contaminants mix with the air and are ventilated out of the room. Then, they are discharged outside the building after appropriate processing (filtration, regeneration). The centrally supplied fresh ventilation air replaces the removed contaminants under the action of atmospheric pressure. If there is no main ventilation, the air balance is restored via air infiltration from neighboring rooms or the outdoor environment.

The **fourth method** used in systems controlling the IAQ is cleaning the air by means of filters or other devices. It is a stage or continuation of the second and third methods. Technically, air cleaning takes place by the separation of the contaminants from the air volume. Neutralization and removal of contaminants take place using mechanical, chemical, photocatalytic or membrane technologies. The methods of separation depend on: (i) dimensions and shape of contaminant particles; (ii) **particle** concentration; (iii) particle electrical and chemical properties. Note also that the separation should take into account the specific character of the actual contaminants.

One can remove gaseous contaminants by means of one of the following five methods:

1. Contaminant absorption. Air washers are used for that purpose (Typically, a solution of H_2O + additional reagents is used in the absorption of gaseous contaminants);
2. Physical adsorption. The most widely used and excellent adsorbent is **activated charcoal**. It is used in combination with other substances to yield better accommodation of chemically active gases;
3. Hemisorption – similar in many ways to physical adsorption;
4. Catalysis. The chemical reaction occurs at the surface of a catalyst, while the gaseous contaminant does not react with the catalyst itself. The gaseous contaminant combines with O_2 or with the supplied chemicals.
5. Catalytic combustion – at lower temperatures. To reduce CO_2 and CO to C and O_2, specialists typically use solid and liquid absorbents.

In other cases odor masking turns to be the best solution. This is done by the introduction of a pleasant odor to mask the unpleasant one, or by mixing two odors.

4.2.2 Building Ventilation and IAQ

A logical issue of how to organize the supply of fresh air to an occupied space faces specialists. Long-standing practice proves that this can be done as follows:

- By control of the "stack effect" of the polluted volume;
- By control of the indoor pressure in the polluted area via air exhaustion or displacement.

While the first approach is applied to higher buildings[27], the second one is universally applied to low and high buildings.

There are two rooms in a building plan, formally called "dirty" and "clean" ones. Considering functionality and human activities, specific contaminants are released in dirty rooms. For instance, vapors and cooking odors are present in kitchens. Specific contaminants are released in bathrooms and toilets too. Office equipment also releases contaminants. Although the emissions are sometimes spontaneous, they are ill-smelling and contain allergens. Their diffusion in other premises is unwanted. Humans and domestic animals release specific emissions in clean rooms with a normal regime of occupation. Their intensity is significantly weaker, but rooms are bound to ventilation.

Dirty and clean rooms are often neighboring, located on one and the same floor, in one and the same part of the building – see Figure 4.26.

The arising problem consists in how to organize simultaneous ventilation and contaminant rarefaction in the two rooms, avoiding at the same time the **transfer of pollutants from the dirty to the clean space.**

The use of total hermetic sealing and encapsulation of individual rooms or local ventilation are expensive solutions, not applied on a mass scale. Typical engineering projects use combined/universal schemes of contaminant removal, including the arrangement of volumes of depressurization and pressurization within the building.

There are three types of building ventilation systems:

a. Exhaustion systems;
b. Displacement systems;
c. Combined systems.

The problems consist in how to design a particular ventilation project – see, for instance, the apartment in Example 4.5.

The basic idea is to organize an area of pressurization with clean fresh air, in order to avoid diffusion of contaminants from a particular site. This can be done by the use of a combined exhaustion – displacement system. At the same time, the installation creates an area of depressurization in the "dirty" rooms. Thus, the contaminated air starts to convect to the dirty premises only, where it is subsequently neutralized and discharged. Figure 4.27 shows two schemes of combined systems of volume ventilation, which differ from each other by the employed method of contaminant removal:

a. Air displacement ventilation where the outdoor air is supplied through one or more intake diffusers by means of mechanical equipment (Item 1, Figure 4.27,a);
b. Combined ventilation where the polluted air is exhausted by a return fan in the ventilation station (Item 1, Figure 4.27,b). Then, intake diffusers (Item 6), direct it to a regenerator and a filter section, and it is finally exhausted through one or more diffusers.

[27] Several techniques are used for that purpose, including ventilation channels, controllable windows and vents located on the fencing.

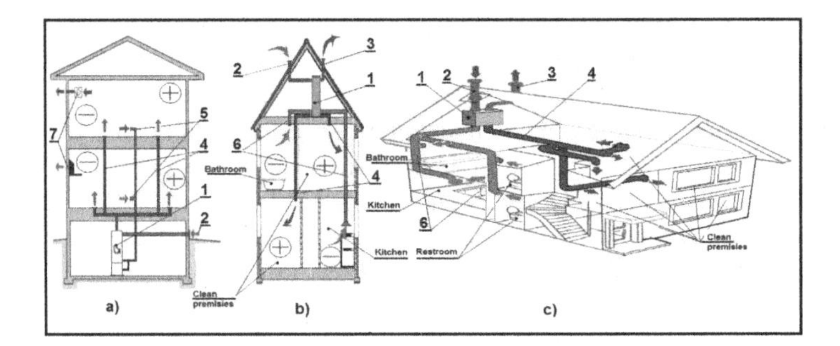

FIGURE 4.27 Combined schemes of space ventilation: a) A displacement ventilation system equipped with local intake diffusers; b) A combined exhaustion-displacement system; c) An axonometric scheme of a combined ventilation system: 1 – Ventilation handler; 2 – Intake confuser for outdoor air; 3 – Terminal for polluted air; 4 – Supply duct-clean air; 5 – Recirculation-indoor air; 6 – Exhaust duct – polluted air; 7 – Vent openings.

The axonometric scheme (Figure 4.27,c) shows that the building design, dividing the volumes into "dirty" and "clean" ones still in the architectural project, is decisive in the successful arrangement of combined ventilation installation. Note that supply and return ducts for fresh and polluted air are positioned in two different parts of the building. This facilitates the ventilation and integration with other building systems (electrical, fire safety, water and gas conveying installations).

The next example illustrates the method of ventilation using outdoor air. It is an extension of Example 4.4 where the necessary flow rates of outdoor air are tabulated and introduced in Table 4.7.

Example 4.6

Design a scheme of a ventilating installation providing air comfort to an apartment, whose scheme is shown in Figure 4.26.

SOLUTION

We use the estimated flow rates of the outdoor air – Table 4.7, which are given in the apartment plan – see Figure 4.28,a). It seems reasonable to use a combined scheme similar to that in Figure 4.27,a), where the displacement system supplies outdoor air prepared in the ventilation handler (see Figure 4.28,b) while polluted air is sucked by the return fan (for instance, Items 1 and 2, Figure 4.28,b) and is discharged out of the building.

The arrangement of the ventilation equipment plan is shown in Figure 4.28,b). The displacement air duct (Items 4 and 6) distributes the air prepared in the ventilation handler in the living room and the bedroom. The nominal flow rate of the ventilation handler amounts to 1.63 m^3/s. The overall dimensions of the air ducts are given in Table 4.8. They are calculated to operate at a velocity of 4 m/s. Air ducts can be installed in a gap over a suspended[28] ceiling or beneath an integrated floor.

[28] Two methods are used in the upper supply of fresh air: – through a perforated suspended ceiling; through diffuser creating the effect of Coanda.

TABLE 4.8
Calculation of air ducts at an air velocity of 4 m/s

Premises	Nominal flow rate $\dot{V}$ - m³/s	Cross section A_0- m²	Overall dimensions m
Living room	0.810	0.2025	0.26 × 0.78
	0.405	0.1012	0.18 × 0.55
Bedroom	0.810	0.2025	0.26 × 0.78
	0.405	0.1012	0.18 × 0.55
Kitchen	0.048	0.012	0.065 × 0.19
Bathroom/toilet	0.216	0.054	0.135 × 0.4
Exit corridor	0.006	0.0015	0.02 × 0.07

Air outflow takes place through diffusers[29], which are regulated such as to operate at a particular air velocity, not causing discomfort[30] to the occupants (about 1–2 m/s), while the actual air flow rate in the inhabited area does not exceed 0.2–0.5 m/s.

The balance of air flow rates in the controlled space of the apartment shows that the inflowing air exceeds by 1.362 m^3/s; the air sucked during the operation of the entire ventilation system. Then, the entire space undergoes pressurization. If the

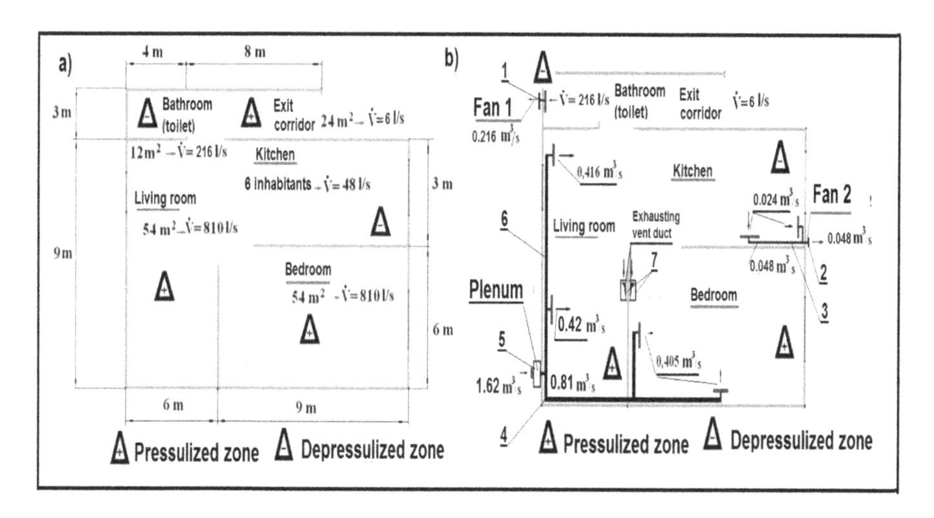

FIGURE 4.28 Design of a scheme of a ventilation installation; a) Task; b) Positioning of the ventilation equipment: 1 – Exhaust fan operating in the toilet; 2 – Exhaust fan operating in the kitchen; 3 – Exhaust duct in the kitchen; 4 – Displacement duct in the bedroom; 5 – Ventilation handler processing the outdoor air; 6 – Displacement duct in the living room; 7 – Exhaust duct used to balance the air pressure.

[29] Similar to those in AVAC systems type "air-air", ventilation installations comprise a wide variety of diffusers, for example: Linear spots, Perforated face, Louvered face or Troffers over lighting fixturs.
[30] It is assumed that a velocity of 0.4 m/s does not create discomfort for most people.

building envelope is not airproof, the indoor air leaves the control space if someone opens doors and/or windows. To reduce the unfavorable effects, one can take the following measures:

- Control of the flow rate of a fan belonging to the supply ventilation handler (Item 5, Figure 4.28) by means of a super-pressure transducer (pressure sensor);
- Design of a ventilation duct and its installation in an apartment in order to balance the air pressure (Item 7, Figure 4.28).

The third rational idea is the use of indoor air in a non-balanced gas device, for instance a fireplace in the living room, a gas radiator-convector or a gas refrigerator.

We discuss in what follows some additional aspects of the involvement of gas devices in building ventilation.

4.2.3 Gas Equipment and Building Ventilation

The above example proves the idea that unbalanced gas devices are compatible with the ventilation system[31], since the flow rate of the outdoor air may exceed that of the exhausted polluted air in combined ventilation installations. Moreover, a device yielding air natural convection, may replace a return fan.

When gas devices are installed in a building, their operation increases the natural stack effect. It can be controlled by an appropriate arrangement of the apertures perforated through the building envelope – see Figure 4.29. Figure 4.29 shows the perforation of a building facade at various levels (Figure 4.29,c and d). Thus, one can provide appropriate lift pressure, convective air lift and indoor air cleaning[32].

Pollutant lift occurs due to the redistribution of pressure acting at different points of the envelope and resulting from a controlled opening of windows.

One can estimate the lift pressure by applying the law of gas dynamic head to the open ventilation line of a building (see Figure 4.29,b) and using Eq. (2.15). Its expanded form reads:

$$-\Delta p_1 - \Delta p_2 - \Delta p_3 + \Delta p_4 = 0$$

where

- Δp_1, Pa – pressure drop at the intake;
- Δp_2, Pa – indoor pressure drop ($\Delta p_2=0$);
- Δp_3, Pa – pressure drop at the outlet;
- Δp_4, Pa – lift head.

[31] Gas devices with a net power of up to 7.5 kW operate without flues while combustion air is supplied through the voids of the envelope. Gas devices with power exceeding 12.7 kW need forced supply of fresh air or one should use dynamically balanced gas devices.

[32] A stack effect occurs only if there is a difference between the densities of the indoor and outdoor air.

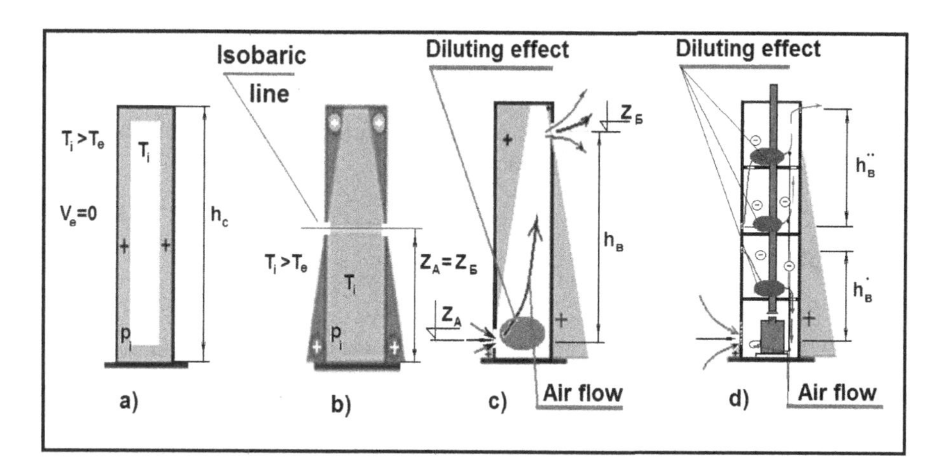

FIGURE 4.29 Stack effect in buildings: mechanisms of natural ventilation – control of pressure acting on the envelope. a) Airproof envelope; b) Envelope opens horizontally; c) Envelope opens at various levels; d) Envelope of a multistory building with open vents at different levels; positioning of an unbalanced gas device in the sub-terrain.

The dynamic lift H is found performing the substitution $\Delta p_4 = +H$, whereas

$$H = \left(p_A - p_B \right) = h_B \, g \left(\rho_A - \rho_B \right) = \Delta p_1 + \Delta p_3 . \text{ Pa} \tag{4.20}$$

Here

- $p_A = Z_A \, \gamma_A = Z_A \, g\rho_A$, Pa – pressure at point A;
- $p_B = Z_B \, \gamma_B = Z_B \, g\rho_B$, Pa – pressure at point B;
- $h_B = Z_A - Z_B$, m – "ventilation height", the difference between the vent levels – see Figure 4.29,c);
- $(\rho_A - \rho_B)$, kg/m^3 – the difference between the specific weight of air at points A and B.

The sum of pressure drops Δp_1 and Δp_3, expressed via the flow rate of the inflowing outdoor air $\dot{V}_B$, the coefficients of intake and outflow resistance – ξ_A and ξ_B, as well as the net areas of the ventilating grids (A_1 and A_3) takes the form:

$$\Delta p_1 + \Delta p_3 = \xi_1 \frac{w_1^2}{2} \rho_1 + \xi_3 \frac{w_3^2}{2} \rho_3 = 0.5 \dot{V}_B^2 \left(\frac{A_3 + A_1}{A_1 A_3} \right) \rho_B, \quad \text{Pa.} \tag{4.21}$$

Substitute Eq. (4.21) in Eq. (4.20) and solve it with respect to the ventilation flow rate $\dot{V}_B$. Then, we find the following expression, which can predict the grid geometry or the needed ventilation height:

$$\dot{V}_B = \sqrt{\frac{h_B \, g \left(\rho_A - \rho_B \right)}{0.5 \left(\frac{A_3 + A_1}{A_1 A_3} \right) \rho_B}}, \text{ m}^3\text{/s} \tag{4.22}$$

Considering multistory buildings, an adequate denivelation of the intake vents is used to assess the dynamic lift H via expression (4.1) – see Figure 4.29,d).

An alternative to the outlined approach is the use of data in Tables P8–P15[33] in order to perform fast calculations. We consider the following four ventilation problems to illustrate the alternative approach.

Example 4.7

A gas flow heater with total thermal power of 15 kW is installed in a garage – see Figure 2.27. The garage area is 36 m^2, and its ventilation is natural. Find the net cross-section of the ventilation grids.

SOLUTION

If only one door is available, only one grid is needed. The specific net area of the ventilation grid A_B^{Tabl} is selected depending on the pure power of the device exceeding 7.0 kW as specified in Table P10.

Considering the pure power 15 kW/1.1=13.64 kW we find the specific area in the second row of Table P10. It amounts to 0.005 m^2/kW. Then, the minimal net area of the grid will be:

$$A_B^{Net} = A_B^{Tabl}\left(Q^{Net} - 7.0\right) = 0.005\left(13.64 - 7.0\right) = 0.033, m^2 \quad (4.23)$$

We may install the grid at any convenient spot of the garage outer wall.

Example 4.8

A gas cooker with total power of 20 kW is installed in the kitchen box of a one-story residential building, where ventilation is assumed to be natural. The horizontal pan of the building is shown in Figure 3.27. Find the net cross-section of the ventilation grids of the kitchen box.

SOLUTION

Since the cooker is a household appliance and has no flues, we select the specific net area of the ventilation grid A_B^{Tabl} from the first row of Table P13, but considering the room volume, only. Figure 3.27 shows that the net volume of the box and the living room is 210 m^3.

We have A_B^{Tabl} = 0 m^2/kW corresponding to a volume of 10 m^3 and the minimal net area of the grid will be zero, since

$$A_B^{Net} = A_B^{Tabl}\left(Q^{Net} - 7.0\right) = 0.0\left(20/1.1 - 7.0\right) = 0.0, m^2.$$

Hence, there is no need for a grid and only one window is needed to ventilate the premises.

[33] Pursuant to BS5440; BS6644; BS6230; IGE/UP/10.

Example 4.9

A gas cooker with a power of 35 kW, type "air-air", is mounted in a hermetically sealed box, where natural ventilation takes place, while the cleaning air directly comes from outside. Find the grid dimensions.

SOLUTION

The specific net area of the ventilation grid A_B^{Tabl} is selected from Table P11 depending on the pure power of the device:

- Intake at the lower level – 0.01 m^2/kW;
- Exhaust at the upper level – 0.005 m^2/kW.

Then, the minimal net areas of the grids will be:

$$A_{B-D}^{Net} = A_B^{Tabl} * Q^{Net} = 0.01 * 35.0/1.1 = 0.32m^2;$$

$$A_{B-Up}^{Net} = A_B^{Tabl} Q^{Net} = 0.005 * 35.0/1.1 = 0.16m^2.$$

Example 4.10

Show how the dimensions of natural ventilation grids vary in the two cases shown in Figure 4.30. Consider the installation of two devices – a convector, type "air-air", with power 12 kW and a flow water heater with maximal power 17 kW.

SOLUTION

Since there are two devices in the rooms, the total thermal power is the power sum of both devices, i.e. the pure thermal power is (12+17)/1.1=26.4 kW.

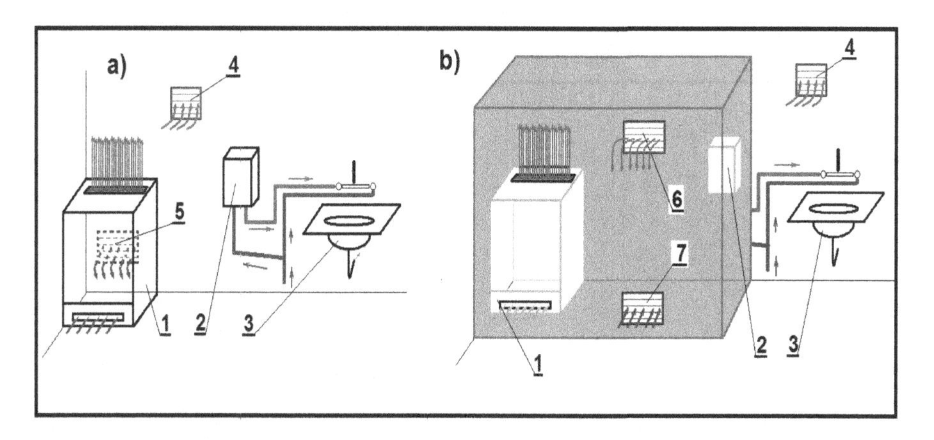

FIGURE 4.30 Natural ventilation of the premises: a) The space is directly ventilated by outdoor air; b) Gas devices are installed in a box, which takes/returns air from/to the room. 1 – Gas convective heater "air-air"; 2 – Gas tankless water heater; 3 – Sink; 4 – Exhaust vent grid for polluted air; 5 – Intake vent grid for outdoor air; 6 – Outflow vent for warm air; 7 – Intake vent grid for indoor air.

Consider the case in Figure 4.30,a) where the space is ventilated by outdoor air. Then, subtract 7.0 kW, which does not yield an increase in the pollution concentration above the admissible limit.

We find the specific area A_B^{Tabl}=0.005 m²/kW in the second row of Table P10, corresponding to a power of 26.4 kW. Then we find the minimal net area of the grid using Eq. (4.22):

$$A_B^{Net} = A_B^{Tabl}\left(Q^{Net} - 7.0\right) = 0.005\left(26.4 - 7.0\right) = 0.097, m^2$$

Consider the second case in Figure 4.30,b), where gas devices are installed in a box and the ventilation air comes from the indoor space. Then, we take the specific area $A_B^{Табл}$ of the ventilation grids (Items 6 and 7) from Table P11, i.e.:

- Box lower level – grid **(Item 7):** A_B^{Net}=26.4×0.02=**0.53**, m²;
- Box upper level – grid **(Item 6):** 50% of A_B^{Net} of the lower level or A_B^{Net}=**0.265**, m².

To ventilate the space, subtract 7.0 kW from the total power, which does not yield an increase of the pollution concentration over the admissible limit. Then, the area of the ventilation grid – **Item 4**, is found via the modified formula (4.22):

$$A_B^{Net} = A_B^{Tabl}\left(Q^{Net} - 7.0\right) = 0.005\left(26.4 - 7.0\right) = 0.097, m^2.$$

The above examples illustrate the application and importance of gas devices in the ventilation of individual rooms.

Yet, ventilation practice proves that gas devices can be included in a general system cleaning the contamination throughout the building. Figure 4.31 shows an example of such an application. Gas devices (Item 2) are the basic components of the ventilation system in both buildings. Except for their principal function of generating heating

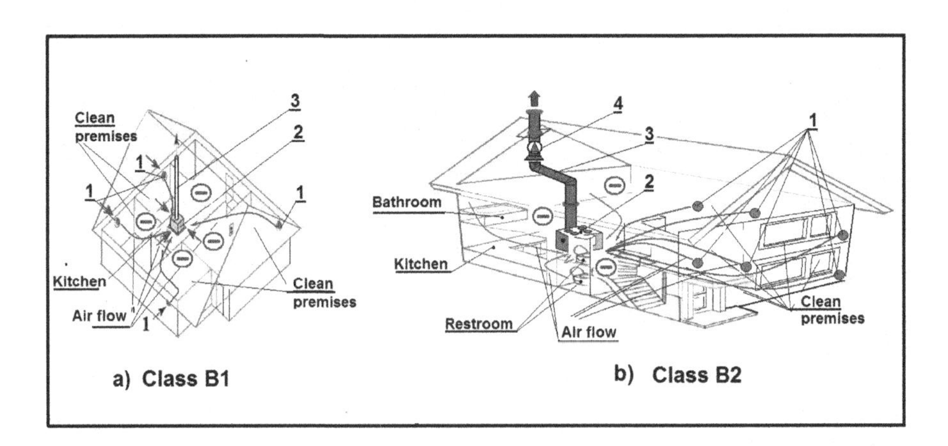

FIGURE 4.31 An exhaust ventilation system of a building; operation of an unbalanced gas device: a) Peripheral clean premises; b) Clean premises are grouped around the frontal facade: 1 – Suction vent; 2 – Unbalanced gas device; 3 – Flue; 4 – Exhaust fan.

or cooking power, they suck indoor air. Thus, the devices create a depressurization area around them and induce dilute air motion throughout the total building space.

The difference between the two schemes consists mainly in the different architectural-planning solutions of the two buildings:

- Consider the first scheme. Note that "dirty" premises are located in the building core where a gas device is installed. During operation, it creates a depressurization area, heating the air in the surrounding space (see Figure 4.31,a). The air in the neighboring premises diffuses through the indoor vents and voids under atmospheric pressure. At the same time, clean air enters the building through the suction vents (Item 1) cleaning the atmosphere along its way. Smoke flues suck the pollutants released by the gas device.
- The "dirty" and clean rooms are grouped in a different manner in the second building, i.e. clean rooms are situated around the frontal facade, usually with south, southeast orientation. Rooms, where contaminants are released, are situated around the back facade. The gas device is installed there too. It is situated in a basement, cellar or box. Gas cookers, whose flue operation is additionally supported by a suction smoke fan, can also be used by the ventilation system. The dilute flow is directed from the vents through the frontal facade (Item 1) to the depressurization area at the back of the building, where the contaminant concentration is high.

Consider the problem of providing clean indoor air in a building where a gas device excites natural indoor convection (as in Figure 4.32,a). Then, one should plot the ventilating flows (see Figure 4.32,b). The plot includes trajectories of air flows. It starts from the intake ventilation grids, whose specification numbers conform to

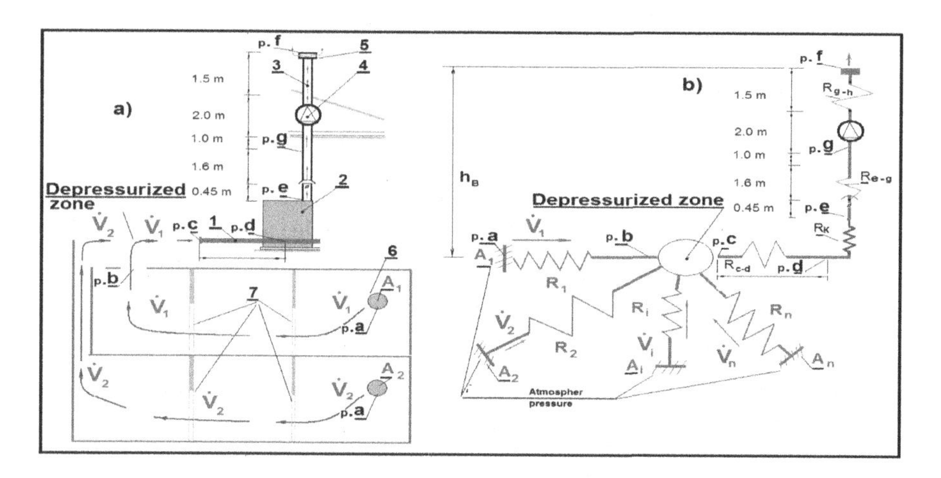

FIGURE 4.32 A gas device in a ventilation system: a) Partial vertical cross-section of the building; b) Plot of the ventilation fluxes: 1 – Suction duct; 2 – Gas device; 3 – Flue; 4 – Suction smoke fan; 5 – Terminal outflow of combustion products & contaminants; 6 – Vents through the building facade. 7 – Local air resistance

those of the facade vents and to the virtual traces of the outside air. The plot ends at the depressurization area where trajectories join and the fuel system of the gas device sucks the air, discharging it in the atmosphere through the chimney.

The number of plot sections (trajectories) of the ventilated air corresponds to that of the ventilating grids (n). Ventilating air flows under the action of a lifting head. It is generated by gravity and is calculated accounting for: (i) the difference between the levels of vents (Item 6) and terminals (Item 5) – h_B (see Figure 4.32); (ii) the difference between densities of the flue gases and the surrounding area $(\rho_e - \rho_{SG})$; (iii) pressure p_{Fun}^{Nom} generated by the smoke suction fan (Item 4), if any.

An assessment of the lifting head is made applying Kirchhoff's second law to an open contour and again using Eq. (2.15). Considering the i^{-th} section in Figure 4.32,b), the total pressure in its open contour has the following expanded form:

$$-\Delta p_1^i - \Delta p_2^i - \Delta p_3^i + \Delta p_4 = 0, \tag{4.24}$$

where: Δp_1^i,Pa – pressure drop in the ventilating grid of the i^{-th} section; – Δp_2^i, Pa – pressure drop in the indoor space ventilated by the i^{-th} section; Δp_3^i, Pa – pressure drop in the gas device and the flue; Δp_4, Pa – lifting pressure head. Performing the corresponding substitution and rearranging the equation with respect to the ventilation flow rate of the i^{-th} section we find the following user-friendly formula:

$$\dot{V}_{B-i} = \sqrt{\frac{h_B\, g\left(\rho_e - \rho_{SG}\right) + p_{Fan}^{Nom}}{0.5\left(\frac{\xi_i^j}{A_i^2} + \sum_{j=1+m} \frac{\xi_i^j}{A_i^2}\right) + \Delta p_{GL}^i}},\ \mathrm{m^3/s}, \tag{4.25}$$

Note the new components in the formula, together with the quantities participating in Eq. (4.22), i.e.:

- P_{Fan}^{Nom}, Pa – nominal pressure head of the smoke suction fan;
- $\sum_{j=1\div m} \frac{\xi_i^j}{A_j^2}$, m^{-4} – coefficient of local resistance and areas of net cross-sections of the indoor vents that the ventilation flow through the i^{-th} vent overcomes;
- Δp_{GL}^i, Pa – pressure drop in the gas device and the flue adjacent to the ventilation flow that passes through the i^{-th} vent.

Treating design issues, one should consider the prescribed flow rates of ventilation, needed for various premises and prescribed in Table 4.6, as well as the dimensions of the premises or the number of occupants. The following Example 4.11 illustrates a method employed in express engineering calculations.

Example 4.11

The building plan in Figure 4.31,a) shows the following clean rooms: a living room (42 m^2); a children's bedroom (20 m^2); a bedroom (25 m^2); a dining room

TABLE 4.9
Specification of the area of net cross-section of the ventilating grids

No	Premises	Needed air flow rate $\dot{V}$ -m³/s	Share of the total flow rate %	Net cross-section of the grid m²
1	Living room (42m²);	0.63	45.0%	0.07
2	Children's bedroom(20m²)	0.30	21.9%	0.033
3	Bedroom (25m²)	0.375	27.2%	0.041
4	Dining room (8 occupants)	0.08	5.9%	0.009
	Total:	1.385 m³/s	100%	0.15

(8 occupants), located in the building periphery. The dirty rooms are grouped around the building core. A gas device, type "air-air", with the pure power of 30 kW, is installed in a box with an open flue.

Find the net areas of the suction ventilating grids.

SOLUTION

We find the total area of the ventilating grid pursuant to the installed power – see Table P10. The specific grid area, corresponding to 30 kW, is taken from the second row, namely $A_B^{Tabl} = 0.005\ m^2/kW$. Then, the minimal net area of the grid will be equal to the product between the specific grid area and the pure power of the gas device:

$$A_B^{Net} = A_B^{Tabl} Q^{Net} = 0.005 \times 30 = 0.15, m^2.$$

The total net area distributes proportionally, pursuant to the needed ventilation flow rates. It is found via formulas 4.17 and 4.18 and using data specified in Table 4.6 (see column 3). The results are inserted in column 5 of Table 4.9. The real rate of the flow passing through the grids is realized via regulation and balance of the entire ventilation system, having erected the building and built the installation.

The examples discussed so far prove that gas devices are entirely compatible with the ventilation system of a building, where they can be successfully integrated.

4.3 SYSTEM FOR VISUAL COMFORT AND GAS APPLIANCES (GAS LAMPS)

4.3.1 Visual Comfort in Buildings

Quite often, the public idea of a comfortable indoor environment puts stress upon the visual comfort in a building only. Visitors like buildings, open to the exterior via glazed envelope areas. Yet, they do not realize the reverse of the medal, i.e. the high cost of energy needed to maintain the visual comfort during a gloomy day and hours of darkness.

Prior to discussing the conditions for visual comfort in a building space, we must clarify the human mechanism of visual perception. Briefly, people see thanks to the light reflected from objects and received by the human eye[34]. The brain transforms the received information into nerve impulses and organizes them in a spacious colored picture, establishing the difference between colors and comparing object illumination. Light may originate from:

- The Sun;
- Conversion of electrical energy;
- Chemical reaction or combustion.

The question often asked is what is visual comfort. It directly concerns visual clarity – if people's vision is clear, clean and there is no eye stress during durable gazing at an object, then people enjoy internal visual comfort. The basic requirement for the availability of visual comfort in an occupied area is the sufficient illumination of the work surfaces.

In other words, visual comfort exists in a room if:

- Occupants are not under pressure or their eyes are not tired of durable reading or gazing at an object;
- There is no indoor strong light contrast due to the brightness of different areas in the visual field;
- The occupants are engaged in continuous visual monitoring of small objects or reading with minimum stress.

If there is no visual comfort in a room, people lose their ability to work and experience fatigue and headache. Upon durable shortage of visual comfort, disabilities and headache, malaise or loss of contract sensitivity and eye damage may occur.

The basic factors affecting the sense of visual comfort include the lack of light; the brightness distribution; the orientation of the luminance; the subjective sensations to colors.

The arrangement of visual comfort is an important engineering-architectural task of building projects.

As practice proves, the use of hybrid illumination systems is appropriate in Southeast Europe, as well as in our country due to its moderate climate. Solar radiation, as well as other available power, "energizes" them – note, for instance, electric or chemical power, power of gaseous energy carriers such as singas, propane-butane or natural gas. The illuminating systems comprise[35]:

- Window lighting system (facade and upper lighting) acting within the building perimeter and using solar light. It is known as Daylight;

---

[34] Hyman eye sensors are sensitive to photons within the range 400–700 nm only.

[35] Systems of artificial illumination of public buildings mostly prevail in countries with equatorial, tropical and subtropical climate and the use of windows as components of natural lighting is not recommended due to eventual deterioration of the building thermal comfort.

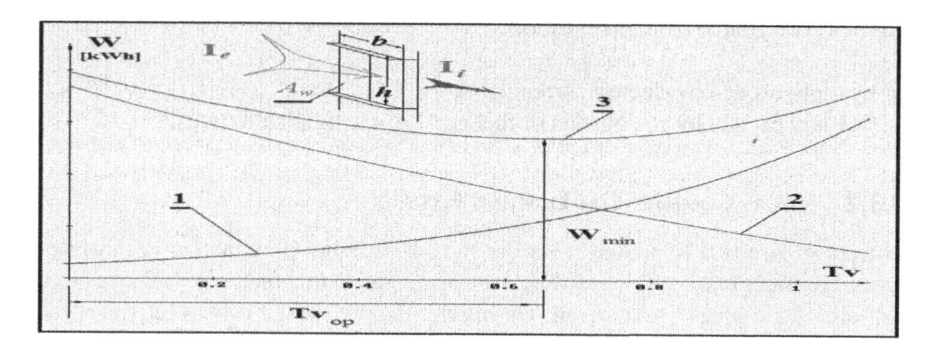

FIGURE 4.33 Effect of the glazed area transparency on energy consumption. 1 – Air conditioning; 2 – Illumination; 3 – Total energy consumption.

- Artificial illuminating bodies (most often, lighting fixtures) are installed in the building core. Illumination is switched on in gloomy days and hours of darkness.

The synthesis of such hybrid lighting systems is quite complex, since one has to meet the requirements of minimal initial investments and the current cover of power expenses.

The basic factor affecting illumination is the glazed area A_w (window overall dimensions h and b, Figure 4.33) and transparency T_v. The larger is A_w the deeper is light penetration into the building and the more successful is the satisfaction of the need for natural lighting. Then, no "artificial" lighting is to be used in that part of the building and overhead expenses for lighting will be lower. On the other hand, the installation of larger windows instead of an erection of solid walls will significantly raise the costs since a glazed envelope is significantly more expensive than a solid one[36]. Besides, the use of larger windows, whose coefficient of thermal conductivity is greater than that of solid walls, yields larger thermal losses (and lower energy gains, respectively).

In this case, the building heating installation should be with greater power, it will be more costly, consuming more energy for the maintenance of thermal comfort. Consider one more factor – energy cost or the ratio between the energy costs for lighting and heating (conditioning)[37].

When the decrease of the overall facade glazed areas is unjustified and extreme, the necessity of additional power for visual comfort increases.

Consider the committed and multidirectional effect of the overall dimensions of glazed components. Their calculation should minimize: (i) the "present value" of the initial investments in the envelope construction, and (ii) future heating and lighting expenses.

---

[36] We do not consider here luxury walls of polished stones.

[37] To solve the optimization problem of calculation of the window overall dimensions we must model the heat exchange between the building and the outdoor environment using energy-simulation codes such as Energy Plus, Energyl0 and ESPR (there are more than 20 commercial codes popular in the field) [41]).

For now, two sources of light exist, i.e.:

- Light excited by electric current;
- Light excited by combustion of fuel gas, i.e. gas lighting fixtures.

4.3.2 Visual Comfort-Gas Lighting Fixtures

The use of fire as a light source set the start of artificial illumination in inhabited areas. Animal grease and wood were used as a light source fuel: olive oil, beeswax, fish oil, whale grease, sesame oil, nut oil etc. People started using fuel gas at the beginning of the 18th century, and the use of fuel gas for that purpose proceeded till the end of the 19th century[38]. Figure 4.34 shows a fragment of a plot illustrating the evolution of artificial light sources into three branches:

- Electric discharge lamps;
- Incandescent lamps with white-hot filaments patented by Thomas Edison in 1879;
- Gas-lighting fixtures.

Pursuant to the principle of operation, electric light sources are:

- Incandescent lamps;
- Fluorescent lamps;
- Highly intensive discharge lamps (mercury or sodium);
- Miscellaneous lamps (halogenous, light-emitting diodes – LED, compact-halogenous, glim lamps).

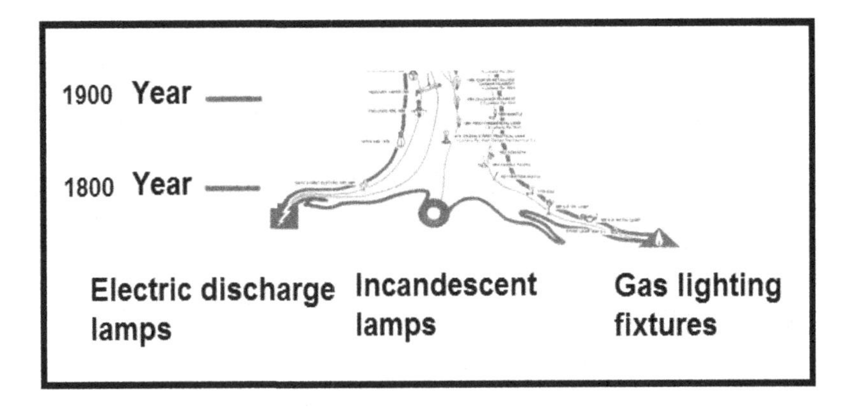

FIGURE 4.34 History of light sources.

[38] As of 1823 a number of small and big cities in Great Britain were equipped with gas lighting fixtures. Their cost was by about 75% lower than that provided by oil lamps or candles, stimulating their development and implementation. Gas lighting fixtures were implemented all over Britain by 1859 and a thousand of plants for gas fuel synthesis started operating. Gas lighting fixtures made reading easy and longer, thus stimulating literacy and education and giving impetus to the Second industrial revolution.

TABLE 4.11
The efficiency of electric light sources

		Lu/w	%	Actual %
1	Lamps with white-hot wire	10 ~ 25	1.5 ~ 3.7	0.67 ~ 1.7
2	Lamps with mercury vapors	20 ~ 63	3.0 ~ 9.2	1.35 ~ 4.1
3	Fluorescent lamps	40 ~ 100	5.9 ~ 14.6	2.6 ~ 6.7
4	Metal-halogenous lamps	50 ~ 110	7.3 ~ 16.1	3.3 ~ 7.2
5	Lamps with sodium vapors – B.H.	50 ~ 140	7.3 ~ 20.5	3.3 ~ 9.2
6	Lamps with sodium vapors – HH	100 ~ 180	14.6~ 26.4	6.6 ~ 11.9
7	Light emitting diodes – LED	80 ~ 180	11.7 ~ 26.4	5.3 ~ 11.9
8	Theoretical luminous flux of от **1 w**	683 Lu/w	100.0%	45.0%

Although the actual power efficiency of electric light sources is less than 12.0%[39] (see Table 4.11), they dominate the market nowadays and are widely used in buildings to achieve visual comfort.

The story of **gas** lighting fixtures is older than that of electric lamps. They were more popular for outdoor and indoor use in towns and suburbs. For now, electric lamps are the main consumer of electric power in buildings despite their inefficiency (the illumination energy occupies the 3rd place in the energy rating of a residential building, while it is 2nd in a business building), and they consume over 20% of

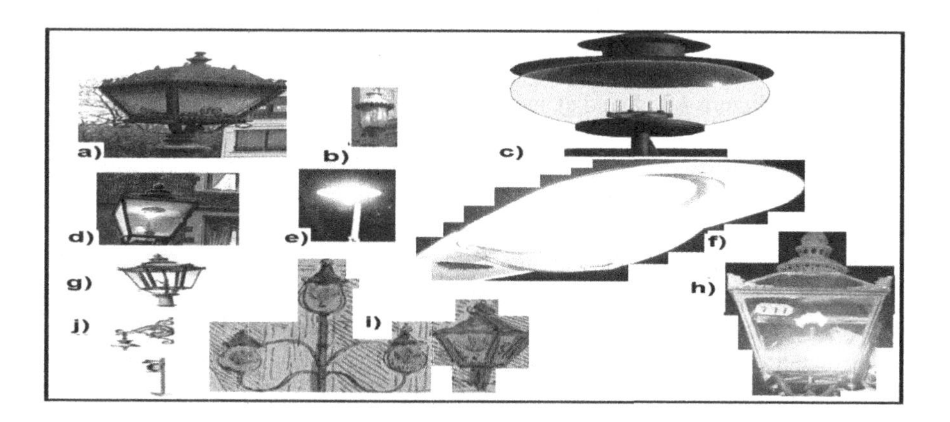

FIGURE 4.35 Gas illuminators as outdoor devices: a) Modern gas illuminator, Rathenow, Germany; b) Illuminator, Baltimore, 1816; c) Reproduction of outdoor gas illuminator, Germany; d) Modern gas illuminator, London, 2008; e) Gas illuminator, Berlin, 2005; f) Modern gas street lamps; g) Wroclaw, Poland; h) Sketch of gas illuminators, London, 1809; i) Gas illuminator, Sweden, 1953; j) Part of a survey of illumination technologies, 1900.

[39] The energy efficiency in column 3 is found per 1 w of electric power. The real energy efficiency of lamps should also account for the efficiency of the electricity-generating technology – see column 4 (it is assumed that the electric power is generated by a gas electric power plant with efficiency coefficient amounting to 45%).

the electric power generated by many national economies till the beginning of the 20th century. Early gas lamps were manually switched on while the automatic switch-on was introduced much later.

The real application of gas lighting fixtures started at the beginning of the 19th century. The devices illuminated the centers of European and American cities – London[40], Baltimore[41], St Petersburg[42] and Paris. Prior to the electricity coming into use, gas lamps illuminated streets as well as prestigious buildings (churches, theatres, residences) and private houses of rich citizens[43].

Although gas-lighting fixtures are significantly more efficient and cheaper than the electric ones, their market share is modest at present. They mainly illuminate camping sites, lighthouses and some historical sites, where a sentimental-romantic effect is sought.

Street gas lighting fixtures are not entirely extinct in some cities and counties – see for instance Boston where 2800 gas lamps still glow. Over 1100 gas lamps glow in the historical quarters of Cincinnati. Gas lamps glow also in the famous French quarter of New Orleans as well as in historical houses throughout the town.

Extended street-illuminating gas networks operate not only in the USA, but also in Europe. For instance, more than 37000 gas lighting fixtures operated within the gas network of Berlin in 2014 (see Figure 35,a and b). About 1500 gas lamps glow in London parks.

While gas-lighting fixtures successfully compete with electric fixtures at outdoor sites such as parks, streets, museums and other public places, indoor gas illumination of residential and business buildings is undeservedly compromised[44].

Hence, gas-lighting fixtures are kept in some old buildings, only, being rehabilitated thanks to their aesthetic qualities – see Figure 4.36.

The improvement of gas lamps by the installation of catalytic grids and flameless radiators (see Figure 4.36,b and e), makes them attractive to buildings erected in the continental or polar climate zones.

New technical designs of gas devices involve their use in a combination with optical fibers and cables. Figure 4.37 shows a lighting system called "optical gas lighting fixture". The main components are a gas converter (Item 12), an optical conductor (Item 9) and illuminators (Items 10 and 11).

The gas converter (Item 12) is a device where flameless fuel combustion takes place and the radiator (Item 1) emits energy at high temperature. The luminous flux regenerates in the concentrator (Item 7) increasing the temperature of the photon gas and the energy density to a level of about 10^5 W/m^2. The optical head (Item 8)

---

[40] F. A. Winsor was the first inventor, who illuminated Pal Mal, London, on January 28, 1807, by a system of gas lamps.

[41] Baltimore was the first American city equipped with street gas illumination – February 7, 1817.

[42] Russia got its first street gas illumination in the autumn of 1819. Gas lamps were installed along a street on Aptekarski Island.

[43] After 1842, r when W. Grove invented his fuel cell, hydrogen fuel was used in the cell. This was the first step to a centralized use of fuel gas.

[44] Gas lighting fixtures release "bad" heat and a typical gas lamp generates luminous flux of 400–500 lm and heat power amounting to 25–75 w. Due to poor combustion, illuminators are blamed as emitters of significant amount of carbon monoxide.

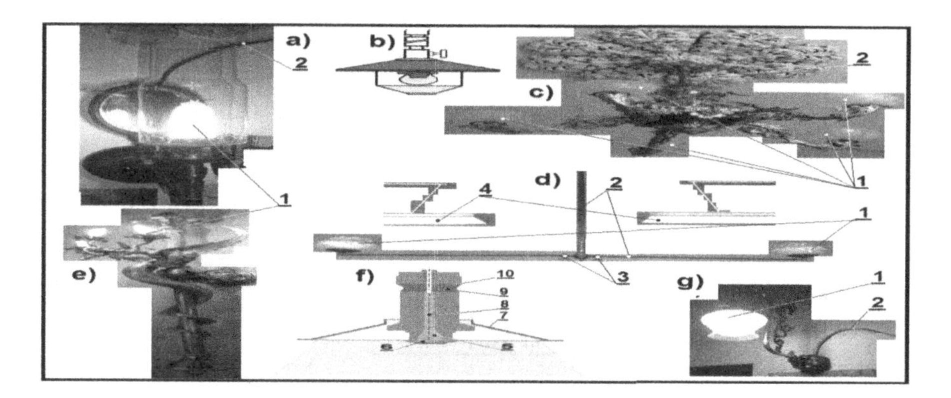

FIGURE 4.36 Indoor gas-lighting fixtures: a) Gas wall-fitting; b) Gas lamp; c) Gas chandelier; d) Gas wall fitting in Minnesota, US; e) Gas lighting fixtures; f) Structure of a gas flameless lamp; g) Gas reading lamp. 1 – Gas lamp; 2 – Gas supply pipe; 3 – Gas valves; 4 – Electric fixture; 5 – Diffuser; 6 – Ceramic head with a catalytic grid; 7 – Reflector; 8 – Mixing pipe; 9 – Fresh air vent; 10 – Gas ejector.

absorbs the concentrated energy flux and the optical conductor (Item 9) transfers it to the indoor lighting fixtures, which glow like white-wire lamps.

The energy efficiency of those gas systems is expected to exceed 30%, which is 3 times greater than that of sodium vapor lamps-HP and LED.

The gas converter of the optical lighting fixture (Item 12) is installed in a box or in the basement like any gas device. It is fireproof and the released combustion products are discharged pursuant to the regular and already discussed methods. The

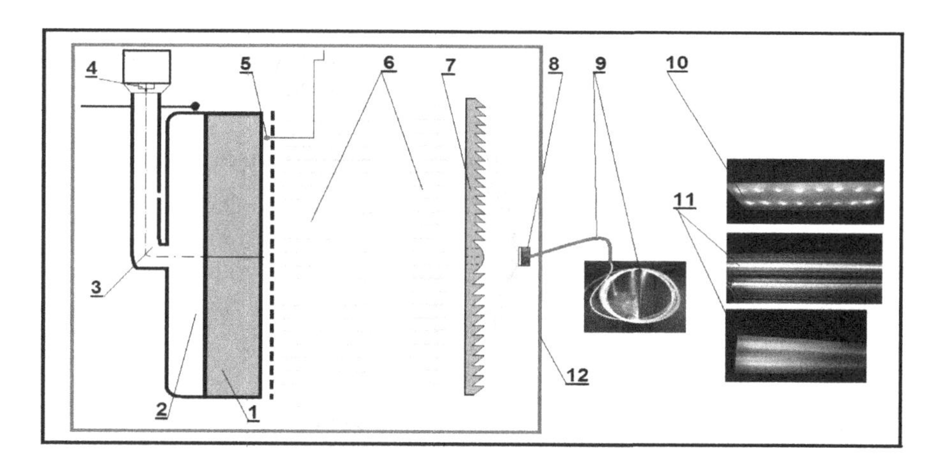

FIGURE 4.37 Optical gas lighting fixture [42]: 1 – Gas radiator; 2 – Distributor; 3 – Mixer; 4 – Ejectors; 5 – Sensors; 6 – Luminous flux; 7 – Light concentrator; 8 – Optical head; 9 – Optical conductor; 10 – Illuminator with optical fibers; 11 – Wall fitting with optical fibers; 12 – Gas converter.

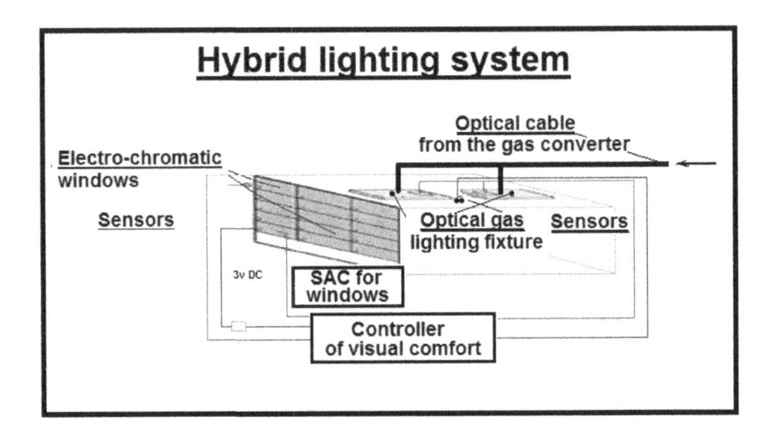

FIGURE 4.38 A hybrid illumination system including electro-chromatic windows and optical gas illuminators.

generated luminous flux runs along the optical conductor not posing any hazard to the building and its inhabitants.

Figure 4.38 shows gas lighting fixtures in a hybrid system together with a system for natural lighting using electro-chromatic windows, which control the passing daylight. The installation operates automatically like a two-position system:

- When luminance exceeds the upper limit of the control area, an electro-chromatic dimming of the windows activates;
- If the luminance is below the lower limit of the control area, an optical gas lighting fixture activates.

The installation of optical conductors (Item 9, Figure 4.37) is similar to that of electric cables.

Due to higher energy efficiency, gas lighting fixtures combined with high-tech optical components outline a perspective trend of illumination technology.

4.3.3 Design Steps in the Calculation of a Visual Control System

The task of achieving indoor visual comfort is an important architectural issue. It is quite often treated using the "Two F-C method"[45]. We shall formally present the succession of steps of visual comfort design. The proposed "algorithm" will be illustrated by the design of an indoor lighting system – see the architectural scheme in Figure 4.39,a).

We propose the following steps to achieve visual comfort:

- **1st step:** Analysis of the visual issues of premises pursuant to their function as prescribed by the technologist of the building[46];

[45] "Two F-C" – finger, floor and ceiling; Figurative meaning – arbitrary solution resulting in the over-calculation of illumination energy and window area, as well as unjustifiable energy over-consumption.
[46] Usually, the chief architect operates also as a technologist of residential and public buildings.

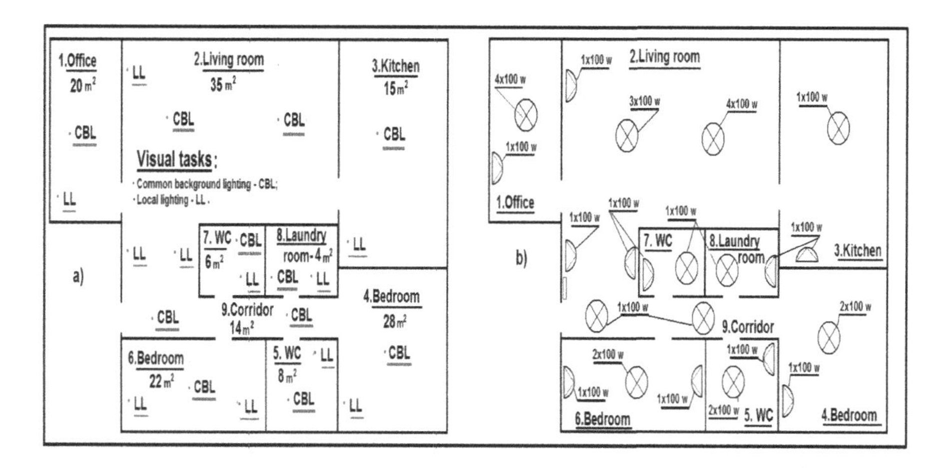

FIGURE 4.39 Visual tasks – 9 CBL and 11 LL: a) Architectural scheme with fixed illumination problems; b) Illuminator location.

- **2nd step:** Selection of lighting fixtures (with respect to type, number and replacement) regarding each lighting task. Specification of the nominal power of each lamp;
- **3d step:** Design of lighting fixtures in lighting groups and circles. Specification of the nominal power of each lighting group[47];
- **4th step:** Design of a scheme of energy supply (electrical or gas scheme) and a scheme of visual comfort control;
- **5th step:** Integration of the lighting system with the main power system of the building;
- **6th step:** Integration of the system of illumination control with the building system of automatic control. Find out how the control of the lighting groups is integrated with the Building BAS.

The lighting fixture type is selected based on investor's requirements and the visual tasks. The number of lighting fixtures depends on various factors including the area of the illuminated rooms, their distribution and location with respect to the control surface, wall reflection capabilities (color, smoothness etc.) and lamp emission as part of the fixture capabilities.

There is a number of methods of calculation of the number of lighting fixtures. One of them recommends the use of the following formula, which is usable and accurate enough:

$$n = \frac{100.E_N.A_{Room}.K_{rez}}{\varphi_L.\eta_{LS}}, \tag{4.26}$$

[47] Designation of all lighting fixtures involved in any illumination task. The devices operate under common supply and control.

where

- ϕ_L, lm – luminous flux generated by the lamps[48];
- A_{Room}, m^2 – area of the illuminated control zone;
- K_{rez} – reserve coefficient (1.3–1.5);
- E_N, lx – standard value of the luminance of the occupied areas[49] (Table 4.10);
- η_{LS}, % – efficiency coefficient of the lighting system[50].

Designers combine gas lighting fixtures into lighting groups in order to optimize the control of their operation. Gas supply takes place through a separate gas pipe network, whose axonometric scheme is similar to that in Figure 4.16.

The following example illustrates the first three steps.

Example 4.12

Design the lighting system of an apartment with a total area of 152 m^2, comprising common background and local artificial light.

SOLUTION

The technological scheme of the arrangement of the apartment equipment is used to define the function of the premises, their area, overall dimensions (l, b and h), wall and ceiling color, their reflection coefficients and eventual arrangement of the household equipment (including kitchen equipment, audio-visual equipment, furniture etc. (see Figure 4.39,a). We shall assume for convenience identical coefficients of wall and ceiling reflection- ρ_{Wall}=0.5 and $\rho_{Ceiling}$=0.7, throughout the building.

We prepare a luminance table based on the available information. Except for the function of the illuminated premises, it contains room areas, normative luminance, calculation indexes and the selected efficiency coefficients of the lighting installation.

The cells of Table 4.12 contain data specified in the architectural sketch in Figure 4.39.

We specify the main lighting issues of each room. Consider, for instance, room 3 (kitchen). Except for the common background lighting (**CBL**), local lighting (**LL**) of the kitchen equipment components also operates. Except for **CBL, LL** of the toilet mirrors is also envisaged.

We should solve 9 problems of CBL and 11 problems of LL. The solutions are presented graphically in Figure 4.39,b).

The efficiency coefficient of the lighting installation (see column 5 of Table 4.12) is selected from Table P18 accounting for the indexes of the illuminated premises calculated via formula (4.27) and the coefficients of wall and ceiling

[48] See Table P16.

[49] See Table P17.

[50] The coefficient is found in Table P18, accounting for the space "index" i_{Room}=l.b/hl+b – Eq. (4.27), wall color and reflection capability.

TABLE 4.12
Table of luminance

No	Premises	Area A_{Room}	Luminance norm E_n	Index i_{Room}	Efficiency coefficient of ELS
		m^2	Lx	-	%
1	Office	20	75	1.71	31.7
2	Living room	35	75	**2.24**	**35.5**
3	Kitchen	15	30	1.44	29.5
4	Bedroom 1	28	30	1.96	33.8
5	Bathroom 1	8	50	1.03	24.2
6	Bedroom 2	22	30	1.78	32.4
7	Bathroom 2	6	50	0.92	23.5
8	Wet space	4	50	0.77	23.0
9	Corridor	14	30	1.19	33.7

reflection – (ρ_{Wall}) and ($\rho_{Ceiling}$). Consider, for instance, the living room (No 2) whose overall dimensions (width b and length l) are 4 and 5 m, respectively, while its height is h=1.3 m (see Figure 4.40). Then, we find the index via expression (4.27), i.e. $i_{Room} = \dfrac{4.5}{1.3(4+5)} = 2.24$.

Refer to Table P18. Account for the above index and assume reflection coefficients ρ_{Wall}=0.5 and $\rho_{Ceiling}$=0.7. Then, we find the efficiency coefficient of the lighting system by means of interpolation. The result is $\eta_{LS} = 35.5$ %, given in Table 4.12. We find the luminous efficiency coefficients of other premises using a similar procedure (see column 6 of Table 4.12).

The number of lamps in the premises calculated by means of expression (4.26) is specified in Table 4.13. We use three types of light sources – an electric lamp with white-hot wire (with a power of 100W – the 5th column), an electric luminescent

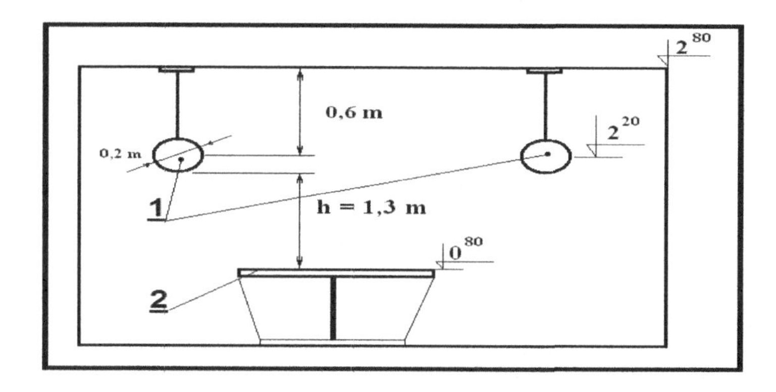

FIGURE 4.40 Vertical cross-section of the illuminated space: 1 – Lighting fixture; 2 – Illuminated control surface.

TABLE 4.13
Number of lamps in the premises

			Lamp type					
			Wire-hot 100 W 1400 lm		Luminescent - 150 W 2050 lm		Gas 75 W 500 lm	
No	Premises	Area	Normative luminance	Number	Normative luminance	Number	Normative luminance	Number
		m^2	Lx	pieces	lx	pieces	Lx	Pieces
1	Office	20	75	4	75	3	75	12
2	Living room	35	75	7	75	5	75	19
3	Kitchen	15	30	2	30	1	30	4
4	Bedroom 1	28	30	2	30	2	30	6
5	Bathroom 1	8	50	2	50	1	50	4
6	Bedroom 2	22	30	2	30	2	30	5
7	Bathroom 2	6	50	1	50	1	50	3
8	Wet room	4	50	1	50	1	50	2
9	Corridor	14	30	1	30	1	30	3

lamp (with power of 150W – the 7th column) and a gas lamp (with power of 75W – the 9th column)[51]. We round data to the nearest integer[52].

The results show that the number of lamps of both types (electric and gas ones), as well as their power and energy consumption, are quite different. This is due to the different ratios between the electric power of the lamp and the individual luminous flux (its value is 1:14 W/lm for electric lamps – white-hot wired and luminescent ones, while its value is 1:6.7 W/lm for gas lamps). The number and power of electric lamps (white-hot wire and luminescent ones) are of one and the same order – 22 and 17 pieces, and 2.2 and 2.55 kW, respectively. Gas lamps are 58 in number[53], and their power amounts to 4.33 kW. During a 5-hour regime of operation (3+2 hours), the daily consumption of primary energy of a lighting system will be 36.6:33.3 kWh_e/kWh_{th}.

Invoices of consumed energy[54] are also in favor of gas lighting fixtures. The example proves that they have higher energy and economic efficiency as compared to electric fixtures.

Based on the calculations summarized in Table 4.13, the lamps of an individual lighting task are arranged into lighting groups controlled by respective controllers and switches. The latter are grouped into electric current contours (see Figure 4.41,

[51] Standard value of the luminance BSS 1786-54 was used in the Example 4.12.

[52] A passing analysis of the results shows that the increased number of gas lighting fixtures is accompanied by a 100% increase of the heat gain as compared to that of the lamps with white hot wire. This is an issue of the thermal comfort and the increase of the capacity of the system of its control.

[53] Regarding a particular project, the large number of necessary gas lighting fixtures is the reason why they are avoided.

[54] At cost ratio higher than 1.7 kWh_e/kWh_{th}.

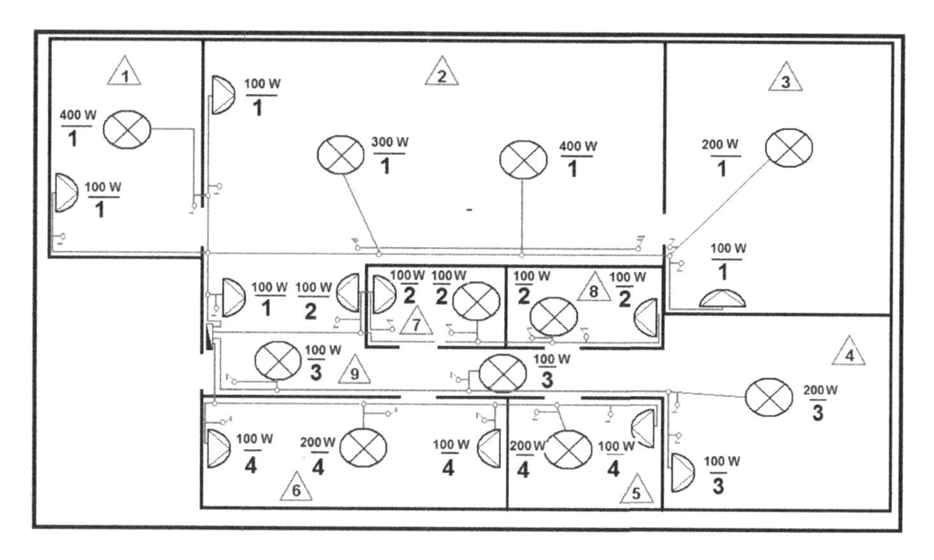

FIGURE 4.41 A group arrangement of lighting fixtures into electric current circles. (The lighting installation consists of four electric current circles)

which shows a contour of white-hot wire lamps while the arrangement of other lamps is identical).

Figure 4.41 offers an arrangement of all **CBL** and **LL** lighting fixtures into four current circles. The used distribution criterions consist in "a convenient technological bonding and a balanced power supply". For instance, the first circle includes CBL and LL lighting fixtures of premises 1, 2 and 3, as well as an LL lighting fixture of the corridor. The fourth circle comprises CBL and LL lighting fixtures of rooms 4 and 5. Finally, we illustrate how particular lighting fixtures are arranged into electric current circles (circulation contours).

The use of optical cables in the near future will imply other criteria of arrangement of lighting fixtures. Yet, the proposed succession of calculation steps will survive.

The manufacture of modern gas lamps varies with respect to their structure and dimensions and depends on their implementation. The power of regular lamps varies from several tenths of watts to 6 kW, while in some cases it may reach 60 kW.

4.4 GAS EQUIPMENT IN A SYSTEM FOR FOOD PREPARATION AND CONSERVATION

4.4.1 A System for Food Preparation and Conservation

Food together with fresh air (oxygen) and water is vitally important for humans. The Food and Agriculture Organization (FAO) of the United Nations recommends a daily personal food portion of about 2400 kcal/day (2.79 kWh/day). Individual consumption of food is different in different countries. The food portion depends on the climate and humans' gender, age and activity.

Food preparation – cooking comprises of more than 50 technologies including dry roasting, barbeque, stewing, blanching, boiling, sousing, steam cooking, sterilization of compotes and canned food, frying, grilling etc. Heating and evaporation of liquids, fats etc. are the basic processes operating during food preparation. Note also that the cooking runs in especially designed cooking appliances.

The basic feature of the cooking technologies is the thermal treatment of food. Hence, food preparation and preservation is one of the most energy-consuming household technologies. The energy consumed in food preparation is the third on the list of all household activities in developed countries such as the USA, Canada, Australia and the EU members. It amounts to 5–11% of the total energy.

Food preparation in modern buildings takes place in separate specialized premises – kitchens equipped with modern devices. Kitchens are treated as "micro-factories" of food preparation and employ innovations in electrical engineering, electronics, mechanics and thermo-technics (gas technics, in particular). Depending on building location, size and type, kitchens are:

- Household kitchens;
- Commercial or business kitchens.

In contrast to household kitchens, commercial kitchens are the logistic units of restaurants, coffee bars, school kitchens, workshops, bakeries and public or kid's kitchens. Food is prepared there on a mass scale – biscuits, cakes, bread and paste products.

The evolution of cooking devices is long, and one can arrange them into two large groups:

- Devices for open vessel cooking (Figure 4.42);
- Devices for cooking in confined space (Figure 4.44).

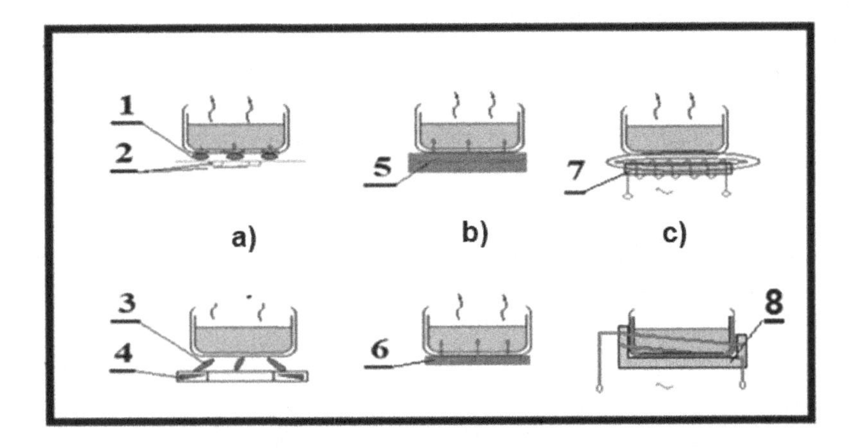

FIGURE 4.42 Types of food thermal treatment in open vessels: a) Discrete heating; b) Hot plate; c) Induction heating: 1 – Briquettes (coals); 2 – Air ducts; 3 – Gas flames; 4 – Gas pipes; 5 – Electric plate; 6 – Radiating ceramic plate; 7 – Inductive heater; 8 – Inductive cup.

Food products mix in open vessels prior to cooking. They are heated under constant pressure (the process is isobaric). The contact between the heat generator and the vessel is of crucial importance for the efficiency of the process.

Figure 4.42 illustrates three methods of contact between the vessel and the heat exchanging surfaces: contact throughout discrete areas (Figure 4.42,a), a surface-to-surface contact (Figure 4.42,b) and a contact throughout the entire volume (Figure 4.42,c). It is seen that heat transfer is different in different cases of contact. Consider, for instance, devices applying discrete heating (Figure 4.42,a). Then, heat transfer takes place via convection and conductivity in the vessel segments. From a thermodynamic point of view, they are the most imperfect ones, since the excitation of a convective flux consumes the main part of thermal energy. Note also that the thermal flux outflows the vessel and scatters into the surrounding environment.

The basic mechanism of heat transfer in devices with a hot plate (Figure 4.42,b) is heat conductivity through the contact area between the plate and the vessel. If the contact is good (i.e. the vessel has a flat, even and clean bottom), heat transfer to the products is effective. However, the device has a comparatively high inertness (big thermal mass), which deteriorates its energy efficiency. The use of less inertial ceramic plates is a successful step toward the improvement of the device efficiency. Faucalt's currents generate heat in the third device – an "inductive heating" type (Figure 4.42,c). An external alternating electromagnetic field induces them in the walls of the cooking vessel.

The popularity of fuel gas usage in residential buildings and everyday life implies the development of a series of open cooking devices, some of them shown in Figure 4.43.

The gas hot-plate is a universal cooker and its structural scheme is shown in Figure 4.43,a). A gas burner consists of a chamber (Item 1), a mixing pipe (Item 2), an ejector (Item 3), main burner ports and retention ports (Item 9) and a crown (Item 8). It does not differ from a device for fuel gas combustion described in Sub-par. 2.

A spring of the regulating cone (Item 5) controls the burner power regulating cone's position, while the control is manual. Two or three burners are installed onto the pan casing[55] – see Figure 4.43,b). The food vessel is positioned on the hot plate or griddle (Item 10).

A gas frying pan (Figure 4.43,f), a gas barbeque (Figure 4.43,c) and a gas grill, type "hot plate" (Figure 4.43,d), are variations of the frying pan. Functionally, the gas frying pan is a complete analog of the regular pan. It thermally processes food in open or closed vessels. The difference lies in the structural arrangement of the fuel chamber, which is ring-shaped and has a central opening facilitating the ejection of secondary combustion air to the flame.

A gas barbeque (Figure 4.43,c) and a gas grill "hot plate" (Figure 4.43,d) bake raw or semi-processed food such as vegetables, meat, sausages etc.

However, baking technologies differ from one another. Gas barbeque baking is contactless under the action of a combined thermal flux. It has radiation and

[55] For cookers.

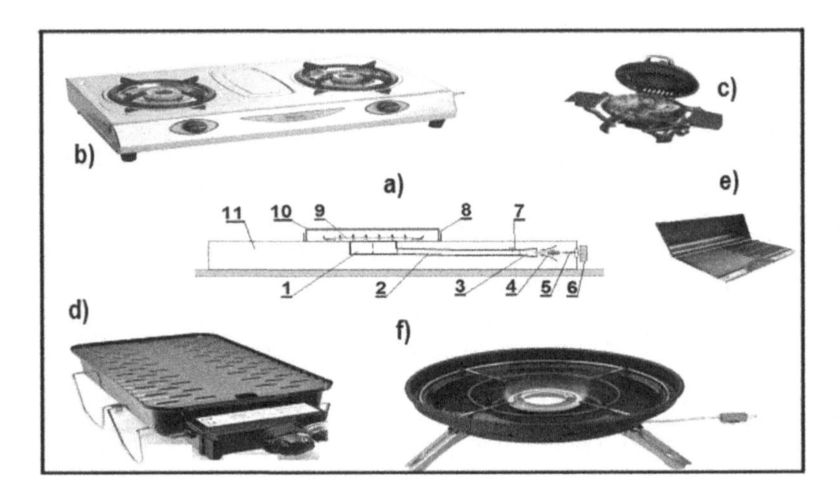

FIGURE 4.43 Open gas devices: a) Structural scheme of a gas hot-plate; b) Gas hot-plate with two burners; c) Gas barbeque; d) Gas grill "hot plate"; e) Plate of a gas stove; f) Gas frying pan: 1 – Chamber; 2 – Mixing pipe; 3 – Ejector; 4 – Primary air port; 5 – Spring of the regulating cone; 6 – Spindle; 7 – Throat restrictor screw; 8 – Crown; 9 – Main burner ports and retention ports; 10 – Hot plate or griddle; 11 – Pan body.

convective components generated by a hot plate, which is far enough from the products. Gas grill baking is of contact type and it runs at a higher temperature. However, all those devices scatter a large part of the generated heat (~70%) in the environments, thus significantly decreasing their energy efficiency. This was practically established still in 1735 when the French architect and designer Francois Cuvillies created the first closed gas stove, "Castrol stove", burning firewood.

Figure 4.44 illustrates three schemes that use confined cooking space (oven), thus saving primary energy and disregarding the effect of the environment. The difference between them lies in the method of heat transfer from the heating device to the cooking oven.

The first scheme is known as a "confined-split oven", and its heating device is installed outside the cooking space. The combustion of an organic fuel in the fuel cell (Item 3) releases heat, and combustion products transfer it through the channels (Item 4, Figure 4.44,a) to the confined space where food is prepared. Having released cooking heat, which passes through the oven walls, the combustion products cool down in the regenerator (Item 9) and exhaust in the surrounding space.

Considering the second and the third schemes, heat-generating devices are installed within the cooking space and the released energy recirculates there. There are different heat generating devices in the two ovens – energy radiators of photons belonging to the visible and infrared spectra (Items 2, 6 and 7). Fans discharge cooking waste outside the cooking space.

Electric stoves designed by the Canadian innovator Tomas Ahearn (1892) became popular after the electrification of cities and buildings. Nowadays, the traditional

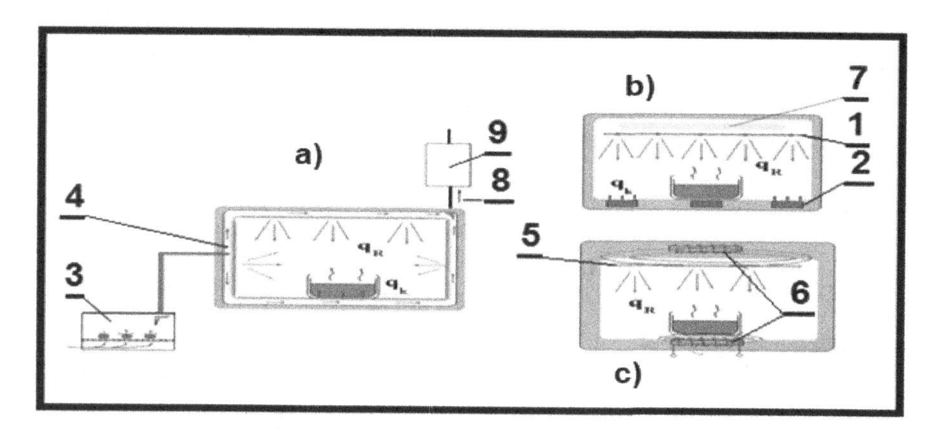

FIGURE 4.44 Thermal schemes of cooking in confined space – oven: a) Isolated-split oven; b) Combine oven; c) Induction oven: 1 – Radiator; 2 – Package heating elements; 3 – Combustion chamber; 4 – Channels for warm gases; 5 – Radiating band;6 – Induction heater; 7 – Energy radiator (photons belonging to the visible and infrared spectra); 8 – Smoke gases; 9 – Regenerating device.

device for food preparation is a combined cooker. Note that a wide variety of gas stoves emerged in the last 40–50 years together with the popular electric stoves – see Figure 4.45. A typical gas stove comprises:

- A unit with gas hot plates located on the stove highest plate;
- A gas oven.

The unit with gas hot pates comprises 3, 4, 5 and even 6 burners arranged in one or two rows.

Gas ovens are of two types (Figure 4.45):

- Ovens with direct heating – Figure 4.45,a);
- Ovens with indirect heating – Figure 4.45,b).

The difference between ovens consists in the method of exploitation of the thermal potential of combustion products. Considering an oven with direct heating, combustion products freely circulate within the oven space and transfer their entropy to the food.

As for the second gas stove, combustion products, having left the burner (Item 2), are directed to the surrounding space through the heat exchanger (Item 5). A fan (Item 6) generates forced convection within the oven transferring the generated heat to the food raw materials. The advantages of the oven are that the food does not absorb smoke gas smell and the internal air gains temperature lower than that of the smoke gases. The efficiency coefficient of gas stoves is $\eta_{eff} = 0.35$,

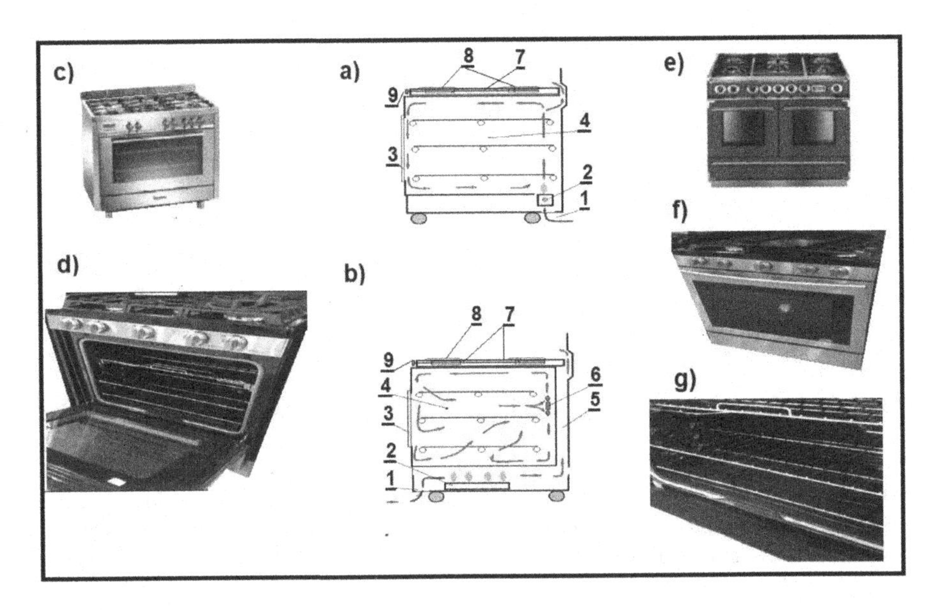

FIGURE 4.45 Combined gas cookers: a) With a directly heated oven; b) With an indirectly heated oven; c); d); e); f); g) Oven view. 1 – Inlet of combustion air; 2 – Oven burner; 3 – Glass panel; 4 – Workspace; 5 – Heat exchanger; 6 – Ventilator; 7 – Hot plate burners; 8 – Hot plates; 9 – Gas regulators.

i.e. it is 2.86 times larger than the theoretically needed one[56]. Yet, the cost of consumed energy is the lowest one – 134 Euro/10t, due to the gas market price, lower than that of electricity.

Note that energy is needed not only for food preparation, but for food preservation too. Food preservation is a serious problem for modern households. Its successful solution may save money and time. Urban life traditions and the convenience of once-a-week shopping in large stores and at lower prices imply the use of household fridges and freezers.

Household fridges are in use for more than 100 years. The first one was manufactured in 1911 by the company General Electric, Fort Wayne, Indiana, USA. For now, fridges are so popular that they consume 6–8% of the total energy of the building.

The traditional classification of fridges follows their principle of operation and outlines three groups:

- Steam-compressor devices;
- Air-compressor devices;
- Absorption devices.

The first two fridges operate pursuant to the inverse cycles of Rankin and Brayton, respectively. Absorption refrigerators operate following the principle of absorption

[56] Note that electricity generation by modern technologies is also inefficient.

and desorption of gases in two-component mixes during their cooling down and heating in cooling cycles. Electric heaters heat two-component cooling mixes (agents) in the classical absorption refrigerators. In the case of gas absorption fridges, heating is done by gas burners.

Figure 4.46 illustrates the structural scheme of a household absorption fridge using a burner and fuel gas to induce a thermal-siphon flow of the cooling agent in the generator (Item 6) – a water-ammonia solution.

The operation of a gas fridge consists in heat "extraction" from products placed in its chamber (Item 1, Figure 4.46). Heat then is transferred through the wall of the pipe serpent of the evaporator (Item 2) to the cooling agent (water-ammonia solution). This activates the ammonia adsorption by the refrigerant mixture and it is called "Strong (saturated) ammonia-water solution or SAWS". The solution density decreases and the hydrodynamic lift head in the piping system increases.

When the refrigerant solution passes through the heating section (Item 6) ammonia evaporates and vapors are directed to the condenser (Item 5). Then, they liquefy, while the released heat is transferred to the environment. The refrigerant solution with decreased ammonia concentration (weak ammonia-water solution – WAWS) continues to circulate in the pipe system to the evaporator (Item 2).

Household gas fridges have a serious market share as components of a food preparation system. They are very popular in gasified urban areas, when fuel gas price is competitive to that of other energy sources (for instance, electricity sources) on the free market.

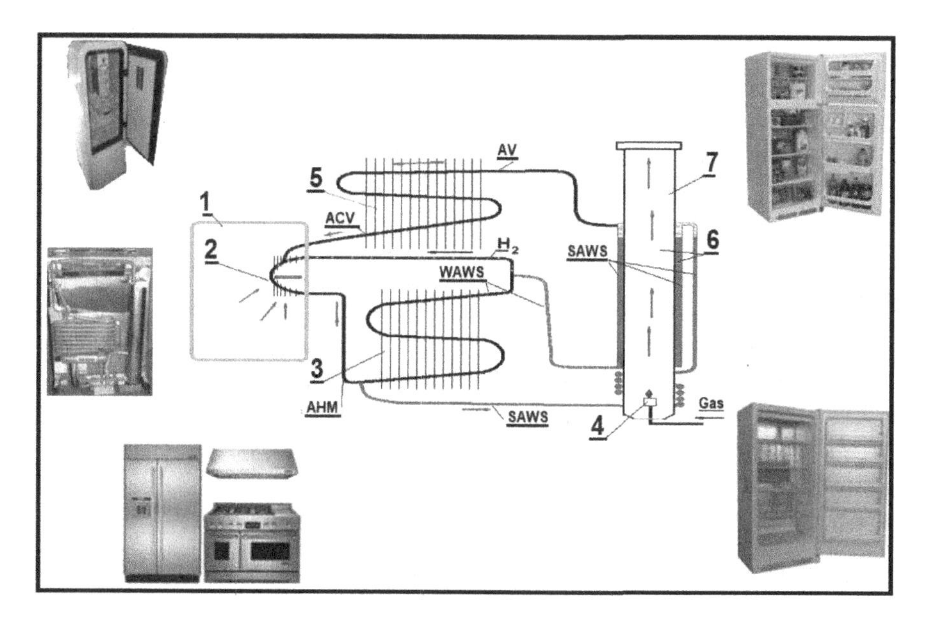

FIGURE 4.46 Gas absorption refrigerator: 1 – Fridge chamber; 2 – Evaporator; 3 – Absorber; 4 – Gas burner; 5 – Condenser; 6 – Gas heating section. WAWS – Weak (scanty) ammonia-water solution; SAWS – Strong (saturated) ammonia-water solution; AV – Ammonia vapors; ACV – Ammonia condensed vapors; AHM – Ammonia hydrogen mix.

FIGURE 4.47 Gas kitchen canopy layout: 1 – Corkboard; 2 – Gas appliances (stoves and hot plates); 3 – Canopy.

4.4.2 Household Kitchens and Gas Appliances

Modern gasified kitchens use various gas appliances together with stoves and hot plates. These are fryers, grills, toasters, wafers and dishwashers and dryers. Figure 4.47 shows a design of gasified household kitchens and their components – gas appliances incorporated into gas stoves, as part of subsequent technical projects.

The availability of such appliances implies the design of a serious logistic system of gas supply, maintenance of air comfort, spill prevention and fire alarm and extinguishment.

A manually operated valve (gas isolating valve) controls the gas supply of outside kitchens. The valve should be close to the emergency exit and easily accessible, thus allowing easy service. To facilitate pipe cleaning, pipes should be installed 0.25 m away from the wall. A flexible pipe and an insolating valve should connect the gas appliance to the main gas pipe.

The kitchen should be ventilated under negative pressure to avoid gas pollution of the air. This is done using a suction installation and canopies installed above the gas appliances. The mechanical ventilation is controlled by a pressurestat maintaining the necessary level of depressurization. Fan switch-off blocks automatically all burners and isolating valves of the appliances.

4.4.3 Commercial Kitchen Arrangement

As discussed, the mission and tasks of commercial kitchens are more important for catering, and their arrangement and logistics are more complex as compared to household kitchens. We outline the main characteristics of commercial kitchens, without delving into details (see Figure 4.48).

Together with rooms for raw materials storage and preparation, the kitchen with its massive equipment and components is an object of a special construction project.

FIGURE 4.48 Commercial kitchen with gas appliances: 1 – Hot plates; 2 – Oven; 3 – Fryer; 4 – Fridge; 5 – Canopy; 6 – Toaster.

Commercial kitchens should conform not only to the designed productivity, but also to the specific clients – restaurants, hospitals, military bases etc.

The general structure of the main kitchen includes:

- Warm kitchen and fire grates;
- Cold kitchen;
- Confectionery kitchen preparing ice cream.

When the building is gasified, the warm kitchen and the fire grates are equipped with gas appliances, which are arranged in cooking units for frying and roasting, fish cooking, food and dish warming up, vessel and utensils washing etc.

To perform basic technological operations such as food and liquid warming and boiling, commercial kitchens use gas boilers with volume amounting to 200 l and power of 40 kW.

Two types of boilers are popular on the market (see Figure 4.49):

- With direct action (Figure 4.49,a and b);
- With indirect action (Figure 4.49,c and d)

The first ones operate under high temperature gradient and they do not process dense liquids, since the latter can burn to the boiler walls. A characteristic feature of the second type of boilers is the indirect warming of food products by means of a buffer heat-conducting medium flowing along an independent circulation contour (water jacket – Item 5). It increases the device thermal inertness thus suppressing the occurring temperature fluctuations. Moreover, the temperature gradient at the vessel bottom decreases, yielding decrease of the probability of fluid burning to vessel's walls and bottom and especially to the internal edges.

There are compact boilers (one-volume ones – Figure 4.49,a and c), as well as boilers, whose allover dimensions are commensurable to those of cookers (for instance,

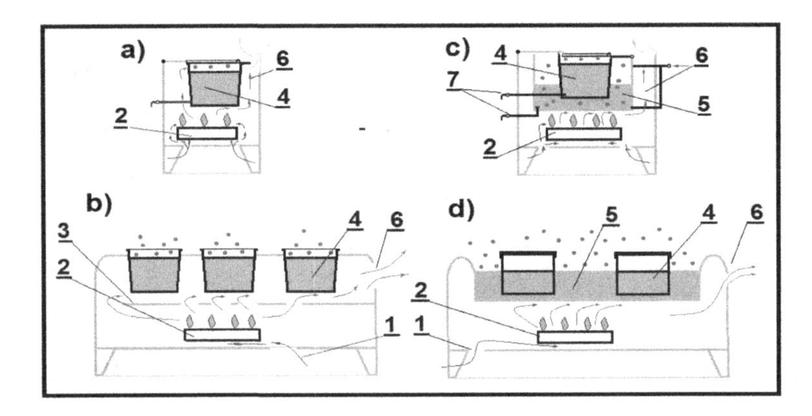

FIGURE 4.49 Boilers a) Single directly heated boiling pan; b) "Bain Marie", serving containers; c) Jacketed bulling pan; d) "Bain Marie" with water jacket: 1 – Intake of combustion air; 2 – Burner; 3 – Separator of warm air; 4 – Serving containers; 5 – Water jacket; 6 – Outlet of combustion products; 7 – Drainage valves.

the multi-volume ones in Figure 4.49,b and d). The choice of their dimensions and capacity depends on the capacity of the hot kitchen and the adopted technological scheme.

Another gas appliances popular in commercial kitchens are those frying raw food such as fish, potatoes etc. known as "fryers". Figure 4.50 shows three types of fryers depending on the bottom structure:

- Flat bottomed (Figure 4.50,a);
- "V" Pan flyer (Figure 4.50,b);
- Tube Fired fryer (Figure 4.50,c).

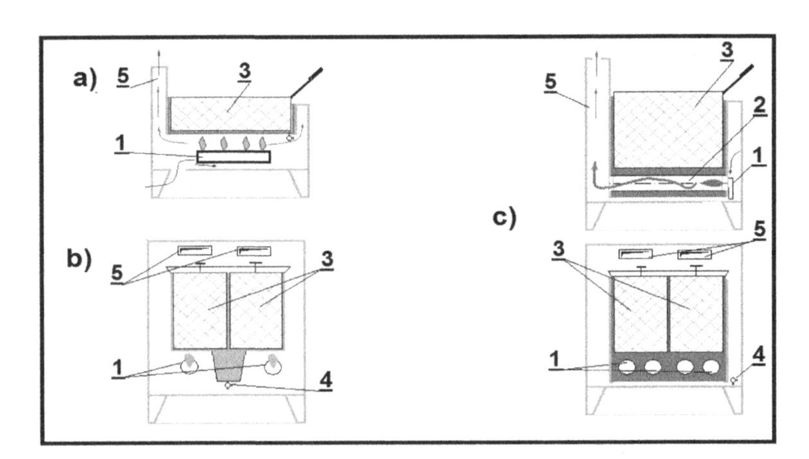

FIGURE 4.50 Gas fryers: a) Flat Bottom Fryer; b) "V" Pan Fryer; c) Tube Fired Fryer. 1 – Burner; 2 – Flame pipe; 3 – Wire basket; 4 – Drainage valves; 5 – Outlet of combustion products.

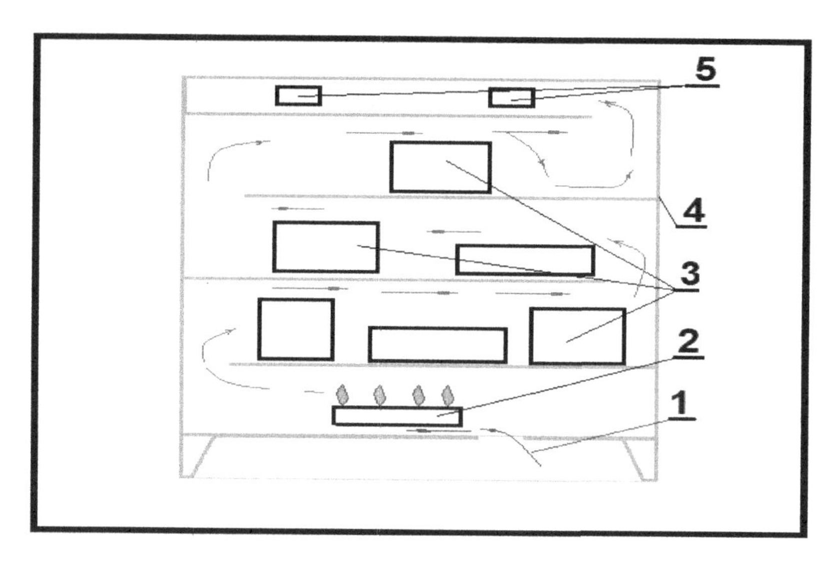

FIGURE 4.51 Cupboard for food heating: 1 – Intake of combustion air; 2 – Burner; 3 – Serving containers ; 4 – Cupboard casing; 5 – Outlet of combustion products.

They have different thermal inertness and duration of operation. Gas fryers whose productivity reaches 20 kg/h are popular on the market. Their volume is 6–10 l, while power is 3.5–18 kW. The temperature is regulated within wide ranges. The devices are equipped with thermal protection cutting the gas supply if the temperature exceeds a specified limit. Their basic components are fabricated from stainless steel and aluminum.

Another gas device used in commercial kitchens is a cupboard heating the ready-cooked food. Figure 4.51 shows a scheme of the appliance. Vessels with cooked food are placed on shelves, the distance between which is 0.15–0.25 m.

A flux of combustion products released by the burner (Item 2) heats them. The burner is located at the lowest cupboard level. The walls of the cupboard (Item 4) are of type "sandwich", fabricated from thin parallel steel sheets with polyurethane sprayed in the gap between them. Their recommended thermal resistance is R50. A thermostat controls the temperature in the cupboard within wide ranges – 0–90°C. The volume of a typical food-heating cupboard is 160 l while its power amounts to 3.5 kW.

There are two variants: stationary and mobile cupboards.

4.4.3.1 Ventilation of Commercial Kitchens[57]

The parameters of the ventilation installation should conform to the effect of the following factors:

- Evaporation due to cooking;
- Food contamination by occupants (cooking personnel and clients);
- Regime of operation and combustion waste released by the equipment.

[57] BS6173.

TABLE 4.14
Recommended suction flow rate of diffusers[60] installed above cooking devices

No	Device	Flow rate m^3/s	No	Device	Flow rate m^3/s
1	Coffee machines in a café	15	5	Lower burning grill	18
2	Boiling pan	18	6	Steam oven фурна	18
3	Oven for dough	18	7	Upper burning grill	27
4	Single oven	18	8	Fryer	18

To overcome the unfavorable effects, one should use a combined ventilations system (suction-pressurizing one), providing 20–40 cycles of air exchange per hour[58]. Kitchen space should undergo significant depressurization so that toxic cooking emissions remain in the cooking area. Hence, diffusers of the suction installation should be placed near the cooking units in conformity with the requirements of the ventilation technology. Except for the correct arrangement, they should be regulated such as to operate under appropriate suction velocity and output. Table 4.14 specifies the recommended flow rates of cooking appliances[59].

Generally, one can assess the flow rate of the supplied air via the method described in Sub-par. 4.2.1 (Eq. (4.19)), considering the admissible concentration of carbon dioxide – 2800 ppm (0.28%).

The efficiency of kitchen cleaning significantly rises when a local canopy ventilation system, installed above the cooking units, joins the central one. The superposition of the canopies and the central ventilation system should produce clean air throughout the building.

4.4.3.2 Design of Local Canopy Ventilation

Local canopy ventilation includes special components called canopies together with air ducts, fans, filters etc. Upon waste release, the canopy encapsulates and removes the emissions, avoiding their diffusion into the kitchen by the formation of aerodynamic traps.

There are various types of canopies employed by various technologies – chemical industrial, chemical-educational, thermal and others. The canopy design should account for the individual characteristics of the processes and the properties of the individual polluters.

Ventilation canopies for commercial kitchens are the basic components of ventilation installations. They are also an object of individual design as part of the kitchen technological project.

[58] The admissible air supply from neighboring premises is considered to be (the rest 85% come from the surrounding medium). Air flow velocity in the area of service and habitation should not exceed 0.25 m/s.

[59] If information on the requirements to the ventilation output lacks, we recommend the following suction velocity: for light meal – 0.25 m/s; for medium meal – 0.4 m/s; for heavy meal – 0.5 m/s.

[60] If no canopies are used, we recommend a minimum flow rate 0.0175 m^3/s – m^2 of exhausting fan.

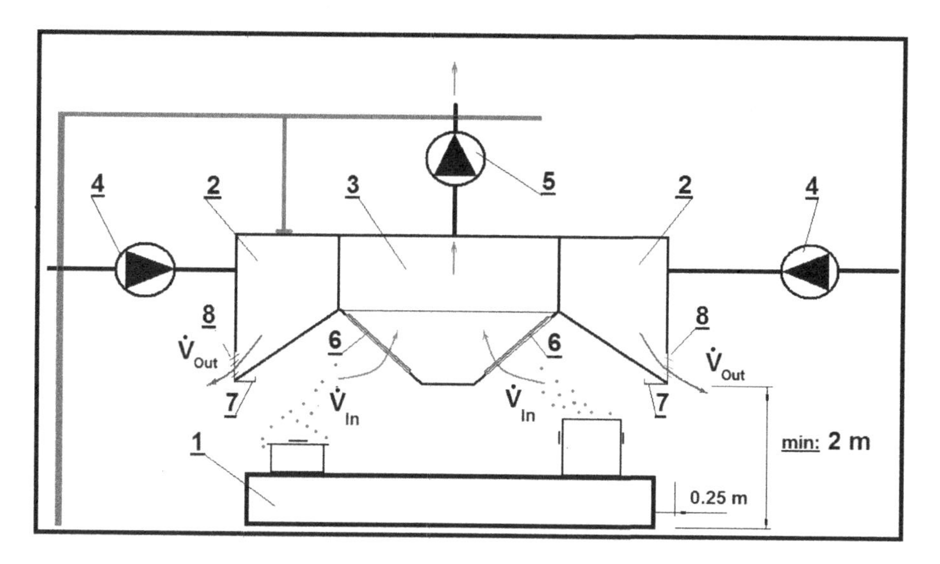

FIGURE 4.52 Scheme of a canopy layout: 1 – Cooking center; 2 – Pressurizing plenum; 3 – Suction section; 4 – Supply fan; 5 – Return fan; 6 – Removable filter; 7 – Condensation gutter; 8 – Movable louvers.

Figure 4.52 shows a general principal scheme of a canopy aggregate for cooking units located at the core of the cooking space. It consists of:

- Pressurizing plenum 2;
- Suction section 3.

It is designed such as to create an aerodynamic "trap" to gas products and vapors released by the operating cooking unit. Depressurization acts within the unit, and the air flowing through the louvers (Item 8) of the pressurizing plenum generates pressurization along its periphery. Released waste mixes with the clean air. It is carried away by the air flow, and a fan (Item 5) discharges it through the suction section (Item 3). The suction section is equipped with removable filter 6 and condensation gutter 7 to catch vapors, which condense at the internal walls of the pressurizing section.

The canopies are fabricated from incombustible and stainless materials including steel and canopy sheets manufactured by specialized and licensed companies.

The recommended dimensions of the installations are specified in Figure 4.52, while the installation of the ventilation line with the canopies is preformed prior to the delivery of the gas appliances, conforming to designer's and manufacturer's instructions.

We conclude that the use of gas appliances in food preparation and storage is economically more profitable due to the low price of blue fuel. Yet, gas devices require strict observation of fire safety and safe work requirement, as well as additional education of work staff and end users.

We shall discuss in the following sub-paragraph the use of gas device in hygiene maintenance systems. Our focus will be on gas water heaters (GWHs), laundries and dishwashers used for household and industrial needs.

4.5 HYGIENE MAINTENANCE SYSTEMS (DOMESTIC HOT WATER SYSTEM.)

4.5.1 Structure of a Hygiene Maintenance System

A hot water installation is an essential part of any household. Hot water is a crucial factor in hygiene maintenance. Effective bathing, furniture cleaning, food and dish washing, laundry etc. seem impossible without hot water and soap, shampoo or various chemicals. Skipping the long and ancient story of the evolution of hot water systems, we discuss in what follows the main installation schemes and components.

The general operational principles of a hygiene maintenance system are set forth based on the methods disclosed in Sub-par. 4.2.1, considering hygiene-deteriorating factors such as gaseous or liquid pollutants, aerosols etc. Similar to the methods of removal of fuel gas pollutants, the basic hygiene-maintaining methods employed in a building include the following techniques:

- Removal or modification of pollution sources located in the occupied areas;
- Use of hot water to dissolve, wash and remove pollutants (supplied by a hot water installation for household needs);
- Local washing;
- Cleaning of waste water.

We focus on hot water systems/installations for domestic needs, considering the incorporation of gas heaters, which are components of the indoor gas network. Due to the distance between the heat generator and the hot water users, we distinguish two classes of devices illustrated in Figure 4.53:

- Heaters close to the user – see Figure 4.53,a);
- Remote heaters – see Figure 4.53,b).

As seen in the first scheme, the heat generator (Item 1) and the end devices (taps, batteries, showers, bath tubs, hydrotherapy tubs, washing machines, dish washers etc. – Item 4) are neighboring ones (Figure 4.53,a). Due to the small distance, water flows in the main pipe under gravity.

As for systems where water is heated by a remote heat generator (see Figure 4.53,b), it flows under the action of a circulation pump (Item 8). To realize an efficient control of the consumption of a primary energy carrier, the system of water heating should be designed as an independent hydraulic circulation contour, comprising heat generator (Item 1), heat accumulator (Item 5) with a heat-exchanging serpent (Item 6) and a circulation pipe network.

The revolutions of the pump motor (Item 8) and pump flow rates should conform to the consumption of hot water by the end users.

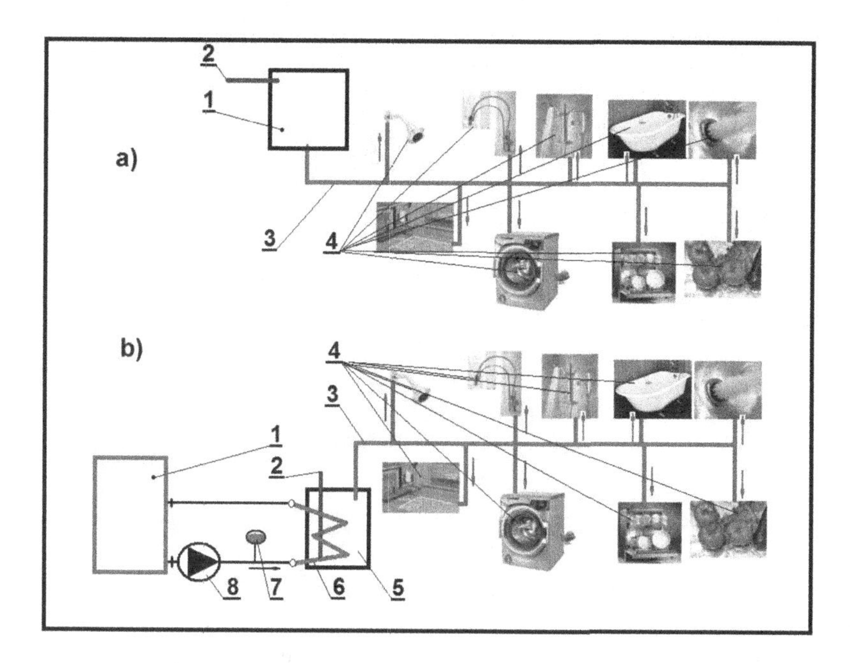

FIGURE 4.53 Typical hot water installations for domestic needs: a) Water heating to the needed temperature; b) Heating in a remote heat generator: 1 – Heat generator; 2 – Cold water intake;3 – Hot water outlet and distributing pipeline; 4 – End users; 5 – Regenerator (heat accumulator); 6 – Heat exchanging serpent; 7 – Membrane expander; 8 – Circulation pump.

Disinfection may take place during washing if water temperature is high enough to kill bacteria, spores or viruses or wash grease. Water heating is programmed in washing machines and it can reach boiling or saturation temperature in the general case.

To heat water for hygiene needs (50–60°C), households usually use boilers, water heaters, flow devices, heat accumulators with heat exchanging serpents etc. Heat is released by the combustion of natural gas, household naphtha, briquettes, coal or fire wood. Heat exchangers or boilers are used in cities with central energy supply, operating with overheated water (130/90°C), electricity or natural gas.

Water heating can take place under atmospheric pressure in open vessels and under isobaric conditions (p=const) or in closed vessel at constant volume (V=const) – so-called isochoric heating. From an energy point of view, isochoric heating is more effective, since the increase of water internal kinetic energy (water temperature) entirely consumes the generated heat. Heating in open boilers (isobaric heating) does not result in effective consumption of thermal energy, since liquid internal potential energy (pressure) and enthalpy rise together with liquid temperature (internal kinetic energy).

A system of hygiene maintenance uses a number of devices heating water under isobaric as well as isochoric conditions – see Figure 4.54. Consider, for instance, the open cauldrons (Figure 4.54,a) and the teapots (Figure 4.54,b and c). Note that heating takes place there under constant atmospheric pressure. As for a capacitive water

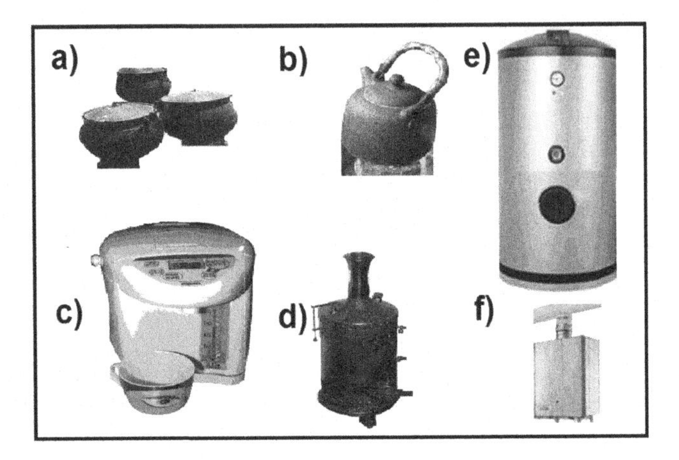

FIGURE 4.54 Water heaters: a) Cauldrons; b) Tea pot; c) Electric tea pot; d) Cauldron with tank; e) Capacitive heater-boiler; f) Tankless – instantaneous water heater.

heater-boiler (Figure 4.54,e), a cauldron with a tank (Figure 4.54,d) and an instantaneous water heater (Figure 4.54,f), heating takes place at a constant volume or under isochoric conditions and it is more effective than the isobaric heating.

The device efficiency determination is a comparatively easy task. Under pressure in a household water heater $(0.66 \le p_{WH} \le 300\text{kPa})$, water specific volume is $0{,}001\,m^3/kg$ and it does not vary significantly. Hence, we may assume that:

$$q' \approx u' \approx i' = c_p t_H = 4{,}19 t_H, \quad kJ/kg^0K. \tag{4.27}$$

The determination of the amount of heat needed to heat water with mass m,kg , is a classical problem. One needs to specify only the respective temperature interval – $\Delta t = (t_2 - t_1), {}^0K$, i.e. water initial and final temperature. There are analytical formulas to perform the calculations:

- In heaters equipped with tanks (capacitive boilers):

$$Q_{Heating} = mc_p\left(t_2 - t_1\right) = 4.186\left(t_2 - t_1\right)m, \quad kJ, \tag{4.28}$$

  (power is defined by the accumulated mass and the needed heating time);
- In instantaneous water heaters we apply the equation of the total flow rate:

$$Q_{Heating} = \dot{m}c_P\left(t_2 - t_1\right), \; kW, \tag{4.29}$$

  where $\dot{m}$, kg/s is the total flow rate of water flowing through the heating device.

Gas devices of a hygiene maintenance system functionally heat water and dry washed materials (clothes etc.). Their share is expected to rise in the next years due to their energy efficiency and drop of the fuel gas price as dictated by the market – see the Third energy package of the EU.

4.5.2 Gas Water Heaters of a Hygiene Maintenance System

GWHs are classified depending on their structure:

- Accumulation type;
- Instantaneous type;
- Water pipe-flow type (with minimal capacity).

According to the method of conveyance of combustion products, the devices are (see Sub-par. 3.3.3.3 for more details):

- With natural or forced draught;
- Unbalanced and dynamically balanced.

Accumulating GWHs are classified depending on the ratio "consumed power/capacity" (CP/C) as:

- GWH for household needs (with CP/C ratio up to 300 W/l);
- GWH for business/industrial needs (with CP/C ratio > 300 W/l).

Accumulating (capacitive) GWHs are classified depending on the type of their system for automatic regulation as: (i) with a proportional system for automatic regulation and (ii) with a two-position regulator (thermostat).

Flow GWHs are non-capacitive devices, volume (accumulating, capacitive) devices and water-pipe flow devices.

The scheme of a tank (accumulator) GWH for household needs is shown in Figure 4.55,a. It belongs to the class of unbalanced devices, since the terminal discharging combustion products is outside the building, and indoor air is sucked to support combustion. The main components of a gas boiler are: burner – Item 4, smoke tube – Item 5, accumulating volume (water tank-jacket) – Item 1 and casing – Item 9.

Figure 4.55,a) illustrates device operation. Cold water enters through the supply pipe 6 the bottom area of the accumulating volume (i.e., the water jacket above the combustion chamber). It accumulates heat and elevates to the top of the volume. Then, being ready for use, it is conveyed through pipe 7 to the end users whose valves are open. Pipe 4 conveys fuel gas to the burner (Item 3) while smoke gases are released in the surrounding atmosphere through smoke tube 5, functioning as a heat exchanger, and through flue 8. Figure 4.55,b) and Figure 4.56 show a view and fittings of an accumulating water heater.

All gas devices are equipped with systems for combustion control, considering water temperature and the needed flow rate. Most of them have thermostat valves installed in the gas-supply pipe, which are normally closed prior to the burner turn-on.

Some gas devices have module burners used for a more adequate control of the temperature of the discharged water. Other devices control water flow rate such that water temperature would not drop below a previously specified value. Most modern

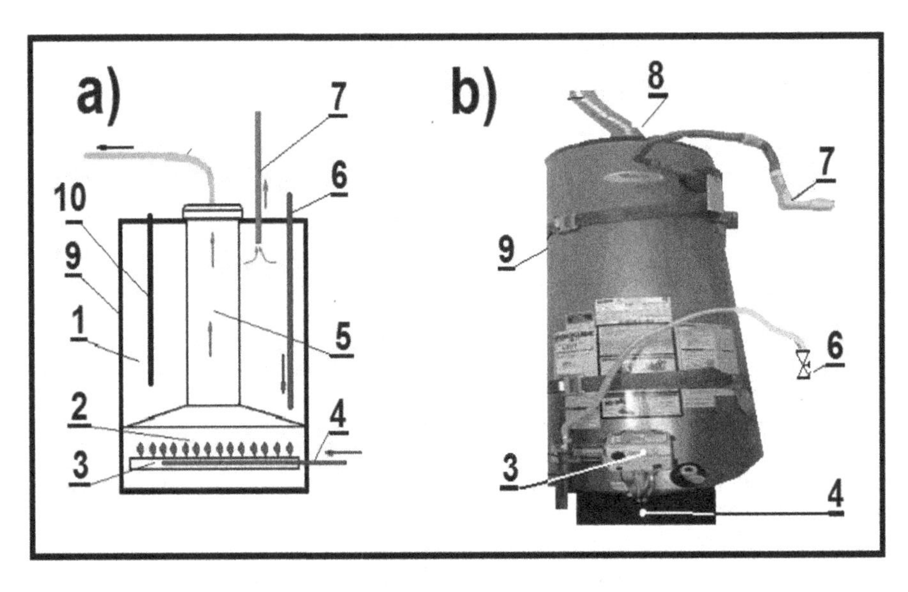

FIGURE 4.55 Gas accumulation water heater: a) Cross-section of an accumulation water heater; b) View of an accumulation water heater: 1 – Accumulation volume; 2 – Combustion chamber; 3 – Gas burner; 4 – Gas supply pipe; 5 – Smoke tube; 6 – Cold water supply pipe; 7 – Pipe conveying hot water; 8 – Flue to chimney; 9 – Casing; 10 – Sacrificial anode.

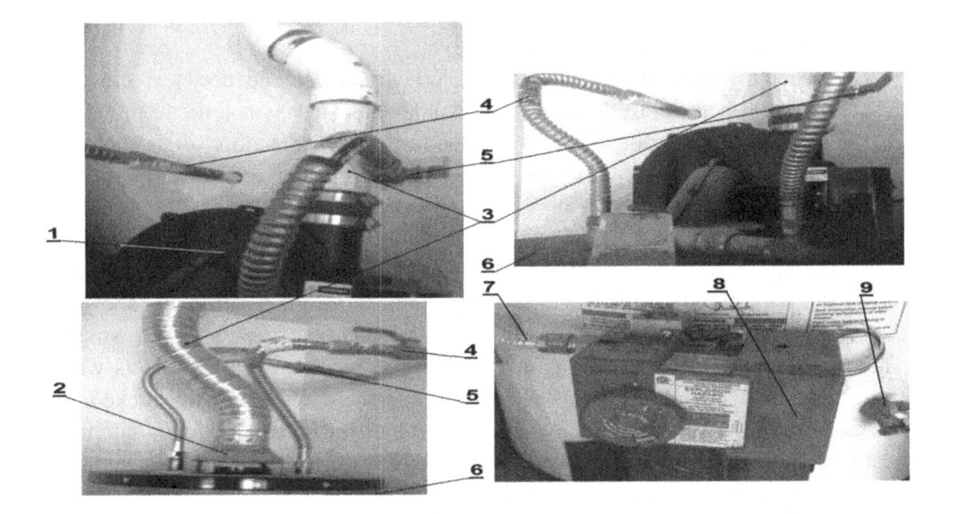

FIGURE 4.56 Fittings of a gas accumulating water heater with capacity 40 l and power 12 kW: Gas pressure: 1000 Pa (max. 3488, min. 1250 Pa): 1 – Smoke fan; 2 – Draught diverter; 3 – Secondary flue; 4 – Cold water feed; 5 – Hot water draw out; 6 – Water accumulator; 7 – Gas connection; 8 – Gas burner; 9 – Drain valve.

devices are equipped with an electronic control of burner ignition minimizing the stationary losses.

We consider the structure and functioning of non-capacitive (tankless) devices as GWHs. For now, they are the most popular ones because of their instant heating action.

Figure 4.57 shows a frontal view – semi cross-section of a water heater of an instantaneous type. Cold water enters the device through pipe 8 connected to the heat exchanging serpent (Item 9) of the primary (Item 12) and condensing (Item 14) heat exchangers. Flowing water is heated there and immediately leaves the water heater through pipe 6.

The instantaneous devices operate only when end users need hot water. The steady losses are negligible, since no water volume is heated in the accumulator in order to maintain device operability. Device thermal capacity is sufficient to cover the timetable of hot water consumption within the entire system, and water heaters are highly efficient. Their efficiency coefficient is of order 0.91–0.95, owing to their specific structure allowing the in-depth exploitation of the energy potential of smoke gases. Their temperature drops to 70–65°C in a condensing heat exchanger, and the heated water extracts the heat of hidden phase transition from water vapors, transforming the latter into a condensate that exhausts through the drainage pipe – Item 13.

Pursuant to the method of installation, flow water heaters operate as dynamically balanced, as well as dynamically unbalanced devices. When the intake (Item 1) and the exhaust (Item 2) terminals are installed on one and the same facade or on

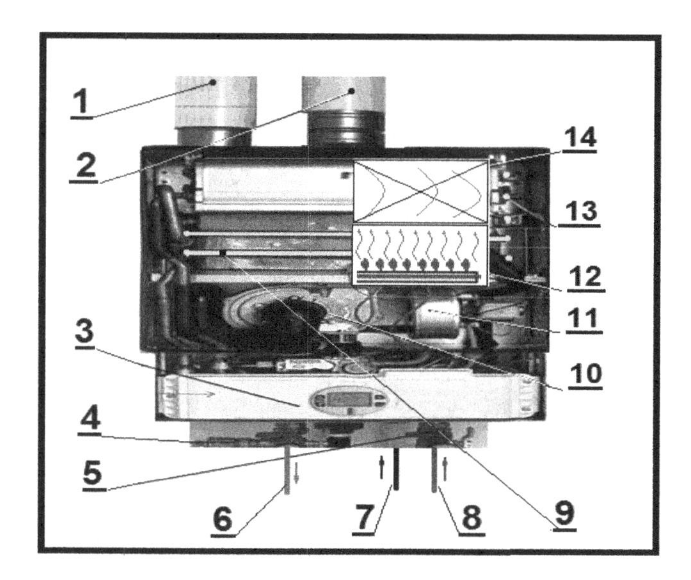

FIGURE 4.57 An instantaneous non-capacitive (tankless) gas water heater: 1 – Intake of combustion air; 2 – Exhaust out of combustion products; 3 – Panel of a system for automatic control; 4 – Hot water outlet; 5 – Cold water inlet; 6 – Hot water outlet; 7 – Fuel gas intake; 8 – Cold water intake pipe; 9 – Heat exchanging serpent; 10 – Fan; 11 – Burner; 12 – Primary heat exchanger; 13 – Pipe draining the condensate; 14 – Condensing heat exchanger.

the building roof, and the wind-generated dynamic pressure is one and the same, their operation is stable. If fresh air is sucked from a room, change of pressure will take place in the combustion chamber at any change of the total pressure within the smoke flue installed outside the building. At the same time, fire extinguishment will gather momentum, while combustion will become unstable and ineffective.

As mentioned, the choice of a heater should be in conformity with the structural concept of the technical-economical task (TET) and the capacity of the employed technology.

4.5.2.1 Rating of Heat Generators in a Hygiene Maintaining System

An important issue that faces a designer of a system for hygiene maintenance is the selection of a heat generator pursuant to the domestic need for capacity and power. Parallel to criteria such as operative reliability, easy control and repair, the decisive argument herein is the life cycle costing of the generator.

Figure 4.58 shows the ranking of different heat generators depending on life cycle costing (including initial investment, fuel gas price and operative costs). As seen, gas heat generators occupy the first four places where the price of the instantaneous gas generator (Item 1) is the lowest one. The highest price is that of electric heaters (Items 6, 7 and 8). The most expensive device is the flow non-capacitive heater (Item 8).

An important characteristic of the devices is their operation in parallel with other heaters that use alternative energy sources, for instance electricity sources and renewable energy sources. The use of such combinations aims at the increase

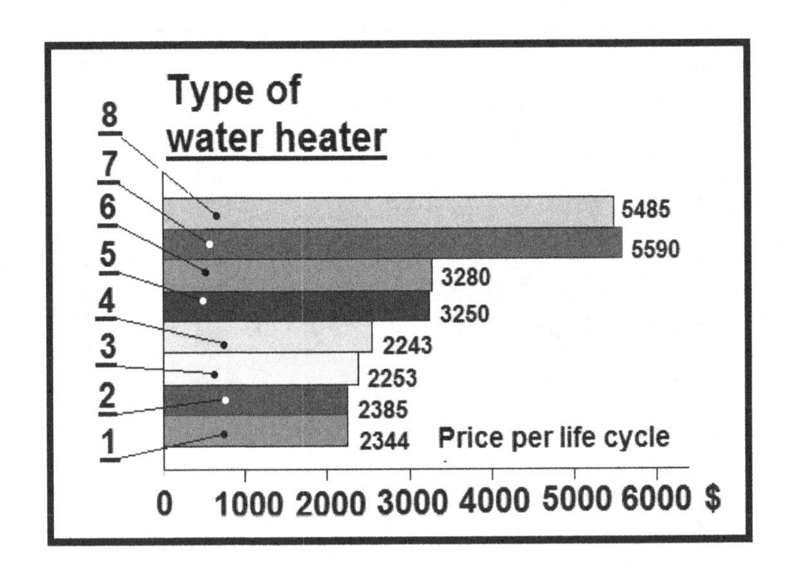

FIGURE 4.58 Price of heat generators participating in a hygiene maintaining system during a life cycle: 1 – Instantaneous gas device; 2 – Gas indirect effective device; 3 – Gas capacitive effective device; 4 – Gas capacitive convective device; 5 – "Solar plus electric" device; 6 – Electric thermal pump; 7 – Conventional electric capacitive device; 8 – Electric flow non-capacitive device.

of the system reliability or economy of primary energy sources. We will consider in what follows some combined schemes using gas heaters together with heaters using solar energy.

Modern systems for hygiene maintenance widely apply bivalent (combined) schemes using two heat generators consuming energy from different sources.

4.5.3 Combined Schemes of Water Heating in Systems for Hygiene Maintenance

Figure 4.58 shows a principal configuration of a hygiene maintaining system and its four modifications, comprising GWHs combined with thermo-solar collectors and thermo-pumps:

- Gas flow and thermo-solar water heater (Figure 4.59,a);
- Gas flow heater and thermo-solar heater with an accumulator (Figure 4.59,b);

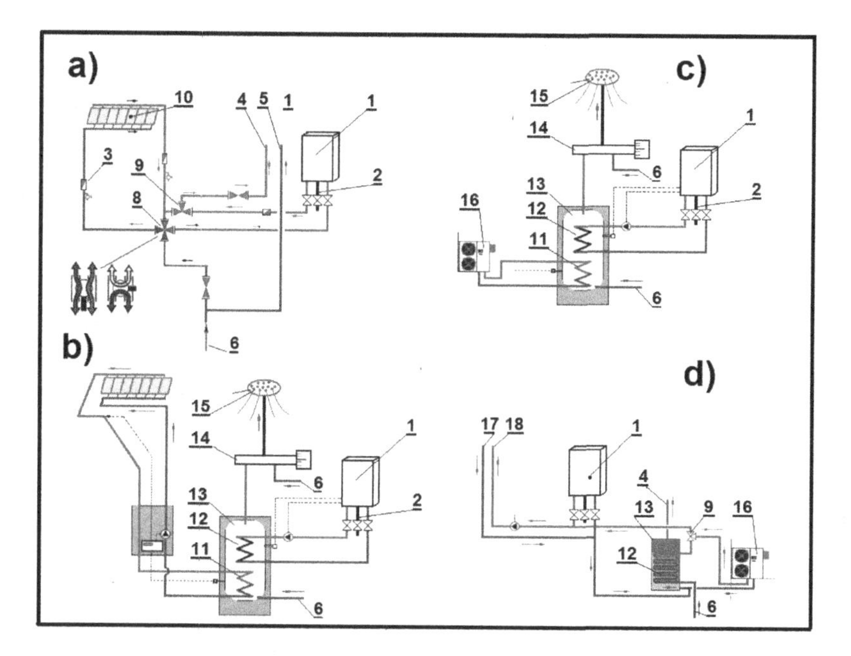

FIGURE 4.59 A combined scheme of the activation of gas water heaters: a) Gas flow water heater and thermo-solar water heater; b) Gas flow water heater and solar water heater with an accumulator; c) Gas flow water heater and a heat pump; d) Gas flow water heater, a heat pump included in heating and hygiene maintaining systems. 1 – Gas instantaneous water heater; 2 – Intake of fuel gas; 3 – Return valve; 4 – Outlet of hot water; 5 – Cold water line; 6 – Cold water inlet; 7 – Operative valves; 8 – Four-way valve; 9 – Three-way mixing valve; 10 – Thermo-solar water heater; 11 – Heat exchanging serpent-solar circle; 12 – Heat exchanging serpent-water heating circle; 13 – Accumulator; 14 – Mixer: 15 – Shower; 16 – Heat pump; 17 – Returning main line of heating installation; 18 – Supplying main line of the heating installation.

- Gas flow water heater and a heat pump (Figure 4.59,c);
- Gas flow water heater and a heat pump in heating and hygiene-maintaining systems (Figure 4.59,d).

The first of them in Figure 4.59,a) has two heat generators – the thermo-solar installation (Item 10) and the gas device (Item 1), which work in parallel controlled by a SAC. The executing device of the latter is a three-way mixing valve (Item 9). When there is not enough solar light and water temperature measured in the solar collector differs from the set point, the three-way valve opens to hot water inflowing from the gas heat generator (Item 1). When the temperature of water inflowing from the thermo-solar water heater is high enough (for instance, 38°C), the three-way valve shuts off, avoiding inflow of fluid heated in the gas heat generator. Hence, the latter is excluded from the scheme and fuel consumption terminates. The scheme is successfully used in areas with tropical climate and one-season supply of hot water.

The second scheme (Figure 4.59,b) is a replica of the first one. A free volume replaces the three-way valve, accumulating and supplying thermal energy and hot water to the end users. A solar heat generator is included in the scheme via an independent hydraulic contour where fluid with low freezing point (antifreeze) circulates. The circulation pump in the solar contour activates when a difference between the temperatures of the solar field accumulator, the collecting line and the distributing line occurs. The solar heat generator operates as a device that heats the water in the accumulating volume to a certain temperature, depending on the amount of absorbed solar energy and on the accumulator capacity. An underestimated thermal capacity yields losses, i.e. unabsorbed solar energy. One should employ an optimization procedure to calculate the solar field, the supplementary power and the accumulator capacity.

The gas device is also connected to the scheme by means of an independent contour, and it operates as a heat generator. The scheme can be modifies if the device is set to operate in a flow regime and without a heat-exchanging serpent – Item 12. Its function is to raise water temperature to a particular value. A system for automatic regulation activates the device when water temperature drops below a fixed value. The scheme can operate successfully in areas with continental climate or throughout the year.

A heat pump – Item 16 replaces the thermo-solar heat generator in the third scheme (Figure 4.59,c). Its function is similar to that of the solar heat generator – it heats the cold water beforehand. A GWH – Item 1, additionally heats water to the specified temperature (about 60°C). Its activation depends on the regime of operation of the hygiene maintenance system.

The type of the used heat pumps depends on the available environmental energy source (water basins, rivers, underground massifs, technological waste heat and even ambient air).

The last scheme in Figure 4.59,d) is that of an integrated system for heating and hygiene maintenance by means of hot water supply from an accumulator – Item 13. The heating installation discharges cooled water to the bottom area of the accumulator. The same area feeds the condenser of the heat pump – Item 16. Water, heated in the heat pump, returns to the accumulator through the three-way valve – Item 13, or enters the distributing main line of the heating installation, where it mixes with water previously heated in the gas heater. Cold water in the pipeline enters the system at the base of the heat exchanging serpent – Item 12 of the accumulator. Heated water leaves the accumulator (Item 13) through the outlet – Item 4 and enters the hygiene maintenance system.

4.5.4 Gas Dryers in Systems for Hygiene Maintenance

We discuss in what follows a gas dryer, which is another component of a system for hygiene maintenance. It is installed in a "wet" space[61], close to a washing machine. Its overall dimensions are about 0.6×0.6×1.2 m, while its power amounts to about 6 kW.

Figure 4.60 shows a typical scheme of the device. As seen, gas dryer does not differ significantly from an electric dryer. Its main component is a rotating drum (Item 5) with volume of about 0.120 m^3, tumbling clothes through heated air to remove moisture after centrifuging. The drum rotates clockwise with a velocity of 50–60 tr/min. It is set in operation by a motor and a driving belt. After several revolutions, it stops and starts rotating anticlockwise. Then, it again stops and starts rotating clockwise, and the cycle periodically repeats. Hot air (50–60°C) enters the drum, removing moisture from clothes for about 45 min (the duration of one technological cycle).

The difference between electric and gas dryers consists in the structure of the air generator of heat only. While electric heaters are components of an electric dryer, here the heat generator comprises a gas burner, an ejector and a smoke gas fan. It uses fuel gas as a primary energy carrier. To avoid damage of clothes, the temperature is controlled within fixed limits by two operative thermostats, whose sensors[62] are

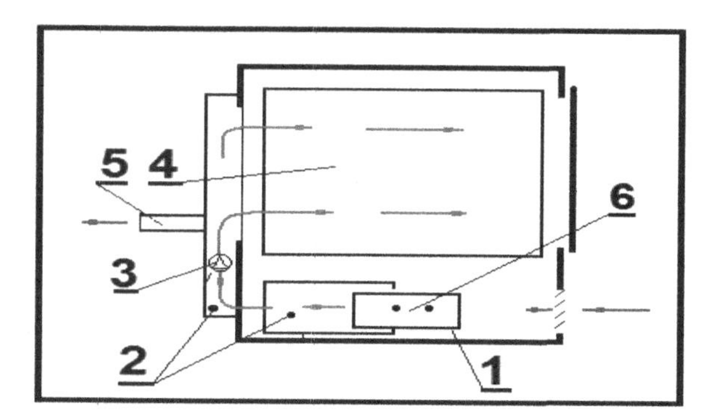

FIGURE 4.60 A household gas dryer: 1 – Ejection chamber and burner; 2 – Air heat generator; 3 – Fan; 4 – Drum; 5 – Outlet for combustion products; 6 – Thermostat.

[61] A dryer should not be installed in a bathroom with a shower or tub. It should not be mounted in a sleeping room or in a small room where the ratio volume/installed power is less that 7 m^3/kW. No gas dryers should be installed in penthouses while LPG dryers should be out of sub-terrains.

[62] A clogged filter decreases the flow rate of the thermal air flux through the device, as well as the device productivity.

mounted in the air duct after the fan – Item 3. There is a second pair of thermostats controlling the temperature. The first of them operates at about 110°C. It is installed on top of the rear of the drum housing. This sensor cuts off the gas supply, but allows reignition as the drum cools down. The location of the second thermostat is identical. It is used to guarantee cut-off of the gas supply when the controlled temperature exceeds 120°C. The thermostat function is to avoid clothes damage during drying.

The gas dryer deteriorates the air in the room where it is installed. Due to its low power, it is classified as a smokeless device, but its operation yields increase of the humidity and the aerosol amount. This is so, since a large amount of cloth fibers are released during drying, joining the air molecules. This yields moisture condensation on outer and non-insulated walls of the room, stimulating the development of plantations of mould and lichen.

An air filter with flow rate capacity of 0.05 m^3/h is installed at the device outlet and it should be periodically cleaned.

The waste mix released after drying contains moisture, aerosols and carbon dioxide. It is discharged through a wall or window vent with diameter of at least 0.075 m, but not less than that recommended by the manufacturer. A smoke flue can convey the waste to the roof, but the waste should not mix with smoke released by other devices.

4.5.5 Calculation of Hygiene Maintenance Systems

Hygiene maintaining devices occupy separate rooms, where special appliances may also operate. Such are for instance:

- Bathrooms;
- Laundry ("wet" rooms);
- Toilets/bathrooms;
- Wash-rooms;
- Kitchens.

where hot and cold water are the work fluids. Hence, the calculation of a hygiene maintenance system of a building is reduced to the use of a standard algorithm calculating a water pipe network. It consists of the following steps:

1. Situation of the devices and the pipe network in the building plan;
2. Specification of the normative consumption of water (hot/cold) expressed in WSFU[63] and concerning each device designated in the sketch;
3. Design of an axonometric (riser) scheme of the pipe network. Calculation of the hydraulic flow rates of the pipe sections using Kirchhoff's second law;
4. Transformation of WSFU into units of water flow rates – m^3/s using the plot in Figure 4.61;
5. Hydraulic calculation of the pipe network.

[63] WSFU (water supply fixture units) – conditional unit of (hot/cold) water consumption by the devices related to the output through a regular water valve (see Table P19 in the Appendix).

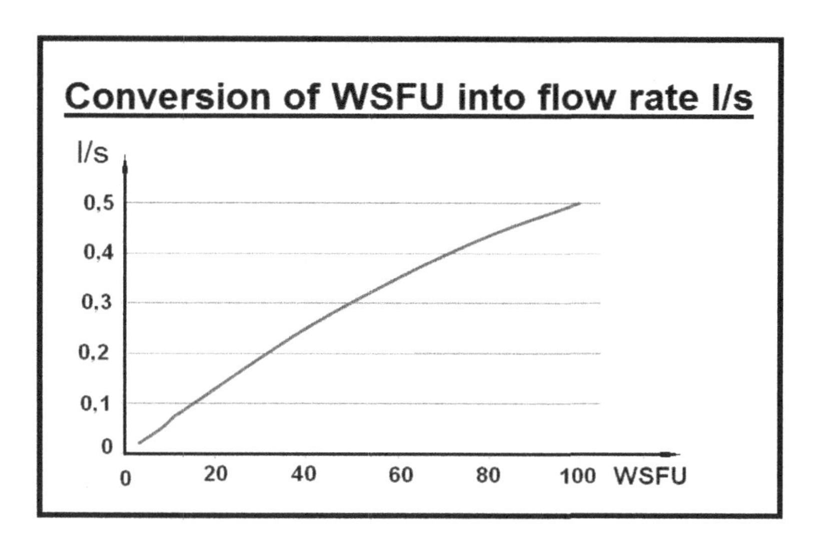

FIGURE 4.61 Plot of the relation between water consumption of the device expressed in WSFU and the equivalent flow output, l/s.

Example 4.13

Calculate the pipe network of a system for hygiene maintenance in a two-storey residential building. The inlet of the main hot water line is 50 m away from the city pipeline (the static pressure amounts to 1.6×105 Pa). The hydraulic loss in the spherical valve is 0.2×105 Pa.

SOLUTION

The connecting pipes are 1.3 m above the floor. Taps and spherical valves are 0.015 m away from each device and 0.022 m away from the showers.

The architectural arrangement of the devices of the hygiene maintaining system is in two premises – a kitchen and a bathroom/toilet/washroom (see Figure 4.62,b), while Figure 4.62,a) shows a partial cross-section of the building.

1. The components of the systems and the pipe network are plotted in the sketch keeping the respective scale (see Figure 4.6,b). We take the length of main lines, risers and connectors from the architectural drawing. The distances are denoted in the axonometric scheme of the pipe network – Figure 4.63. We chose a scheme with upper supply consisting of three risers (these are three vertical main pipes joined by the horizontal branches to the devices). To guarantee the reliability of water supply, we consider the installation of a 250-liter water tank in the attic or on the roof.
2. We record the water amount (measured in WSFU) needed by each end user in compliance with data in Table P19.
3. Using the plot in Figure 4.61, we transform the prognostic water consumption measured in WSFU into prognostic flow rate measured in ***l/s***, inserting the new data in the sketch (Figure 4.63,a).

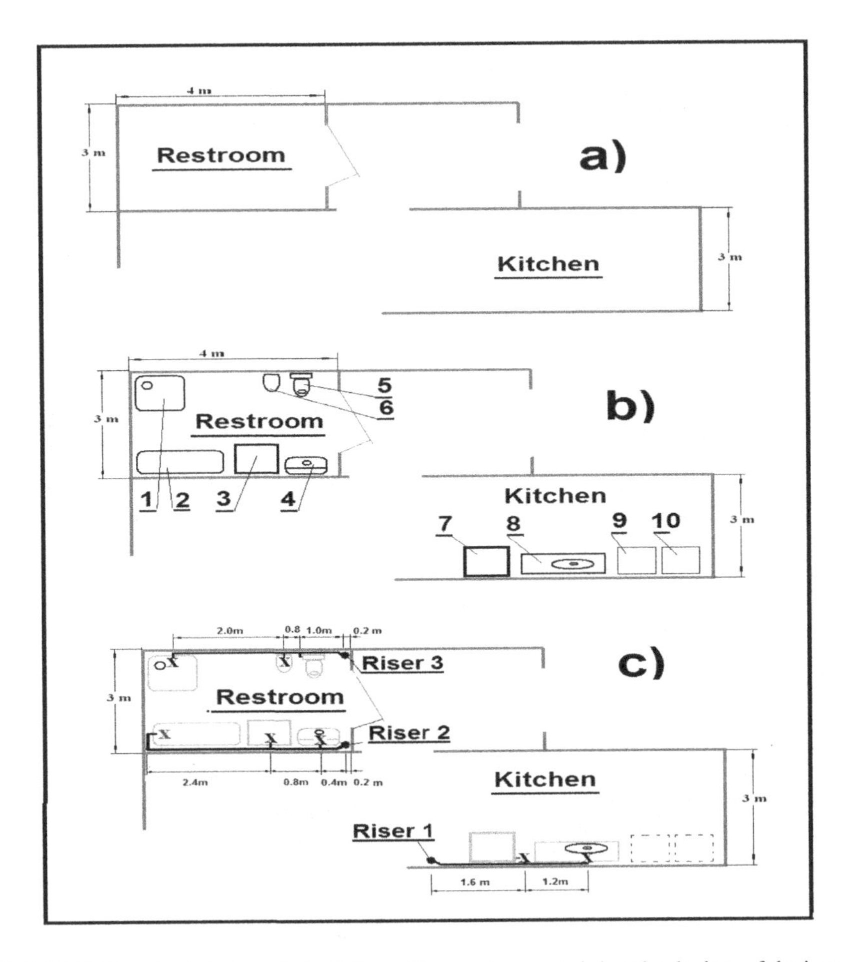

FIGURE 4.62 Partial view of a building with premises containing the devices of the installation – first and second floor: a) Architectural sketch; b) Scheme of device location; c) Scheme of pipe network for water supply: 1 – Shower cell; 2 – Bath tube; 3 – Laundry; 4 – Sinks; 5 – Toilet; 6 – Bidet; 7 – Washing machine; 8 – Kitchen sink; 9 – Gas cooker; 10 – Fridge.

Consider, for instance, a certain amount of toilet water – 3 WSFU. Then, we transform 3 WSFU into 0.025 ***l/s***, 11 WSFU – into 0.075 ***l/s*** and 14 WSFU – into 0.1 ***l/s***. To make the sketch clearer, we use the 1st column of the matrix table:

WSFU	
$\dot{V}$ *l/s*	-

3	-
0.025 *l/s*	-

We insert the needed amount of water and the prognostic flow rate of each device in the matrix table, following the notations in Figure 4.63,b).

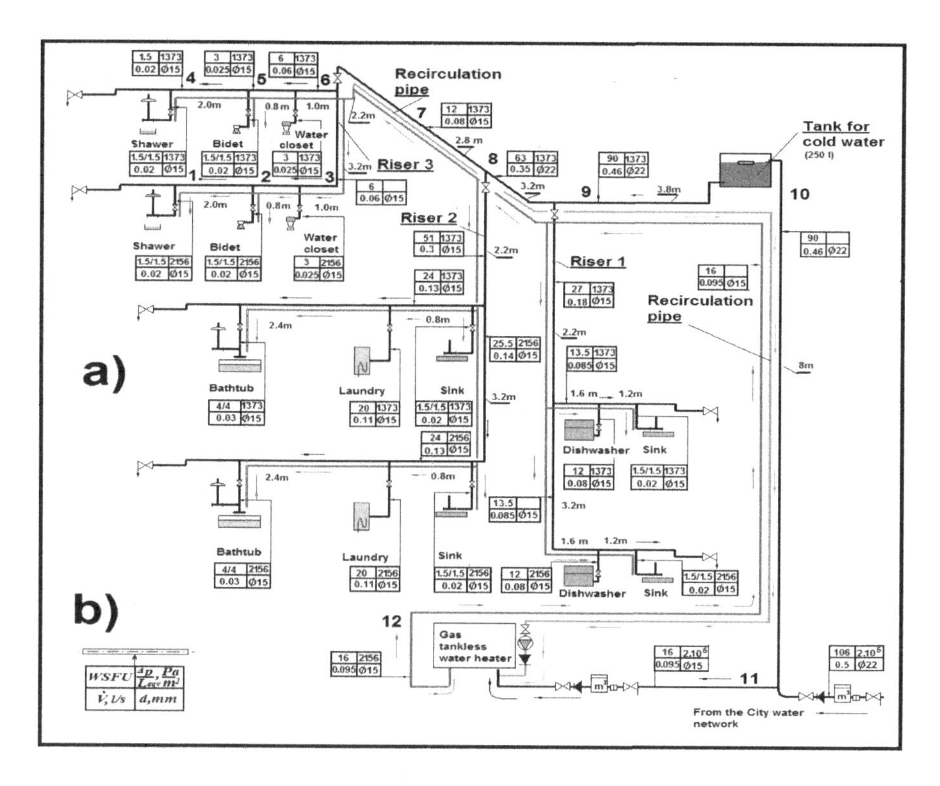

FIGURE 4.63 A riser scheme of the cold/hot water pipe network.

4. Based on Kirchhoff's second law, we summate all WSFU along a direction opposite to that of water flow in the network.
   The results for the respective pipe sections are ordered in a matrix table inserted in the riser scheme – Figure 4.63,a).
5. We find the indexes[64] of each circulation contour;
   5.1 Determination of the water head:
   - First floor – 6.1 m (see Figure 4.62,c), and Figure 4.63,a);
   - Second floor – 2.9 m;

   5.2 Determination of the equivalent length of the pipe sections ($L_{\text{екв}}$) in (m);
   We multiplying the measured length of the pipe section (L_M) by a coefficient reducing the local resistance to the linear one by means of Eq. (2.22) – see Table 2.4 in Sub-par. 2.22 for details.
   5.2.1 Second floor
   The remotest water user is connected to the hydraulic contour of the third riser – a shower-cell on the second floor (see Figure 4.62,b and Figure 4.63,a)

$$L_M = 3.8 + 3.2 + 2.8 + 2.2 + 1.0 + 0.8 + 2.0 = 15.8 \text{ m}$$

$$L_{eqv} = 1.3 \times 15.8 = 20.53 \text{ m}$$

[64] The index of the circulation contour is the least value of the ratio "hydrostatic head/equivalent length" calculated via the formula $IC_i^H = H_i / l_{eqv}^i$.

5.2.2 First floor (see Figure 4.62,c and Figure 4.63,a).
The remotest water user is connected to the hydraulic contour of the third riser – shower-cell on the first floor (see Figure 4.62,c and Figure 4.63,a)

$$L_M = 3.8 + 3.2 + 2.8 + 2.2 + 1.0 + 0.8 + 2.0 + 3.2 = 19.0 \text{ m}$$

$$L_{eqv} = 1.3 \cdot 19 = 24.7 \text{ m}$$

The following table gives the calculated indexes of the water contours of the first and second floors:

Circulation contour	Equivalent length l_{eqv}, m	Hydraulic head H_i, m	Contour index – IC_i^H
First floor (to the shower-cell)	27.7 m	6.1 m	0.22
Second floor (to the shower cell)	20.53 m	2.9 m	0.14

Using these values, we calculate the available hydraulic pressure head $\Delta\bar{P}_i$, belonging to the water devices of the first $\Delta\bar{P}_1$ and second $\Delta\bar{P}_2$ floors, pursuant to the expression:

$$\Delta\bar{P}_i = 9807 * IC_i^H, \quad \text{Pa/m} \tag{4.30}$$

Consider the water contour of the first floor. Then, we have the following available specific hydraulic head – $\Delta\bar{P}_1$=0.22×9807=2156.0 Pa/m. As for the second floor, we have $\Delta\bar{P}_2$=0.14×9807=1373.0 Pa/m. We calculate the pipes using these values.

6. Calculation of the pipe diameters considering the specific pressure head

We must solve the so-called **inverse problem.** Knowing the specific pressure head, find an appropriate pipe diameter guaranteeing the necessary water flow rate. We use Table P21[65]. For instance, if $\Delta\bar{P}_1$=2500 Pa/m, the mass flow rate of pipes with diameter Ø15will will be about 0.268 kg/s, while that of pipes with diameter Ø42 will be about 4.26 kg/s.

Using the table, we calculate pipe sections №1 – №6 denoted in the schemes of the water contours feeding devices in the first and second floors (Figure 4.63,a). This, however, is done, using the already calculated specific hydraulic drop $\Delta\bar{P}_1$=2156.0 Pa/m in the first floor contour, as well as that in the second floor contour – $\Delta\bar{P}_2$=1373.0 Pa/m.

№	$\Delta\bar{P}_i$ Pa/m	$\dot{V}$ l/s	d mm	№	$\Delta\bar{P}_i$ Pa/m	$\dot{V}$ l/s	D mm
1	2156.0	0.02	15	4	1373.0	0.02	15
2	2156.0	0.025	15	5	1373.0	0.025	15
3	2156.0	0.06	15	6	1373.0	0.25	15

[65] The table contains data on the output capability of copper water-supply pipes with nominal diameters within ranges Ø15–Ø45, considering a particular value of the available specific hydraulic drop of pressure. Source:CIBSE Guide (1986).

Attention!

We have additionally increased the diameters of the horizontal distributing pipe (second floor section) to decrease the probability of noise occurrence in the pipe system.

The calculated diameters of all pipe sections are summarized in the riser scheme shown in Figure 4.63.

4.5.5.1 Water Heater

We select a water heater pursuant to the "calculation" flow rate and power.

The calculation flow rate is found by summation of the hot water flow rates needed by the end users, adopting a maximal coefficient of simultaneousness (n=1). The total need for hot water/hot water flow rate amounts to 16 WSFU/0.095 l/s (95×10^{-6} m^3/s) in Example 4.16.

We calculate the needed net power of the water heater using expression (4.26). It takes the following form in Example (4.13):

$$Q_{Heating} = 0.095 * 10^{-6} * 995 * 4186 * \left(65 - 5 \right) = 23740.0, \quad W$$

The total thermal power is:

$$Q_{WH} = k * Q_{Heating} / \eta_{WH} = 1.2 * 2374.0/_{0.96} = 2967.6, \quad W.$$

Here k is a resource coefficient while η is the coefficient of device efficiency.

We select a GWH with the following parameters: total thermal power – 3 kW, nominal flow rate – 120×10^{-6} m^3/s.

4.5.5.2 Recirculation Pump

We selected it pursuant to the flow rate and head. Its function is to recirculate the cooled water in the distributing riser through the water heater. Yet, considering Example 4.13, a pump with head of 0.06 MPa and flow rate of 0.003 m^3/s will successfully operate in this case.

4.6 GAS HEAT PUMPS IN ENGINEERING INSTALLATIONS OF BUILDINGS

4.6.1 Heat Pumps Principle of Action

(Heat pumps as increasing thermal transformers.) Classification

Heat pumps (HP) are devices regenerating low-potential thermal energy or energy waste extracted from a cold source (**CS**) by means of a special treatment of the working fluid in the control volume of a thermodynamic system (TDS – Figure 4.64)[66]. The treatment allows the subsequent use of the regenerated energy in various installations – in thermal comfort or hygiene maintaining, industrial or utilization installations etc.

[66] Lord Kelvin theoretically described the principles of a heat pump in 1852, P. Rittinger designed and built the first HP in 1855–1857.

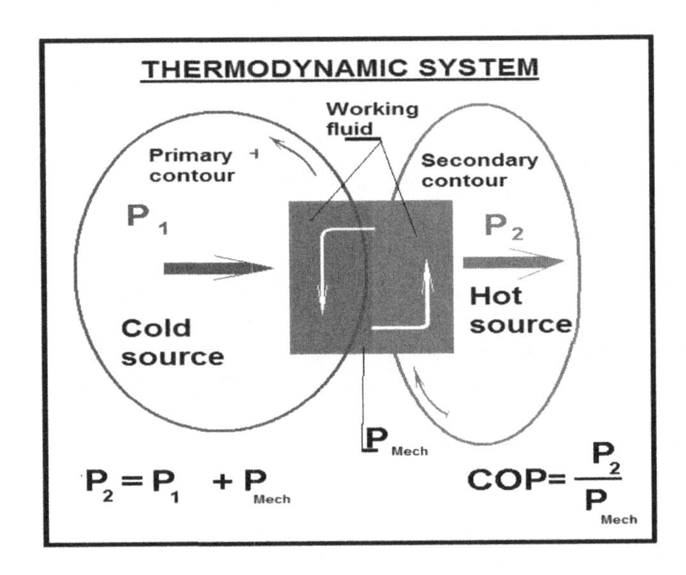

FIGURE 4.64 A heat pump as an energy regenerator (increasing transformer).

CSs of low-potential energy can be:

- The environment (i.e., the surrounding air, soil around or beneath a building, water reservoirs – lakes, rivers);
- Certain industrial technologies;
- Certain natural phenomena (solar light, sea tides, atmospheric air motion).

A classical example of a cold low-energy source is the Geneva Lake[67] (a natural accumulator of heat): – it accumulates heat during the summer and water temperature rises to 10–12°C; it releases heat during the winter, but water temperature does not drop below 3–2°C).

The principle of operation of a heat pump is illustrated in Figure 4.64. The working fluid of a TDS is charged in a **CS**, and the gained power $\mathbf{P_1}$ is its representative characteristic. Then, an external mechanical force[68] compresses the working fluid. Thus, work is done whose intensity is measured via the **mechanical power –** P_{Mech}. The TDS parameters, i.e. operational temperature and pressure, increase together with the increase of energy, while the sum P_1+P_{Mech} presents the newly gained energy per second in the TDS.

The relation below expresses the transformation running in a TDS:

$$\mathbf{C.O.P. = P_2/P_{Mech} = (P_1 + P_{Mech})/P_{Mech}}, \tag{4.31}$$

which is called a coefficient of performance.

[67] During the last 30-ies Swiss engineers designed the first economically profitable industrial HP operating with low potential heat accumulated in the Geneva Lake during the summer.

[68] Specialized devices known as compressors are used (these are piston, turbine, scroll, wafer etc. compressors).

Here

- P_1, kW – power "extracted" from the **CS** (incoming);
- P_2, kW – power introduced into the hot source (outgoing);
- P_{Mech}, kW – power of compression (introduced into the compressor).

(quantity C.O.P.=$1+P_1/P_{Mech} > 1.0$ is experimentally found and recorded in the HP catalogues). Theoretically, C.O.P. may reach 14.0, but in rare cases, only (C.O.P. > 3.5[69]).

The defining expression (4.31) is modified as

$$\mathbf{P_{Mech} = P_2 / C.O.P.}, \quad (4.32)$$

which is often used in the analysis of the energy schemes of heat pumps, as in Sub-par. 3.2.1.

The **thermal energy** regenerated in a heat pump passes to the hot source (thermal accumulator) or directly to the utilization installations thanks to the "mediation" of the working fluid.

As seen, an HP structure comprises two types of physical devices (see Figure 4.64):

- Devices regenerating low-potential thermal energy;
- Devices facilitating the HP contact with cold and hot energy sources.

The physical contact of an HP with hot and cold energy sources takes place applying two techniques:

- Direct contact (**DxHP**[70]);
- Indirect contact (**InDxHP**).

Under direct contact, the working fluid approaches the areas of energy utilization (occupied areas in buildings, transport vehicles or industrial sites) where it releases energy (after a depressurizations through expansion or throttling) thus doing useful work.

Under indirect contact, the working fluid does not leave the heat pump and its throttling takes place in the HP closed scheme, without mass exchange with the hot source (**HS**). The **HP-HS** energy exchange runs in specialized heat exchangers (finned-tube coils, plate heat exchangers or heat pipes). Combined with the heat pump, they form a monolithic structure.

From a theoretical point of view, regeneration of the low-potential thermal energy takes place, decreasing the TDS entropy. At a technical level, this results in the decrease of the Lagrange multipliers and shift of the orbitals of the energy

[69] C.O.P.=7.0 in the heat pump EcoQue manufactured by Sanyo and used in a hygiene maintenance system. The working fluid/gas is CO_2

[70] Direct-exchange heat pump.

carriers to the short-wave zone of the spectrum[41]. The regeneration mechanisms include:

- Mechanical compression of the TDS;
- Chemical compression (during adsorption of binary working bodies);
- Concentration of the energy orbitals of the energy carriers in the short-wave zone of the TDS (by the introduction of external energy or use of physical concentrators).

A traditional HP operates following the first two mechanisms. Hence, the market offers three types of heat pumps operating under counter clockwise thermodynamic cycles[71], which compete with each other:

- Vapor-compressor heat pumps mechanically compressing the TDS;
- Air-compressor heat pumps mechanically compressing the TDS (turbo-compression);
- Absorption heat pumps applying chemical; compression (absorption-desorption).

During the second half of the 20th century, leading technological companies stimulated the development of various energy-regenerating devices with the aim of decreasing the cost of their products. Hence, the development and improvement of heat pumps became an object of special interest. Various types of heat pumps are popular nowadays, and they are classified as follows:

Pursuant to the source of low-potential energy, heat pumps are:

- Air HP;
- Geothermal HP;
- Water HP;
- Utilizing HP.

Pursuant to the areas of application, heat pumps are:

- Building HP;
- Transport HP;
- Industrial HP.

Pursuant to the number of controlled volumes (zones), heat pumps are:

- Single-zone HP;
- Multi-zone HP.

---

[71] A thermodynamic cycle (TDC) consists of successive thermodynamic transformations of the working fluid, which starts at and returns to an initial thermodynamic state (p_n,T_n,v_n). Pursuant to the direction of travel over intermediate thermodynamic states, a TDC can be a direct (clockwise) or indirect (counter clockwise) cycle. Thermal engines operate under a direct cycle, while heat pumps – under an indirect cycle.

We will focus our attention on steam-compressor heat pumps in what follows. They operate in residential, public and business buildings where ambient air is the **CS** of low potential energy. A number of arguments are in support of those devices:

- Attainment of higher efficiency, i.e. higher C.O.P.;
- Larger market offer and circulation;
- Greater technological experience of their manufacture and hence, higher quality;
- Lower noise level.

4.6.2 Gas Vapor-Compression Heat Pumps

Working fluids of vapor-compression heat pumps can be liquids easily evaporating under low temperature, such as freon, type R22a, R134a, R123, R407C, R410A and CO_2 in some cases. The running processes are described by periodically repeating closed contours of the type "**Rankine counter clockwise cycle**".

The thermodynamic cycle of vapor-compression heat pumps can dissociate into four elementary thermodynamic processes of the working fluid, presented by the P-V diagram plotted in Figure 4.65,a):

- Isothermal heating: p.A-p.1;
- Adiabatic compression: p.1-p.B;
- Isothermal cooling: p.B-p.2;
- Adiabatic expansion: p.2-p.A.

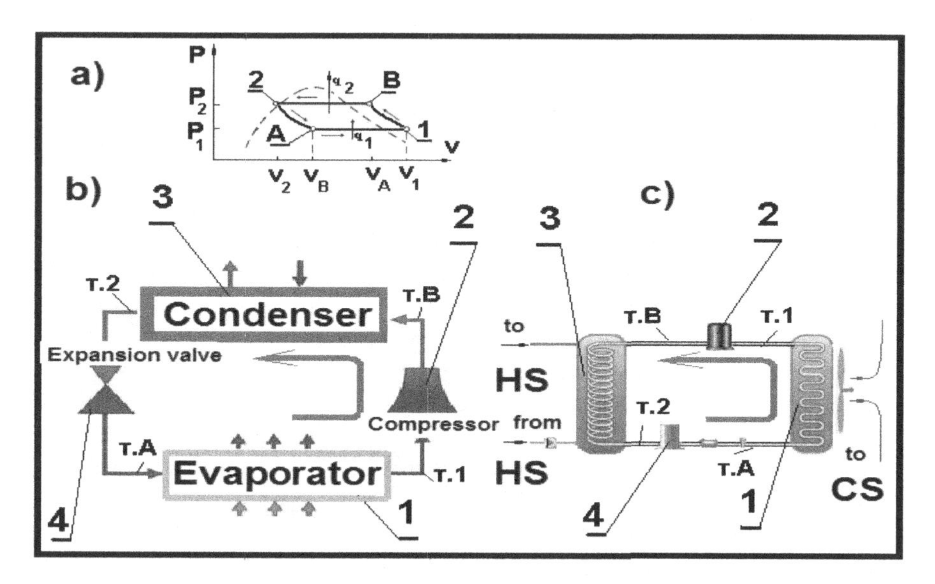

FIGURE 4.65 Heat pump with indirect contact with a hot and a cold energy sources: a) Rankine counter clockwise cycle; b) Structural scheme of an HP; c) Scheme with indirect connection between hot and cold energy sources via external, auxiliary contours and drives – pumps and fans.

The processes run in four different physical devices, which are the main components of the heat pump shown in Figure 4.65,b) and c) and called:

- Evaporator – Item 1;
- Compressor – Item 2;
- Condenser – Item 3;
- Throttling (expansion) valve – Item 4.

In fact, processes of regeneration of the thermal energy described in Sub-par. 2.2, run in each device. The working fluid contacts the cold energy source (the process starts at p.A – see Figure 4.65,a), while the sequence is as follows:

- Contact and extraction of low potential energy + $\mathbf{q_1}$ from the cold energy source in the evaporator (Item 1). The working fluid evaporates and transforms into ***cold vapor*** sucked by the compressor (Item 2);
- Regeneration of the low-potential energy by compression of the TDS in the compressor (Item 2). Here the TDS energy increases by the mechanical work L_{Mech} done by the compressor. The working fluid transforms into a superheated steam;
- Contact and release of regenerated energy – $\mathbf{q_2=q_1+L_{Mech}}$ to the hot energy source in the condenser (Item 3). The working fluid (as superheated steam) liquefies and becomes a boiling liquid under high pressure;
- Expansion of the working fluid via throttling in the valve (Item 4). Pressure in the TDS drops from p_1 to p_2, while the working fluid (i.e., a cooled liquid corresponding to a state denoted by p.A of the P-V diagram in Figure 4.65,a), is ready to repeat the cycle.

Heat pumps shown in Figure 4.65,b) and c) belong to the class of devices operating under **indirect contact** of the TDS with cold and hot energy sources. The working fluid in the pumps (slightly boiling liquid) flows in a closed pipe system within the control volume of their components that form a compact and monolithic structure. Connections to the hot and cold energy sources take place through auxiliary external contours (Figure 4.65,c). Quite often, heat exchanging devices with large overall dimensions are used in indirect HP. Such are the devices shown in Figure 4.66 and known as chillers or heat-pump stations. Heat exchange in those HP deteriorates and the energy efficiency drops due to various structural and exploitation issues. This is not the case in installations with HP, where the working body contacts directly **HS and CS** (i.e. the HP is of DX type). Systems maintaining the thermal comfort in residential or administrative buildings use these installations – see, for instance, the arrangement in Figure 4.67.

Two heat-pump installations with direct evaporation of the working body (DX) operate with significant efficiency and flexibility in a regime of simultaneous heating/cooling of inhabited areas, satisfying the respective energy demand. They are known as "VRF"[72], meaning "installations with variable refrigerant flow". The working

[72] Variable refrigerant flow.

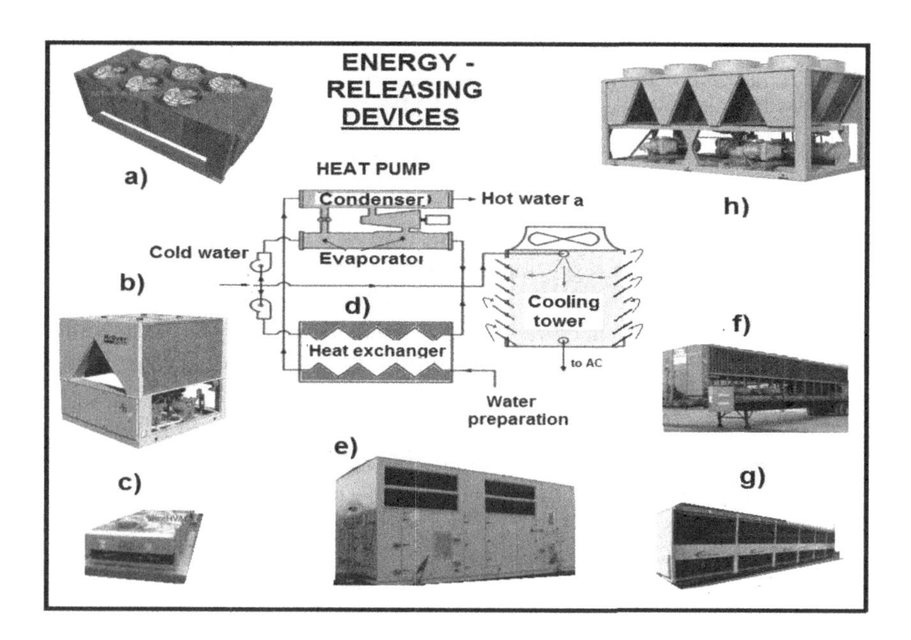

FIGURE 4.66 Heat exchanging devices with a heat pump connected to cold and hot sources.

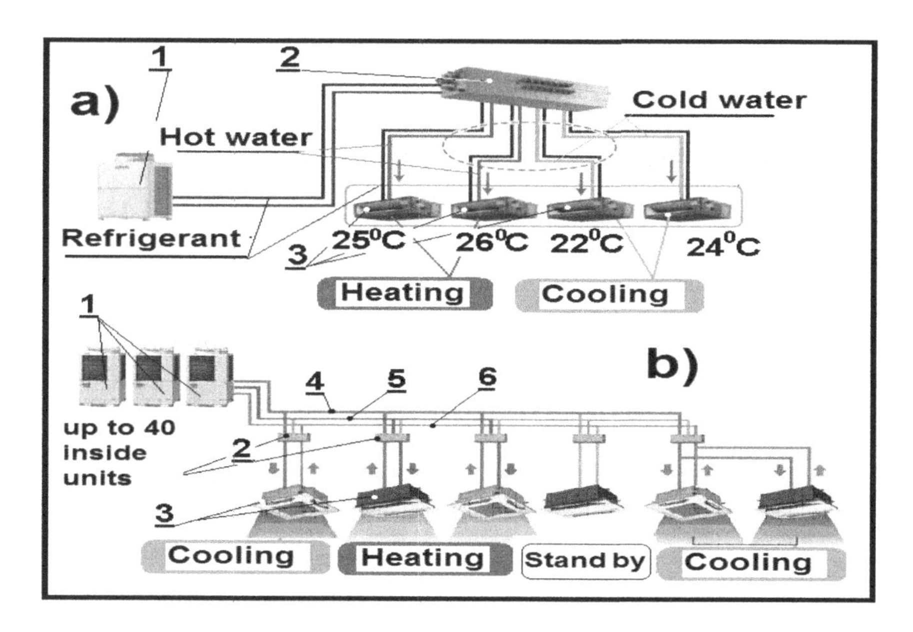

FIGURE 4.67 Gas heat pump with direct evaporation (used in a multi-zone building) – a VRV system: a) Two-pipe system; b) Three-pipe system: 1 – Outdoor unit; 2 – HBC controller; 3 – Indoor units-cassettes; 4 – Liquid pipe (medium temperature and pressure); 5 – Discharge pipe (high gas temperature and pressure); 6 – Suction pipe (high temperature and pressure).

fluid of the HP (the refrigerant) circulates in two-pipe or three-pipe systems (Items 4, 5 and 6) installed in the building and the indoor units. Note also the more effective contact between the working fluid and the cold/hot energy source[73].

4.6.3 Operation of a Gas Heat Pump

As clarified in Sub-par. 4.6.2, the mechanical work done during a TDS compression is a key factor in the regeneration of thermal energy in a vapor-compression HP. Hence, the financial equivalent of the mechanical work done L_{Mech} is of crucial importance in the estimation of the benefit of using a heat pump technology in real economy. The energy needed for the operation of the heat pump can come from different sources:

- Environment – coming from renewable energy sources;
- Centralized systems;
- Decentralized systems.

It is historically proved that the energy from flowing water and wind is the cheapest one. Yet, the exploitation of these sources in urbanized areas is limited. At the same time, various limitations were overcome in the era of cheap hydroelectric power generation. In the absence of a reasonable alternative, specialists preferred and included inverter and all-season heat pumps with high C.O.P. in the engineering solutions.

At present, when new and higher ecology norms and requirements to the environment preservation prevail, specialists look for alternative solutions – for instance, the use of natural gas in primary mechanical drives.

As clarified in Sub-par. 2.5, the use of natural gas as a primary energy carrier yields a significant decrease of green house gas emission. In case of a decentralized generation of electric power by means of micro HPS, the CO_2 and green house gases emissions drop 3 times. Suppose that HP are operated by gas internal combustion engines (ICE) instead of electric motors[74]. Then the following questions naturally arise: (i) What would be the economy of using primary energy carriers and (ii) Would the emissions of green house gases actually drop.

The efficiency coefficient of typical ICE operating under the direct cycle of Otto-Langen or Diesel is within ranges 28–30%.

A characteristic feature of a gas ICE is the high compression degree (not lower than 18) attained due to the fuel low octane number. This urged designers to modify the classical ICE. Other modifications of the thermal-mechanical technology of Otto-Langen were made to increase its mechanical efficiency. Atkinson and later

[73] In fact, so-called "indoor bodies" (Item 3), diffusers or convertors reversely operate as evaporators or condensers, depending on the comfort requirements in the occupied area. Controller (Item 2) is functionally intended to let through the working fluid flow rate in order to regulate the thermal comfort in admissible limits.

[74] It is considered apriori that the use of fuel gas by ICE yields decrease of CO_2 emissions by 25%, and NO_x emissions by 75%. It is known that the combustion of 1 m^3 of natural gas emits 1.058 m^3 of carbon dioxide, 2.019 m^3 of water vapors and 8.777 m^3 of nitrogen oxides.

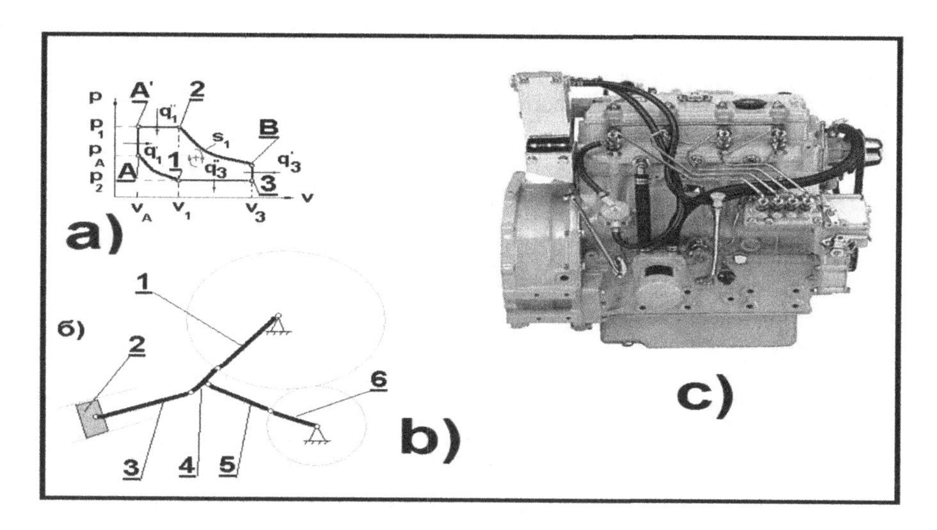

FIGURE 4.68 Three-stroke Miller's engine (1947): a) Direct cycle of Miller; b) Crank-gear; c) Miller's gas engine: 1 – Knee; 2 – Piston; 3 – Rod; 4 and 5 – Rockers; 6 – Knee.

Miller in 1947 introduced two thermodynamic innovations to increase the work done by gases after gas-air mix combustion.

Figure 4.68,a) and b) illustrates the thermodynamic cycle of Miller and the innovative mechanism of its realization in a gas engine. Improvement of the mechanical efficiency is attained by injection of additional fuel $\mathbf{q_1''}$ in the combustion chambers using a high-pressure compressor (as in the Diesel engine). The injection takes place at the moment prior to the adiabatic expansion of the TDS and the prolongation of the contact with the **CS**, adding the isobaric compression of the gas mixture (along the line p.3→p.1).

Consider the start of the engine operation at **p.**1-suction of the gas-air mixture. The crank-gear has two rockers (Items 5 and 6, Figure 4.68,b) and it performs an additional inverse semi-drive during compression. Thus, part of the gas-air mix returns to the suction tract. Just then the distribution mechanism closes the suction valves and the mix compresses and ignites at p.A. Mix pressure rises to the isochore $\mathbf{v_A=const}$. The TDS reaches p.A'. Prior to TDS adiabatic expansion along the line 2-B, a new portion is injected into the chamber – additional fuel $\mathbf{q_1''}$. After fuel burning, the TDS performs additional isobaric expansion along line A'-2. This "innovation" of Miller's engine significantly increases the positive work done; graphically presented by the area under line A'-2-B of the diagram. Reaching p.B, release valves open and the cylinders discharge combustion gases (TDS interaction with the "cold" source). Pressure drops along the isochore B-3, and then compression follows along the isobar 3-1[75].

[75] The technical details of the control of those processes fall beyond the scope of the present study. See.http://youtu.be/xUsxE5_ttsM;http://youtu.be/8haoPu1cLhg; http://youtu.be/gmwpljA8sqQ;http://youtu.be/CNsJOsnuIMA; http://youtu.be/pGaISFg_ZIw for more details.

The prolongation of the contact between the TDS and the "cold" source results in the release of additional heat $q_2^{'}$ under isobaric cooling (3-1), while the specific volume changes from v_3 to v_1, and the work done to compress the gas-air mix significantly decreases.

We may systemize the operation of Miller's engine adopting five stages:

- Suction: p.3→p.3';
- Partial release: p.3'→p.1;
- Compression: p.1→p.A;
- Ignition and expansion: p.A→p.A'→p.2→p.B;
- Gas discharge: p.B→p.3.

Compare the diagrams of the mechanical work (drawn in P-V coordinates) done under Otto-Langen and Diesel cycles, on the one hand, and that done under Miller's cycle, on the other hand. It is obvious that the **work of TDS expansion** done under Miller's cycle significantly exceeds that done under Otto-Langen and Diesel cycles. Yet, **the work of TDS compression** done under Miller's cycle is significantly smaller than that done in regular carburetor and diesel engines. Hence, it follows that Miller's gas engine is more efficient than its predecessors and its efficiency coefficient attains the impressive 40–43%, while the financial prime cost of the mechanical work done drops significantly.

This is the argument of designers of a number of heat pumps to consider a compressor drive in their projects based on Miller's gas engine.

4.6.4 Heat Pumps Driven by Natural Gas and Available on the Market – An Investment Point of View

Although investors in various projects are encouraged to stress upon issues of energy efficiency and ecology stability, the main grounds are the initial investment and the time of its reimbursement, i.e. investor's profit, and its efficiency.

Analyzing the financial efficiency, **an investor should be convinced** that buying a heat pump instead of a water heater or a boiler would result in investments efficiency. There are well grounded reasons in this respect – eventual use of free energy extracted from technological waste, energy supplied by renewable energy sources or by the environment, energy regenerated for secondary use etc. Thus, one can avoid the use of expensive fossil energy carriers. At the same time, investors should answer the question of to what extent the initial investments in a gas heat pump are economically more efficient than the investments in an electric pump.

For now, the technical market offers a number of building gas HP[76], which can be an object of investors' interest. Note that some HP technical characteristics are given

[76] "NNCP 096J, NNCP 120J, NNCP 144J and 168J" produce of YANMAR; "K 18" produce of ROBUR INC; "Polo 100" produce of TEDOM; "40HT and 40HS" produce of LOCHINVAR; "aroTHERM 5, 8, 11 and 15" produce of VAILLANT; "Vitodens -050 W, -100 W, -290 W, -224 F, -200" produce of VIESSMANN; "NextAire" produce of INTELLICHOICE ENERGY; "Remeha Fusion GAHP" produce of REMEHA; "GYAQ 8ANV1, 10ANV1, 13ANV1, 16ANV1, 20ANV1 and 25ANV1" produce of DAIKIN; "AXGP224E1, AXGP280E1, AXGP335E1 AXGP450E1 and AXGP560E1" produce of AISIN SEIKI; "SGP EW240M2G2W" produce of SANYO; "ECO G HP" produce of PANASONIC.

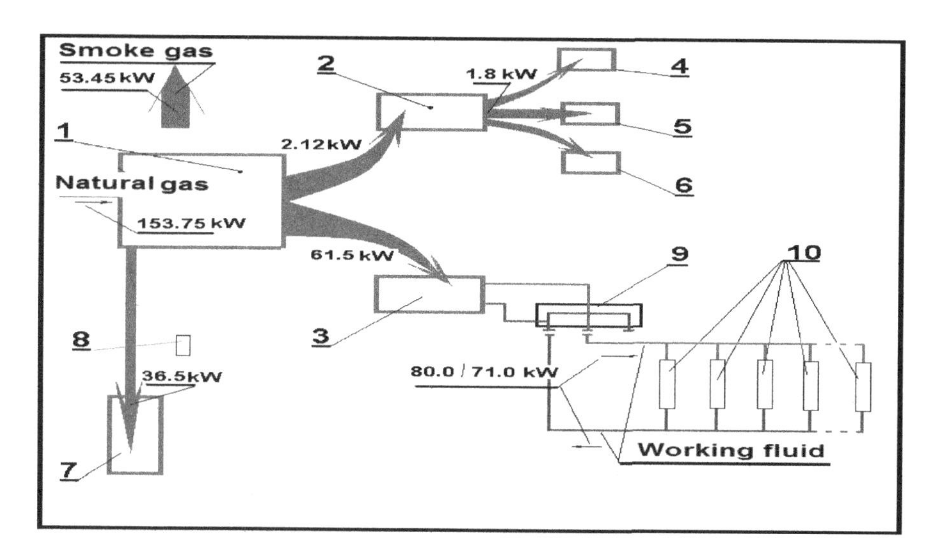

FIGURE 4.69 Energy scheme of gas HP type ECO G HP25 Panasonic: 1 – Gas ICE; 2 – Cogenerator; 3 – Compressor; 4 and 5 – Fans; 6 – Pump; 7 – Recuperating heat exchanger; 8 – Three-way valve; 9 – Controller; 10 – Fan-convectors (indoor units).

in Table P26, where HP^{-ies} are juxtaposed to electric pumps. We give here the characteristics of a gas heat pump, model **ECO G HP 25** and its "twin sister" – electrical HP ECOi 26 HP, produce of Panasonic. Its structural components are as follows (see Figure 4.69):

- **Engine with internal combustion**: 6-cylinder gas ICE operating under Miller's cycle;
- **Utilizer** of heat waste with power 36.5 kW_{th} – heat waste is released during cooling of the ICE belonging to the domestic hot water system;
- **Generator** with power 1.8 kW_e, coupled with an HP feeding the cooling ventilators and the recirculation pump.

The HP, type **ECO G**, has additional energy-efficient capabilities and components such as:

- New propellers of the cooling fans whose energy efficiency increases by 30%;
- Heat exchanger type "L", whose heating surface increases by 25%;
- Decrease of stop-and-start losses;
- Increase of the HP efficiency under small loads;
- Stable operation during winter at a temperature of down to minus 20°C (253 K).

As a whole, a gas driven HP, type **ECO G**, is by 30% more efficient than a heat pump with identical dimensions fed by electricity. This conclusion is drawn based

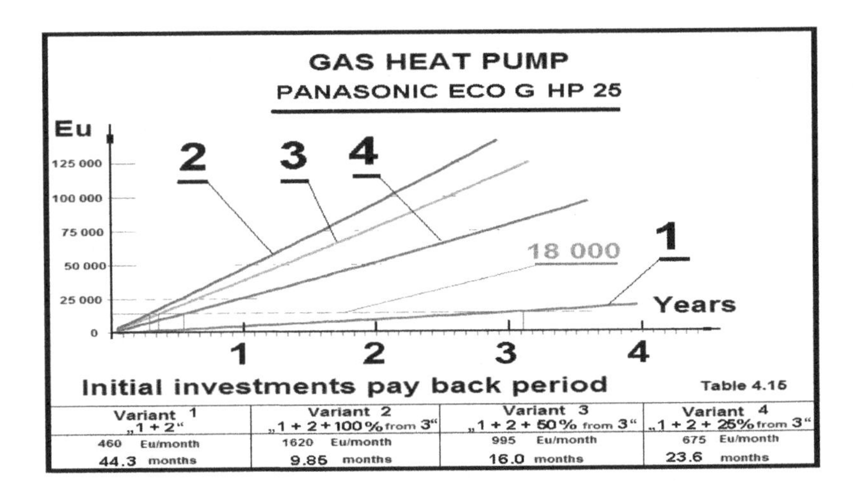

Variant 1 „1 + 2“	Variant 2 „1 + 2 + 100 % from 3“	Variant 3 „1 + 2 + 50 % from 3“	Variant 4 „1 + 2 + 25% from 3“
460 Eu/month	1620 Eu/month	995 Eu/month	675 Eu/month
44.3 months	9.85 months	16.0 months	23.6 months

FIGURE 4.70 Dynamics of the return back of the initial investments in the purchase of a gas heat pump.

on catalogue data of Panasonic. The difference between the initial investments of £14442[77] (31904 lv) is in favor of the electric heat pump.

The rising cost of a gas HP and the reimbursement time are analyzed considering the **monthly savings** with respect to three articles:

- Operation of the gas engine of the heat pump;
- Cogeneration;
- Utilization of smoke gas energy.

The calculation results in Table 4.15 (see Figure 4.70) show that reimbursement time of initial investments is 10–44 months[78] – an optimistic prognosis encouraging the investors to purchase a gas HP.

As a whole, the reimbursement time in heat-pump technologies is about 10 years. Hence, investors need encouragement from the state via purposeful credits or direct subsidy.

4.7 NECESSITY OF INCREASING THE SHARE OF GAS DEVICES IN A BUILDING

During the winter of 2017, the national energy system set a 20-years record of demand for electric power amounting to 7679 MW_e (10.01.2017).The power demand was covered by setting in operation of the cold reserve of six thermal electric power

[77] A check up in the price list of Panasonic for 2017 shows that the **initial price** of the model ECO G 25 HP is **£31489**, while that of ECOi 26 HP – the equivalent model with an electric drive, is **£17047.**

[78] The four studied versions are: – sum of the work done by the heat engine and cogeneration (1+2, without utiliztion); -sum of three types of savings (1+2+3); -sum of the three types of savings accounting for the effect of a utilization system operating with 50% of its capacity; -sum of the three types of savings accounting for the effect of a utilization system operating with 25% of its capacity.

stations (TEPS)[79] in this extreme situation. This was done parallel to the operation of condensing TEPS (a share amounting to 47.8%), Kozlodui nuclear power station (a share amounting to 28%), hydroelectric power stations (a share amounting to 8%) and combined TEPS (a share amounting to 6%). It was found at the same time that national photovoltaic power stations, with their total power amounting to 1200 MW, covered **zero (0)%** of the power demand. Note that the renewable energy sources and photovoltaics, in particular, did not attribute to the successful management of that "stress test". Hence, energy experts concluded that the Republic of Bulgaria needed a **new energy strategy** in the period 2020-2030-2050, when household electric power should be replaced by power from alternative energy sources.

Our analysis shows that the increased use of natural gas in the household is an attractive perspective. Today share of household fuel gas consumption amounts to 5% of the total gas consumption, which is quite unsatisfactory and does not correspond to the situation in other European states. We discuss in the present paragraph the capabilities of indoor gas installations and their components to attribute to the intensification of fuel gas consumption. Gas supply on large scale implies reduction of centrally supplied electricity and thermal power by more than 35%.

The correct combination of used primary energy carriers is an object of state policy and a long-term planning. The new energy strategy of the government in the period 2020–2030 should reflect it. The main challenge facing that strategy is the offer of fair prices of fuel gas to end users. The implementation of the Third Energy Package of the EU may contribute to the achievement of that goal, setting up a realistic energy market that would guarantee free access to end users.

4.7.1 Gas Micro-Cogenerating Systems in Domestic Autonomous Energy Centers

One of the perspective options of using natural gas pursuant to the EU Third Energy Package is the use of micro fuel cells in building cogenerating centers (mHPS), as discussed in Sub-par. 3.3.1. There are numerous reasons in support of such advance of local energy practices, i.e.:

1. The ecology issues of coal and electric power stations grow stronger.
2. Most of the electric power stations in our country operate after Rankine's cycle, whose efficiency coefficient is low (32–34%);
3. Power line connecting electric power stations to end users, power substations (HH→BH; BH→CH; CH→HH) and indoor networks lose additionally 12–16% of the useful electric power. As a whole, 75–78% of primary energy released by energy sources are lost during central transfer, produced by TEPS, NEPS, WEPS, PhEPS and so on;
4. The energy cost rises due to the additional taxes imposed in the form of public obligations assumed to use renewable energy sources;

[79] Two of the cold reserve plants (TEPS Ruse and TEPS Bobov dol) were set in operation with difficulty due to the frozen coal and logistic devices – band transporters, screws etc.

5. New domestic gas mHPS were developed and successfully implemented during the recent 10–15 years. They have high efficiency coefficient, reaching 95%. Their installation yields a significant economy of energy carriers and drop of greenhouse gas emissions.

Conditionally, one can classify domestic mHPS into two groups:

- Systems consisting of gas micro HPS (see Figure 2.13,a);
- Systems comprising cogenerators with heat pumps (see Figure 2.13,b).

To successfully design micro-cogenerating systems, one must coordinate their operation with the established regime of building occupation. The following mHPS are considered in investment – engineering projects of individual residential houses or apartments:

- Systems concerning a "two-shift" regime of building occupation (periods of inhabitants' presence/absence);
- Systems concerning a continuous regime of building occupation (inhabitants' persistent presence).

Consider houses with a "two-shift" occupation. Then, analysis proves that mHPS with optimally calculated electric and thermal accumulators are the most appropriate ones. Note also that they can have accumulators that operate synchronously with the external network, using it as an active accumulator. Residential buildings with continuous regime of occupation have mHPS that operate in a co-energizing regime with the outdoor network.

We discuss in the present paragraph typical schemes of incorporation of mHPS in individual residential buildings or apartments belonging to a condominium – see the incorporation sequence here:

- Microsystems covering the main load of the electric grid (EG);
- Microsystems covering the entire load of the electric and heating installations;
- Microsystems covering the entire load using renewable energy sources.

The integration of the mHPS into the regular building installation is an essential part of any installation project. We propose in what follows two variants of integration of micro-cogenerating systems (mHPS):

a. Integration into a network for a semi-automatic switch-over to a central system – Figure 4.71,a);
b. Integration into a system for parallel (simultaneous) operation with the central system – Figure 4.71,b).

Considering the **first variant**, the domestic mHPS and the central power supply system operate independently (Figure 4.71,a). The electric energy is supplied by an

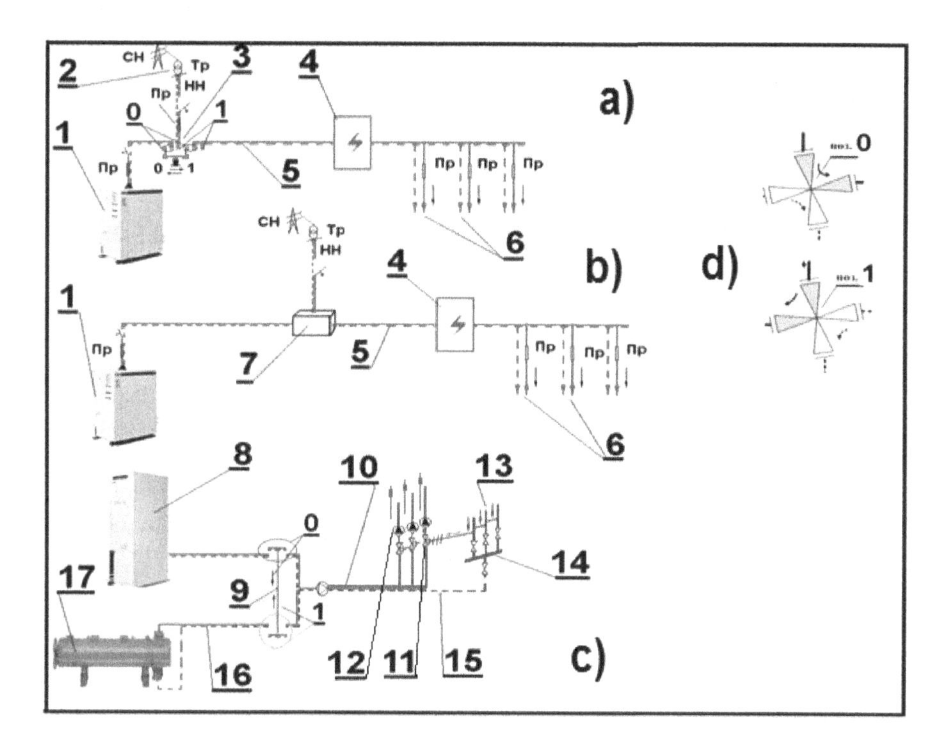

FIGURE 4.71 Integration of a gas mHPS with the electric and heat installations. a) mHPS with a fuel cell operating in a regime of semiautomatic switch-over to the central system; b) Fuel cell receiving a constant electric load in a co-energizing regime with the central system; c) Fuel cell – an alternative heat source; d) Switch of the heat source with a quatro-valve. 1 – mHPS body; 2 – Power substation CH→HH; 3 – Switch-over panel; 4 – Apartment distributing panel; 5 – Connecting cable; 6 – Current contours of the end electricity users; 7 – Interconnector (ASHREA2001, ASME Standard PTC50); 8 – Body of the heat accumulator; 9 – Switch (quarto-valve); 10 – Distributing main line; 11 – Three-way valves; 12 – Circulation pumps; 13 – Gathering lines; 14 –Gathering device; 15 – Returning main line; 16 – Main lines to the power station and the network heat exchanger; 17 – Network heat exchanger.

mHPS during most of the day. Yet, it automatically switches over to the central supply system during peak hours.

The key element in the proposed scheme is the **switch-over panel** – Item 3. Its implementation enables the building electric installation to operate after two regimes:

- Supply from an mHPS (position 0 of the switch)

or

- Central energy supply (position 1 of the switch).

The regime switch (Item 3) is itself an electric device similar to those turning on the emergency lighting. Its overall dimensions are such as to correspond to the actual installation, and it is turned on manually or by means of a timer.

Considering the **second variant** (Figure 4.71,b), the mHPS and the central systems operate in parallel, controlled by a system for automatic tracing control. Switch 3 shown in Figure 4.71,a), is replaced in this case by an interconnector with tracing control (Item 7, Figure 4.71,b), whose functions have been clarified in Sub-par. **2.5.2**.

Setting of the tracing control system is such that when the mHPS balance is positive (i.e. the power demand of domestic electric devices is less than the power provided by the mHPS), the excess of generated energy "returns" to the central system. If the mHPS balance is negative, the central system "buys" the energy shortage.

Having integrated the mHPS interconnector with the electric installation, one should undertake the inevitable next step of connecting the utilization block (UB) to the thermal comfort and hygiene maintaining systems (see Figure 4.71,c). It is advisable to include a two-position switch (Item 9) or a quatro-valve (see Figure 4.71,d) in the integrated thermal scheme. This enables the mHPS UB to operate following two regimes:

- Regime of supply by an mHPS (position **0** of the switch); or
- Regime of supply by the central system (position **1** of the switch).

An illustration of the mHPS integration with indoor building systems is shown in Figure 4.72. The indoor electric installation supplies the following end users: a system for visual comfort (including the illuminators – Item 5; a system for food preparation and storage (fridge – Item 8); a system for entertainment (a TV set – Item 6); a hygiene maintaining system (washing machine – Item 7; energy center components (circulation and recirculation pumps – Items 11 and 12, electric accumulator – Item 3).

As adopted in the installation practice, end users are arranged in separate current contours. In determining the composition of a certain current contour, despite the type of supplied end users and their total power, the selected method of their control is also adopted as a criterion (i.e. manual or automatic control, two-position or tracing control, analogous or digital control).

The integration of the mHPS interconnector with the indoor electrical installation is realized in the connecting panel – Item 19, where the switch or SAC are installed. The electric cables are installed depending on the building construction and architectural design and pursuant to one of the following classical methods: in rigid conduits or thin-wall conduits (EMT) and in flexible metal conduits and greenfield flexible conduits or BX. Cables are driven through thin metal pipes with d_p=0.5÷4" fixed by clamps, through rigid pipelines fabricated from galvanized steel or plastics with d_p=0.5÷6". Cables can be installed in flexible metal pipelines; in PVC pipes with strengthened walls and d_p=40 and 80 mm or in wall shafts with d_p=3÷8"; in cable trays and in a cellular floor. Cables are running in bus ducts, in cassettes or under floor ducts in a raised floor and wire ways.

The axonometric (riser) scheme of integration of the mHPS utilizing block with the heating and hot water installations for domestic hot water supply (DHWS) is

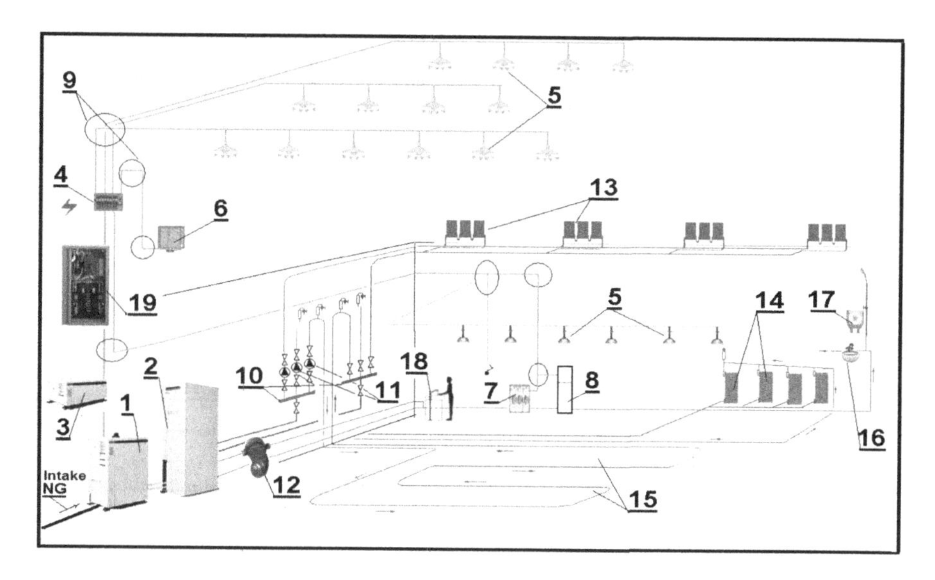

FIGURE 4.72 Integration of mHPS with the building installations – electric, heating or domestic hot water installations: 1 – mHPS body; 2 – Body of the utilization block; 3 – Electric accumulator; 4 – Apartment distributing panel; 5 – Electric lighting fixtures; 6 – TV set; 7 – Washing machine; 8 – Fridge; 9 – Distributing panels; 10 – Distributing tubes; 11 – Circulation pumps; 12 – Recirculation pump; 13 – Heaters; 14 – Heaters; 15 – Heating serpents; 16 – Sink; 17 – Bath-tube; 18 – Kitchen sink; 19 – Connecting electric main **switch board** (panel).

illustrated in the same figure. The installations belong to the systems of thermal comfort and hygiene maintenance, respectively. The distribution of the heat carrier exiting a single-zone UB (Item 2) is performed by a distributer and collector (Item 10, Figure 4.72). Circulation pumps (Item 11), included in each circulation contour, are installed after those devices. Three-way mixing valves (Item 11, Figure 4.71) regulate heat release. To provide fast supply of hot water to end users and save water, a recirculation pump controlled by a temperature sensor is installed along the (DHWS) contour at investor's will.

A second example of mHPS integration with indoor installations is illustrated in Figure 4.73, where a serpent belonging to the utilizing block of a gas mHPS is installed in the air tract of the conditioning/ventilation installation. Based on that scheme, one may consider two variants of connection between devices:

- Direct connection between the utilizing block (Item 2) and the heating section of the central power station (Item 5);
- Connection through a flow GWH (Item 4).

The second variant needs the inclusion of a buster-pump (Item 3) in the scheme due to the increase of the hydraulic resistance against the heat carrier flow through the heater.

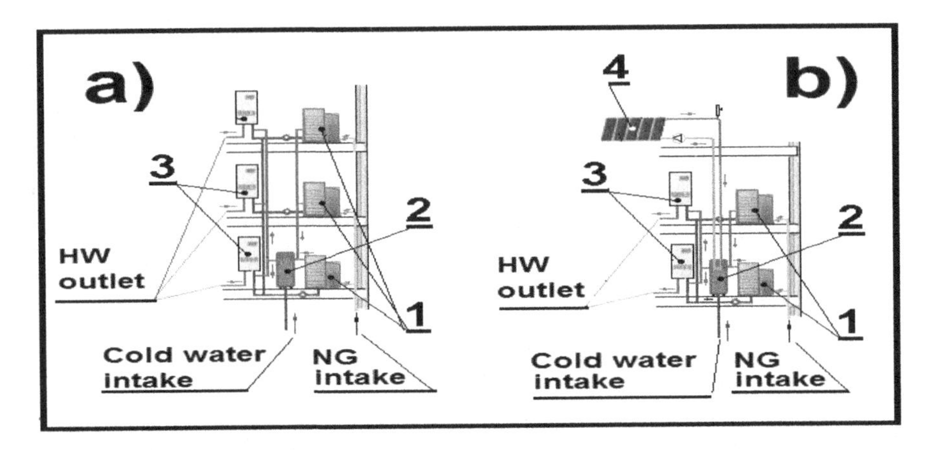

FIGURE 4.73 Joint operation of a set of mHPS, flow water heaters and a heat accumulator within systems maintaining the thermal comfort and the hygiene in a building: a) mHPS bodies connected in parallel; b) Combination between an mHPS and solar collectors: 1 – mHPS body; 2 – Heat accumulator; 3 – Additional flow water heaters; 4 – Solar collectors.

The utilizing blocks of the gas mHPS can be used in individual apartments of multi-storey residential buildings combined with a common heat accumulator – Item 2 (Figure 4.73). As an illustration of these arguments, Figure 4.73 shows two schemes of connection of an mHPS UB (Item 1) to the common heat accumulator positioned in the building sub-terrain:

a. mHPS bodies connected in parallel (Figure 4.73,a);
b. a combination between an mHPS and thermo-solar collectors (Figure 4.73,b).

Considering the first scheme, the utilizing blocks of the three micro-cogeneration cells are connected in parallel, and they collect hot water in the common heat accumulator installed in the building basemet. If necessary, hot water flowing out of the accumulator can be additionally heated in the flow water heaters (Item 3).

The second scheme (Figure 4.73,b) is a modification of the scheme of parallel connection and it offers an option of joint usage of mHPS and solar collectors.

The mHPS can be integrated with an air-conditioning or with a ventilation system how it is shown on Figure 4.74 (the two modifications are different with the additional GWH – Item 4 in Figure 4.74,b).

Example 4.14

Calculation of the infrastructural components of a fuel cell:

We give a numerical example, illustrating the use of catalogue data in current design practice. The problem consists in the calculation of the necessary

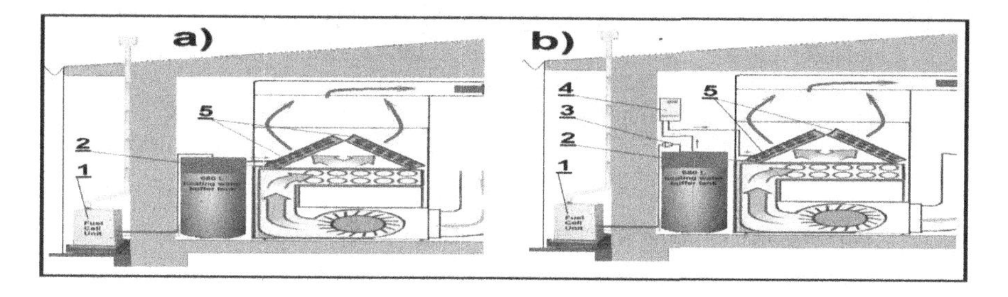

FIGURE 4.74 Integration of the mHPS with the ventilation and air-conditioning installations: a) Aggregation of the mHPS with the air-conditioning handler; b) Connection of the mHPS with a handler with additional gas instantaneous water heater: 1 – mHPS body; 2 – Utilization block of an mHPS; 3 – Buster-pump; 4 – Additional gas flow water heater; 5 – Heating sections of the air-conditioning handler.

infrastructural components connecting the above shown mHPS with the energy center of an office building. Initial data: η_e=0.45; η_{tot}=0.9; w_{NG}=5 m/s; w_{HW}=3 m/s, while current density is 5 A/m2.

SOLUTION

1. **Assessment of fuel cell total power and consumption:**

   Total power: $Q_{tot}=P_e/\eta_e=200000/0.45=444\ 444.4$ w

   Gas consumption:

Total power: J/s	444 444.4
Gas flow rate: m³/s	0.01279
Gas flow rate: m³/day	1105
Gas flow rate: m³/month	33151
Gas flow rate: m³/year	397 820

   Electric parameters:

Electric power: J/s	200 000
Amperage: A	525.5

   Thermal parameters:

Thermal power: J/s	244 444.4
Output of the circulating fluid: m³/s	0.003

2. **Calculation of the main lines**

   Gas main pipeline:$d_{pG}=(4V_{NG}/\pi/w_{NG})^{0.5}=(4\times12.79\times10^{-3}/3.1415/5.0)^{0.5}=0.057$ m

   Electric main cable:$d_e=(4A_e/\pi)^{0.5}=(4\times525.5/3.1415/5.0)^{0.5}=11.56$ mm=0.012 m

   Hot water main pipeline:$d_{pHW}=(4\ V_{HW}/\pi/w_{HW})^{0.5}=(4\times0{,}003/3.1415/3)^{0.5}=0.035$ m

   Gas reducing valve - / 103422 Pa

3. **Calculation of the components of the electric installation**

   Sockets: (for 525 A);

   Fuses: (for 525 A);

   Disconnectors: (for 525 A);

   Returning power station: (360 V × 11 kV for 200 kW_e);

   Returning main line: (11 kV for 200 kW_e I=10.5 A);

4. Calculation of the components of the heating installation

Circulation pumps: pursuant to the circulation contours;
Mixing valves: pursuant to the circulation contours;
Intermediate heat exchangers: pursuant to the circulation contours.

4.7.2 Importance of Gas Micro-Cogenerating Systems for the National Energy Strategy

The large scale use of micro-cogenerating systems (mHPS) in households unveils challenges to national economy of saving primary energy carriers and decreasing the emissions of greenhouse gases.

Assume, for instance, that a number of **200.10³** fuel cells[80] with individual power of 1.5 kW_e and 0.6 kW_{th} operate. They free main power equivalent to 420 MW (3.0×10^5 kW_e + 1.2×10^5 kW_{th}) and normally consumed by the **energy mix**, i.e equivalent to the power of a 500-MW block of a nuclear power plant. The potential profits are as follows:

- Decrease of the risk of network overload – see overload having occurred during the winter of 2017;
- Saving of primary energy otherwise consumed by ineffective central manufacturing and transfer technologies (41.39+4.97 PJ/annually in this case) – see the scheme in Figure 4.75;

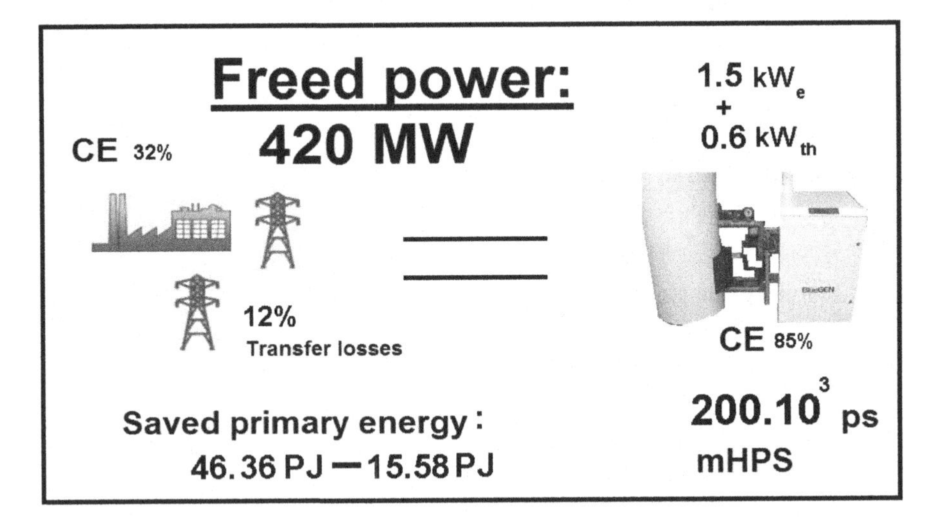

FIGURE 4.75 The installation of domestic mHPS in residential buildings frees energy from a thermal electric power station (TEPS) and saves energy otherwise generated and supplied to end users.

[80] The figure is prognostic.

- Three-times decrease of natural gas consumption (46.39 PJ:15.58 PJ) when only these **domestic mHPS (<u>200.10</u>^{3}** in number) operate, yet supplying the end users with the same amount of energy (electric and thermal), delivered by the central energy system.

As a whole, our example involving a number of **<u>200.10</u>^{3} domestic mHPS**, may yield annual economy of primary energy carriers (natural gas) amounting to 0.8 Gm^3. This amount is consumed by a TEPS with capacity of 30,78 PJ, and it is about 27% of the annual natural gas consumption at present. One should add the significant decrease of greenhouse gas emissions[81] to the economy of energy supplying main electric appliances and domestic water heating. On annual basis, the emissions of CO_2, water vapors and NO_x will be reduced by 0.76 Gm^3, 1.45 Gm^3 and 6.3 Gm^3, respectively, in the above example.

Yet, the investors face a **motivation issue** in the large scale implementation of mHPS in households. Regarding the present status quo and the annual consumption of natural gas by an individual fuel cell of about 2.10^3 m^3/year, one may save net 1120.0 Eu/year (subtracting the expenses of 1400 Eu for electricity and 570 Eu for thermal energy). This money will remain in the budget of the end user. It is found that the user will reimburse its investments during the 10th year of the fuel cell exploitation. This time s commensurable with the life cycle of the device.

Hence, the route to commercialization and mass development of household mHPS (with power of up to 5 kW$_e$) passes through a **purposeful state policy of subsidizing**[82] the initial investments. The aim is to stimulate investors' interest in the mHPS purchase and installation.

The national governments can generate funds in support of the commercialization and mass manufacture of a number of **<u>200.10</u>^{3} domestic** mHPS (12500 Eu per piece – 2.5.10^6 Eu in total) by a redirection of 25% of the expected investments in the "Belene" nuclear power plant. This will help households intending to install mHPS in conformity with the good governing practices.

[81] A local gas mHPS cuts about 65% of emissions of CO_2, water vapors and NO_x.

[82] The program of the Japanese government for commercialization of the installation of household fuel cells successfully started in 2005, where 3000 fuel cells were installed. 120000 devices were installed in 2009 and 240000 – in 2012. The analysis of the dynamics of market fuel cell prices of the Japanese program ENE FARM showed that prices dropped twice in 6 years (from 140000 to 70000 J), while the number of installed devices increased 55 times. This process is controlled by the Japanese government. It granted 14000 J to the any end user having purchased an mHPS in 2009. The Japanese government subsidized a single mHPS, type SOFC, with $4300 and an mHPS, type PEFC – with $3800 in 2014. At the same time, the total amount of money (the total budget) granted by the Japanese government increased 3.28 times (from $61 to $200 million). This corresponds to the end user's multiply growing interest to these systems.

5 Conclusions and Acknowledgment

This book was written in the period 2014–2017 and expanded in 2019 for lecturers in the introductory academic course "Gas Engineering Installations". The course is part of the M.Sc. program of the Department of Civil Engineering at the European Polytechnic University in our country. The book can be useful to students of all specialties at the worldwide University of Architecture, Civil Engineering and Geodesy. It can also be useful for students of the technical universities, studying thermal energy, heat transfer and related phenomena, intending to further qualify as designers of indoor gas installations. Studious readers can gain additional knowledge on the design of gas networks in specialized literature – see, for instance, [13] and [25].

In view of the visionary material of **Par. 4**, the book may also be useful for all regular and postgraduate students in architecture. The paragraph addresses the structure and organization of various building systems, including the logistics of electricity, gas and water supply.

I am grateful to Prof. Dr. Arch. Jordan Radev for his kind help in reading through the final material and for his introductory remarks, as well as to Associated Professor Robert Kazandgiev for the improved English translation of the book.

I express my deepest thanks to my wife, 1st Degree Assist. Prof., Mech. Eng. Zita S. Dimitroff for her support and understanding and for the technical edition, which significantly improved the quality of the book.

Appendix

TABLE P1
Coefficients of internal and external resistance

Device	Internal Resistance R_I	Flue Diameter D, m	External Resistance K_0
Devices with gas stoves:			
- sliding rule 0.1 m	2.5	0.1	2.5
- sliding rule 0.125 m	1.0	0.125	1.0
- sliding rule 0.15 m	0.48	0.15	0.48
Gas stove	3.0		
Gas stove with a boiler	2.0		

TABLE P2
Coefficient of resistance of chimney components

Components	Internal Dimension m	Resistance Coefficient K_e
Pipes (per 1 m length)	0.1	0.78
	0.125	0.25
	0.15	0.12
Brick chimney (per 1 m length)	0.213×0.213	0.02
Dismountable blocks (per 1 m length)	0.317×0.63	0.35
	0.231×0.65	0.65
	0.197×0.67	0.85
	0.200×0.75	0.6
	0.183×0.90	0.45
	0.149×0.102	0.6
Knee 90°	0.1	1.22
	0.125	0.5
	0.15	0.24
Knee 135°	0.1	0.61
	0.125	0.25
	0.15	0.12
	0.197×0.67	0.3
	0.231×0.65	0.22
	0.317×0.63	0.13
Roof terminal	0.1	2.5
	0.125	1.0
	0.15	0.48
Non-roof terminal	0.1	0.6
	0.125	0.25
	0.15	0.12
Hollow blocks (per 1 block)	Any dimension	0.3
Passable clock	Any dimension	0.5

TABLE P3
Pipe characteristic K [18]

V v/s l 10^6	1.5 m	3.0 m	4.6 m	6.1 m	7.6 m	9.1 m	12.2 m	15.2 m	22.8 m
39 m^3/s	46	65	79	92	102	112	130	145	177
79	92	130	158	184	204	224	260	290	354
118	138	195	237	276	306	336	390	435	531
158	184	260	316	368	408	448	520	580	708
197	230	325	395	460	510	560	650	725	885
236	276	390	474	552	612	672	780	870	1062
276	322	455	553	644	714	784	910	1015	1230
315	368	520	632	736	816	896	1040	1160	1416
355	414	585	711	828	918	1008	1170	1305	1593
394	460	650	790	920	1020	1120	1300	1450	1770
473	552	780	948	1104	1224	1344	1560	1760	2124
552	644	910	1106	1288	1428	1568	1820	2030	2478
630	736	1040	1264	1472	1632	1792	2080	2320	2832
709	828	1160	1422	1656	1836	2016	2340	2610	3186
780	920	1300	1580	1840	2040	2240	2600	2900	3540
985	1150	1625	1975	2300	2550	2800	3250	3625	4425
1182	1380	1950	2370	2760	3060	3360	3900	4350	5310
1379	1610	2275	2765	3220	3570	3920	4550	5075	6195
1576	1840	2600	3160	3680	4080	4480	5200	5800	7080
1773	2070	2925	3555	4140	4590	5040	5850	6525	7965
1970	2300	3250	3950	4600	5100	5600	6500	7250	8850
2364	2700	3900	4740	5520	6120	6720	7800	8700	10620
2758	3220	4550	5530	6440	7140	7840	9100	10150	12300
3152	3680	5200	6320	7360	8160	8960	10400	11600	14160
3546	4140	5850	7110	8280	9180	10080	11700	13050	15930
3940	4600	6500	7900	9200	10200	11200	13000	14500	17700

TABLE P4
Diameters of gas pipes conforming to the characteristic K [18]

Pipe Characteristic *K*	0-1170	1171-2650	2651-5330	5331-11700	11701-18100	18101-36700	36701-60200	60201-109500	109501-162300
Diameter	½"	¾"	1"	1 ¼"	1 ½"	2"	2 ½"	3"	3 ½"

TABLE P5
Coefficient of simultaneousness of 4 and more gas devices k_0 [16]

	Single Stove		Tankless Water Heater and Stove		Capacitive Water Heater and Stove	
	4 Burners	2 Burners	4 Burners	2 Burners	4 Burners	2 Burners
4	0.35	0.59	0.31	0.325	0.34	0.44
5	0.29	0.48	0.28	0.29	0.287	0.38
6	0.28	0.41	0.26	0.27	0.274	0.34
7	0.27	0.38	0.25	0.26	0.263	0.3
8	0.265	0.32	0.24	0.25	0.257	0.28
9	0.258	0.289	0.23	0.24	0.249	0.26
10	0.254	0.263	0.22	0.23	0.243	0.25
15	0.24	0.242	0.19	0.2	0.223	0.228
20	0.235	0.23	0.181	0.19	0.217	0.222
25	0.233	0.221	0.178	0.185	0.215	0.219
30	0.231	0.218	0.176	0.184	0.213	0.216
35	0.229	0.215	0.174	0.183	0.211	0.213
40	0.227	0.213	0.172	0.180	0.209	0.211
45	0.225	0.212	0.171	0.179	0.203	0.208
50	0.223	0.211	0.170	0.178	0.203	0.205
60	0.22	0.207	0.166	0.175	0.202	0.202
70	0.217	0.205	0.164	0.174	0.199	0.199
80	0.214	0.204	0.163	0.172	0.197	0.198
90	0.212	0.203	0.161	0.171	0.195	0.196
100	0.21	0.202	0.16	0.17	0.193	0.195
400	0.18	0.17	0.13	0.14	0.15	0.152

TABLE P6
Coefficient of local resistance ζ of the most popular pipe fittings [15]

No	Type of Local Resistance	Coefficient of Local Resistance ζ
1	Instant narrowing	0.35
2	Tees	
	• transition:	1.0
	• deflection:	1.5
3	Crossing	
	• transition:	2.0
	• deflection:	3.0
4	Knee with diameter:	
	• 15	2.2
	• 20	2.1
	• 25	2.0
	• 30	1.6
	• ≥50	1.1
5	Stoppage cock with diameter:	
	• 15	4.0
	• ≥20	2.0
6	Actuator	0.5

Losses in gas regulating and measuring panels are assumed to be 100 Pa.

TABLE P7
Gas consumption of appliances [16]

No	Appliance	Power kW	Volume Flow Rate (Natural Gas 36 MJ/m³)	Volume Flow Rate (Propane 92 MJ/m³)
1	Cooker:			
	• with two burners	6.5–7.2	0.19×10^{-6} m³/s	0.075×10^{-6} m³/s
	• with three burners	8.5–9.65	0.25×10^{-6} m³/s	0.098×10^{-6} m³/s
	• with four burners	11.2–12.0	0.32×10^{-6} m³/s	0.125×10^{-6} m³/s
2	Water heater:			
	• tankless:	21.0–29.0	0.69×10^{-6} m³/s	0.27×10^{-6} m³/s
	• capacitive 80 l:	7.0	0.195×10^{-6} m³/s	0.077×10^{-6} m³/s
	• capacitive 120 l:	14.0	0.39×10^{-6} m³/s	0.15×10^{-6} m³/s

TABLE P8
Selection of gas pipe diameter considering specific pressure drop $\Delta\bar{P} = 0.0654\ \dot{V}^2/d^5(0.000015/d + 763.33 * 10^{-6} d / \dot{V})^{0.25}$, pa/m (density – 0.73 Kg/m^3; kinematic viscosity – $14.3\times10^{-6}\ M^2/s$; relative roughness – 0.000015 m)

$d \times 10^3$ m \ $\dot{V}$	50 Pa/m $\times10^3$ m^3/s	100 Pa/m $\times10^3$ m^3/s	150 Pa/m $\times10^3$ m^3/s	200 Pa/m $\times10^3$ m^3/s
6	0.12	0.165	0.21	0.28
8	0.26	0.37	0.56	0.61
10	0.48	0.7	0.85	1.0
15	1.4	2.0	2.6	3.0
20	3.0	4.4	5.6	6.6
25	6.0	8.0	10.0	12.0
32	11.0	16.0	19.0	23.0
40	19.0	28.0	39.0	42.0
50	37.0	52.0	63.0	72.0
65	70.0	110.0	130.0	151.0

TABLE P8A
Selection of gas pipe diameter considering specific pressure drop (density – 0.73 Kg/m³; kinematic viscosity – 14.3 10^{-6} m²/s; relative roughness – 0.000015 m)

$\dot{V} \times 10^3$ m³/s \ d, m	50 Pa/m ×10³ m	100 Pa/m ×10³ m	150 Pa/m ×10³ m	200 Pa/m ×10³ m	250 Pa/m ×10³ M	300 Pa/m ×10³ m
0.1	8.0	7.5	7.0	6.5	6.0	5.5
0.2	12.0	10.5	10.0	9.5	9.0	8.5
0.3	15.0	12.0	11.0	10.5	10.0	9.5
0.4	16.5	14.0	13.0	12.5	12.0	11.0
0.5	18.0	16.0	15.0	13.0	12.5	12.0
0.6	19.0	17.0	16.0	13.5	13.0	12.5
0.7	20.0	18.0	16.5	15.0	14.5	14.0
0.8	21.5	19.0	17.5	16.0	15.0	14.5
0.9	23.0	20.0	18.0	17.0	16.0	15.0
1.0	23.5	20.5	19.0	18.0	17.0	16.5
1.3	26.0	22.0	20.5	20.0	19.5	18.5
1.6	29.0	24.0	22.0	21.0	21.0	20.0
2.0	31.0	27.0	24.0	23.0	21.5	21.0
3.0	36.0	31.0	29.0	28.0	26.0	25.0
4.0	41.0	35.0	32.0	31.0	28.5	28.0
5.0	44.0	39.0	36.0	33.0	32.0	30.5
6.0	48.0	41.0	38.0	36.0	35.0	33.0
7.0	51.0	44.0	40.5	39.0	37.0	35.0
8.0	53.0	47.0	42.0	41.0	39.5	38.0
9.0	56.0	49.0	45.0	42.0	40.5	39.5
10.0	58.0	51.0	47.0	44.0	42.0	41.0
20.0	76.0	67.0	62.0	58.0	56.0	53.0
30.0	91.0	79.0	72.5	69.0	65.0	63.0
40.0	110.0	89.0	82.0	77.0	74.0	71.0
50.0	151.0	97.0	90.0	86.0	81.0	78.0
60.0	120.0	104.0	96.0	91.0	87.0	83.0
70.0	129.0	111.0	103.0	97.0	92.0	89.0
80.0	135.0	117.0	108.0	102.0	98.0	94.0
90.0	141.0	123.0	113.0	107.0	102.0	99.0
100.0	147.0	128.0	118.0	111.0	107.0	103.0
120.0	158.0	138.0	127.0	120.0	115.0	110.0
140.0	168.0	146.0	135.0	123.0	122.0	117.0
160.0	177.0	155.0	143.0	127.0	129.0	123.0
180.0	186.0	162.0	149.0	134.0	135.0	130.0
200.0	203.0	169.0	156.0	141.0	140.0	136.0

TABLE P9
The flow rate of outdoor air necessary for the ventilation of residential buildings

Room	Unknown	Number of Occupants	Known	Number of Occupants	
	Rate Frequency of the Exchange s^{-1}		Non-smokers m^3/s- for person	Smokers m^3/s- for person	Minimum m^3/s- for person
Living room nursery	1				
Corridor	<2	3	0.017	0.023	0.012
Library	3–4	6	0.011	0.014	0.007
Kitchen (common)	10 (>20)	9	0.008	0.01	0.005
Bathroom	6	12	0.008	0.008	0.004
Toilet	6–10				

TABLE P10
Dimensions of the ventilation grids in rooms differing from boiler rooms

Device Type (Pure Power)	Recommended Norms for the Grids m² за kW	
Open flue <7 kW	No requirement for additional ventilation	
Open flue >7–70 kW <	0.005	
Indirect air heaters, overheated radiating heaters, cauldrons with natural and forced convection >70–1800 kW <	No requirements for additional ventilation if the rate frequency of the exchange is 0.5 per hour.	0.002
Gas devices in hermetic rooms	No requirements for additional ventilation	
Decorative fireplaces	0.1 for devices <20 kW (If two devices are installed in the room, the limit is 0.235)	

TABLE P11
Dimensions of the ventilation grids for gas boilers mounted in encapsulated or sealed rooms

Device Type	For Low Lever m² per kW		For High Level m² per kW	
(pure power)	Outside	Inside	Outside	Inside
Open flue <70 kW	0.01	0.02	50% of that of HH	50% of that of HH
Open flue >70 kW	0.01	No installation is allowed	50% of that of HH	No installation is allowed
Gas boilers in hermetic boxes	0.005	0.01	100% of that of HH	100% of that of HH

TABLE P12
Dimensions of the ventilation grids for natural ventilation of balanced gas boilers and devices

Device Type	Grid Dimensions m² per kW	Method of Ventilation
<70 kW	0.0075	With air duct at a low level
	0.0125	With air duct at a high level
>70–500 kW <	0.01	Vent allowed at a high level, only
>500–1800 kW <	0.008	No vent at low level

TABLE P13
Dimension for the ventilation grids for natural ventilation of household flue-less gas devices (directly outside)

Device Type	Pure Power	Room Volume	Minimal Grid Dimension	Window that Can Be Open
		<5 m³	0.1 m²/kW	Yes
Household stove with an oven	No restrictions	5–10	0.05 m²/kW (if a street door is available 0.0)	Yes
		>10	0.0 m²/kW	Yes
Tankless water heater	11 kW	<5 m³	No installation is allowed	Yes
		>5–10	0.1 m²/kW	Yes
		10–20	0.05 m²/kW	Yes
		>20 m³	0.0 m²/kW	Yes
Dryer	N/A	>10	0.1 m²/kW	Yes
	N/A	>10	0.0 m²/kW	Yes
Fridge	N/A		0.0 m²/kW	No
Natural gas heater	45 W/m²		0.055 m²/kW to 2.7 kW plus 0.1 m²	Yes
	90 W/m²		0.0275 m²/kW to 5.4 kW plus 0.1 m²	Yes
Liquefied gas heater	45 W/m²		0.0275 m²/kW to 1.8 kW plus 0.05 m²	Yes
	90 W/m²		0.0137 m²/kW to 3.6 kW plus 0.05 m²	Yes

TABLE P14
Dimensions of ventilation grids for natural ventilation of heaters directly operating in boiler sections

Pure Power kW	Low Level m^2/kW	High Level m^2/kW
<70	0.005	The same as that for low level
>70	0.0025 до 70 kW plus 0.35 m^2	The same as that for low level

TABLE P15
The flow rate of the mechanical ventilation of heaters directly operating in boiler sections

Device Type	Minimal Flow Rate of the Confusor-low Level m^3/h	Minimal Flow Rate of the Suction Confusor-low Level m^3/h
	2.4 за kW	5–10% lower than that of the low level confusor

TABLE P16
Characteristics of hot-wire lamps

Electric Power, W	Light Flux, lm	Energy Efficiency, lm/W
25	230	9.2
40	415	10.4
60	715	11.9
100	**1400**	14.0
150	2050	13.7
200	2880	17.4

TABLE P17
Standard values of luminance

	Room Type	Standard Luminance, lx
1	Living room	100 (200)
2	Office	200 (400)
3	Library	300 (300)
4	Dining room	100 (200)
5	Bedroom	100 (100)
6	Home workshop	300 (600)
7	Washroom, bathroom, toilet, pool	75–100 (100)

TABLE P18
The lighting systems efficiency coefficients

	ρ_d					ρ_{laf}			
$\rho_{Seiling}$	0.5		0.7		0.3	0.5		0.7	
ρ_{Wall}	0.3	0.5	0.5	0.7	0.3	0.3	0.5	0.5	0.7
i_{Room}	η, %								
1.0	19	22	24	31	17	19	22	24	31
1.1	19	23	25	32	18	20	23	25	33
1.25	21	24	28	35	19	21	25	27	35
1.5	23	26	30	36	21	23	27	30	37
1.75	25	29	32	39	23	26	30	33	39
2.0	27	30	34	40	24	27	31	34	41
2.25	28	31	36	42	26	29	32	36	42
2.5	29	33	37	43	27	30	34	37	44
3.0	31	35	39	45	29	32	36	40	45
3.5	33	37	41	47	31	34	38	42	47
4.0	35	38	43	48	32	35	39	44	49
5.0	37	40	46	49	33	36	41	46	50

TABLE P19
Consumption of cold and hot water in WSFU

	Consumer	Cold Water	Hot Water	Total Amount
1	Sinks:			
	- public kitchens:	3	3	4
	- private kitchens:	1.5	1.5	2
2	baths:			
	- public	3	3	4
	- private	1.5	1.5	2
3	Water cooler	0.25	-	0.25
4	Wash rooms:			
	- **public**	1.5	1.5	2
	- **private**	0.75	0.75	1
5	Urinals			
	Siphon:			
	- **public**	5	-	5
6	Toilets with flushing cisterns:			
	- **public**	5	-	5
	- **private**	3	-	3
7	Toilets with siphons:			
	- **public**	10	-	10
	- **private**	6	-	6
8	Toilets under pressure:			
	- **public**	2	-	2
	- **private**	2	-	2
9	Vacuum toilets:			
	- **public**	0.3	-	0.3
	- **private**	0.3	-	0.3
10	Bath complex	4	4	8

TABLE P20
Correlation between WSFU and l/s

Devices Using Head Reservoirs		Devices Using Siphon Valves	
WSFU	l/s	WSFU	l/s
6	0.315	-	-
10	0.5	10	1.7
15	0.7	15	1.96
20	0.88	20	2.2
25	1.07	25	2.4
30	1.26	30	2.59
40	1.58	40	2.96
50	1.83	50	3.22
60	2.08	60	3.47
80	2.46	80	3.91
100	2.78	100	4.29
120	3.1	120	4.67
140	3.34	140	4.92
160	3.6	160	5.24
180	3.85	180	5.49
200	4.1	200	5.74
225	4.42	225	5.99
250	4.73	250	6.31
300	5.36	300	6.04
400	6.62	400	7.88
500	7.88	500	8.83
750	10.72	750	11.03
1000	13.25	1000	13.75
1250	15.14	1250	15.14
1500	17.03	1500	17.03

TABLE P21
Source: CIBSE guide (1986) flow rate in copper pipes with different diameters

	Ø15	Ø22	Ø28	Ø355	Ø42
1000 Pa/m	0.16 kg/s	0.429 kg/s	0.933 kg/s	1.6 kg/s	2.58 kg/s
1500 Pa/m w=1.5 m/s	0.201 kg/s	0.537 kg/s	1.17 kg/s	2.0 kg/s	3.2 kg/s w=3.0 m/s
2000 Pa/m	0.236 kg/s	0.63 kg/s	1.37 kg/s	2.34 kg/s	3.77 kg/s
2500 Pa/m w=2.0 m/s	0.268 kg/s	0.712 kg/s	1.54 kg/s	2.66 kg/s	4.26 kg/s w=4.0 m/s
3000 Pa/m	0.296 kg/s	0.787 kg/s	1.71 kg/s	2.92 kg/s	4.7 kg/s

TABLE P22
Gas constant R of different gases

	Water	Freon R22	Freon R407C	Freon R410C
Gas constant R J/kgK	462	96.17	96.47	114.6
C_p kJ/kgK	4.186	1.209	1.46	1.519

TABLE P23
Solid-oxide combustion cell "BLOOMENERGY"

Fuel:	Natural gas
Pressure:	15 psich
Outgoing power (AC)	100 kW
Electric efficiency (net AC):	>50%
Connections:	480V; 60 Hz; 4-wire cable; 3 steps
Physical weight:	10 t
Dimensions:	5690 × 2133 × 2058 mm
Emissions:	NO_x < 0.07 lbs/MW-hrm; S_{ox}- none; CO <0.10 lbs/MW-hr; VOCs < 0.02 lbs/MW-hr

TABLE P24
Pressure at some characteristic inlet points of the national gas-transferring network in bulgaria

Inlet Point	Pressue, MPa
Gas transferring network (GTN) Varna	0.6÷1.6
GTN Novi Pazar	0.6÷2.5
GTN Shumen	0.6
GTN Razgrad	0.6
AGTN Targovishte	0.6÷2.5
AGTN Russe West	0.6
AGTN Russe East	0.6÷1.0
GTN Lovech Ловеч	1.0
AGTN Sevlievo	0.6
GTN Pleven	0.6
GTN Sofia 1	0.5÷0.6
GTN Sofia 2	0.6
GTN Sofia 3	0.6÷1.0
GTN Sofia 4	0.6÷1.2
GTN Pernik	0.5÷0.6
GTN Burgas	0.65÷2.5
AGTN Straldja	0.6÷1.2
AGTN Yambol	0.6
AGTN Sliven	0.6
GTN Dimitrovgrad	0.5÷2.6
GTN TEPP North Plovdiv	0.6÷2.5
GTN South Plovdiv	0.6
AGTN Pazardjik	1.2÷2.5
AGTN Stamboliyski	0.2÷0.6

TABLE P25

Specific gas-dynamic losses in copper pipes pursuant to DIN EN 1057

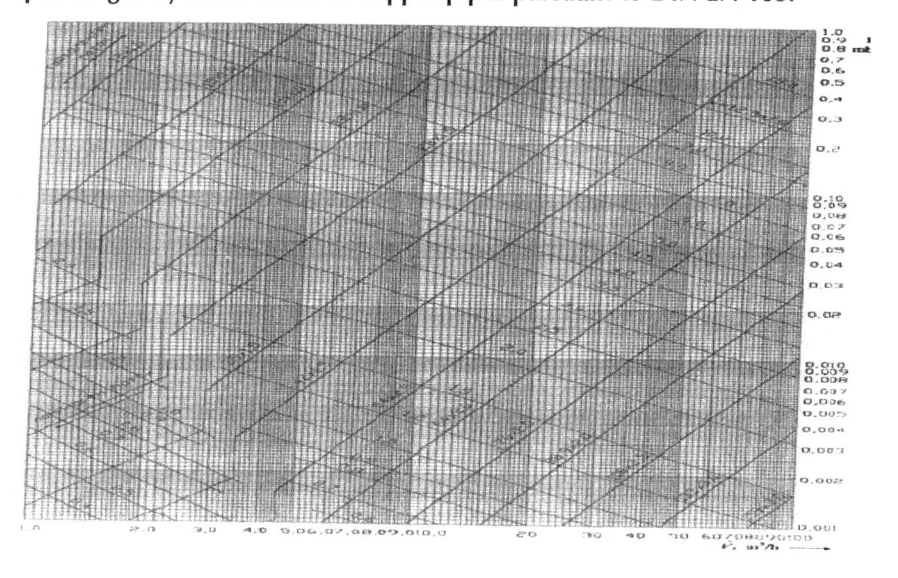

TABLE P26

Specific gas-dynamic losses in steel pipes pursuant to DIN EN 2440

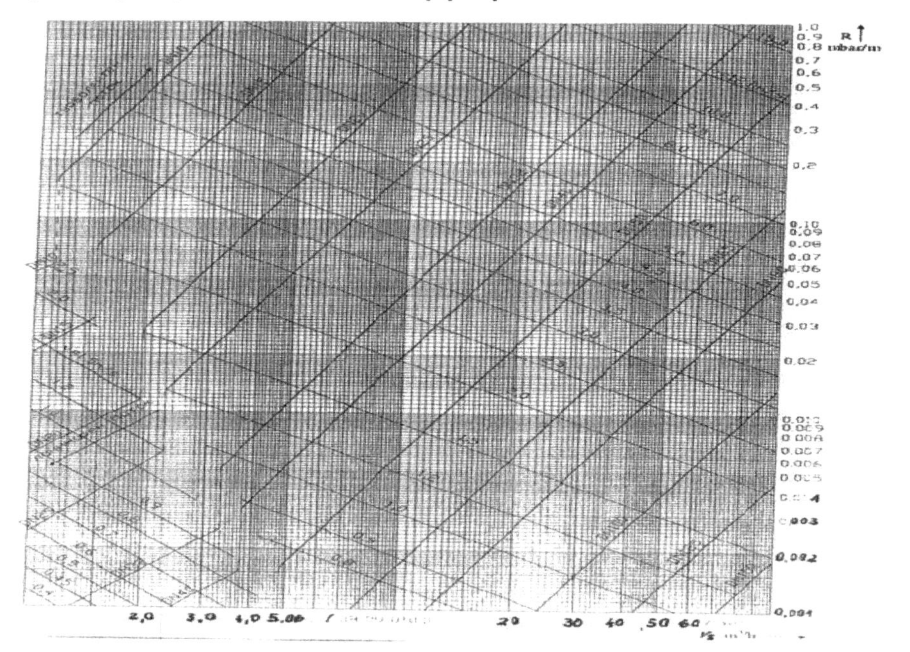

TABLE P27
Overall dimensions of gas regulators and gas-measuring panels [25]

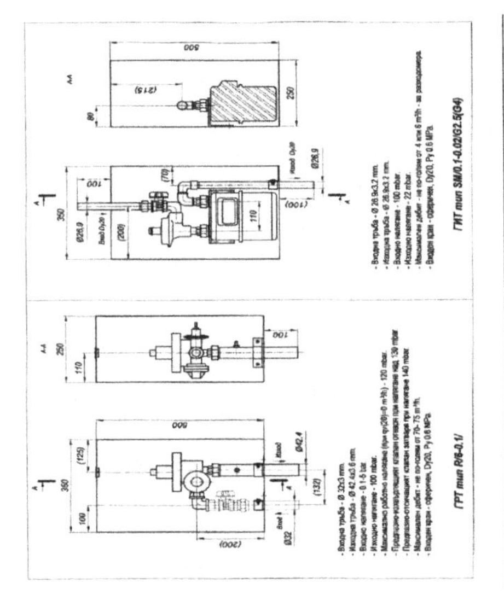

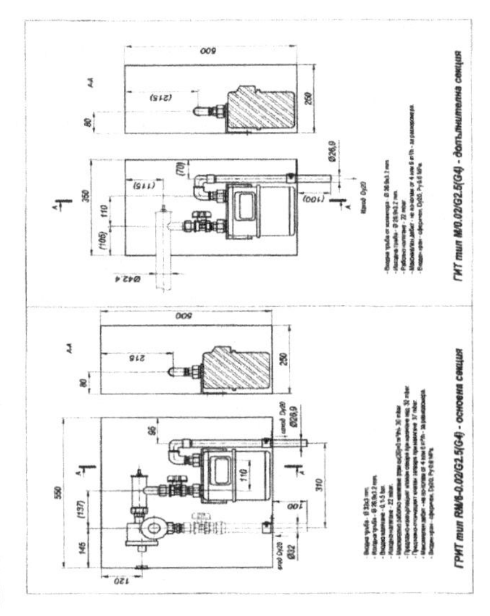

TABLE P28
Gas devices – BDS EN 437A1[37]

Device Class	Subclasses of Devices and Used Gas Groups	Nominal Pressure kPa
Class I	G110; G112	0.8
Class II	**H:** G20; G21; G222; G23	2.0
	L: G25; G26; G27	2.5
	E: G20; G21; G222; G231	2.0
Class III	**3B/P:** G30; G31; G32	2.5–5.0
	P: G31; G32	3.7
	B: G30; G31; G32	2.9

References

1. Territory Management Act (TMA) (Bulgarian).
2. Regulation of the arrangement and safe exploitation of supplying and distributing gas pipes networks and devices, installations and natural gas appliances. Approved by DCM No 171 from 16.07.2004 r.Publ. OG, No 67 from 2 August 2004. Modified, OG, No 78/2005, No 32/2006, 40/2006, No 93/2006, No 46/2007, No 79/2008, No 32/2009, No 5/2010, No 7/2011, No 99/2011, No 103/2012, No 24/2013, No 43/2014, No 50/2014, No 88/2014.
3. Regulation No 6 issued by MRDPW 25.11.2004 (Bulgarian).
4. Regulation for the access to gas supplying and/or gas distributing networks, SCER, Official Gazette No 67, 2004, Published 27.08.2004 (Bulgarian).
5. Regulation for the essential requirements to and the estimation of the gas device correspondence. Approved by DCM No 250 from 5.11.2003, publ. OG, No 100/2003, modifies 24/2006, No 40/2006 (Bulgarian).
6. BDS EN 30-1-1.
7. Regulation for the essential requirements to and the estimation of the correspondence of devices under pressure (RERECDUP) approved by Decree No 204 of CM from 2002 (Bulgarian).
8. Law for technical requirements to products (LTRP), Art. 7 (Bulgarian).
9. Dimitrov A.V. (2013), A building system under ecology stability. A monograph for conferring the scientific degree "Professor" on "Engineering installation in buildings", EPU (Bulgarian).
10. Zokoley S.V. (1984), Architectural design and exploitation of building sites. Relations to the environment, M.; Stroiizdat (Russian).
11. Dimitrov A.V. (1989), Characteristic features of the computer hydraulic calculations of water heating installations. Proc. VNSVU "Gen. Blagoy Ivanov", Sofia (Bulgarian).
12. Dimitrov A.V., S.D. Stamov, Z.S. Budakova (1983), Accuracy of the analytical expressions of the coefficient of hydraulic losses in heat carriers. Proc. of Conf. of TU'83-Sofia, 1983 (Bulgarian).
13. Nikolov G.K. (1993), Transport and preservation of oil and gas. MGU, Sofia (Bulgarian). Nikolov G.K. (2007), Distribution and usage of natural gas. Euconomics, ISBN 978-954-92023-1-1 (Bulgarian).
14. Stoyanov A. (1945), Technical mechanics, Part 2, Fund "Scientific Tasks" VTU (Bulgarian).
15. Idelchik I.E. (1979), Reference book on hydraulic resistance, M. Mashinostroenie (Russian).
16. Designer's reference book (1972), Heating, water conduit, sewerage, Part 8. Gas supply, M. Stroiizdat (Russian).
17. A textbook on HHNiH (1971), under the editorship KKuzov K., Tehnika, Sofia (Bulgarian).
18. Philadelphia Gas Works (1960), Specifications for the installation of gas piping and domestic gas appliances. Division of UGIC.
19. Treloar R.D. (2010), Gas Installation Technology, Second edition, Wiley-Blackwell.
20. Tolley's Domestic Gas Installation Practice (2006), Gas Service Technology, Volume 2, Fourth edition, Edited by Saxon F., Newnes, Elsevier.
21. Ministry of economics, energy and tourism: http://www.mi.government.bg/ (Bulgarian).
22. Bulgargas: http://www.bulgargaz.bg/ (Bulgarian).
23. Bulgartransgas: http://www.bulgartransgaz.bg/ (Bulgarian).
24. Energy strategy of Republic of Bulgaria till 2020 (Bulgarian).

25. Novakov Z. (2015), A course on the design of building gas installations, Branch normal for the design of building gas installations, Module DGI, Iss. 9: http://www.overgas.bg// (Bulgarian).
26. Ministry of regional development and public utilities: http://www.mrrb.government.bg/ (Bulgarian).
27. Bloomenergy Co: http://www.bloomenergy.com
28. Janis R.R., W.K.Y. Tao (2009), Mechanical and Electrical Systems in Buildings, Fourth edition, Pearson, Prentice Hall.
29. Grandzik W.T., A.G. Kwo, B. Stein, J.S. Reynolds (2010), Mechanical and Electrical Equipment for Buildings, Eleventh edition, John Wiley.
30. Mechanical and Electrical Systems for Construction Managers (2010), Edited by Gosse J.F., ATP Ltd.
31. Tao W.K.J. (2001), Mechanical and Electrical Systems in Buildings, Prentice Hall.
32. Hoboken N.J. (2006), Mechanical and Electrical Equipment for Buildings, John Wiley.
33. Dagstino F.R. (2005), Mechanical and Electrical Systems in Construction and Architecture, Pearson, Prentice Hall.
34. Dimitrov A.V. (2014), Manual for the Homework on Technical Installation Preparation (Step by Step), EPU, under edition.
35. Dimitrov A.V. (2013), 501 Questions and Multiple Choices in Eng. Installations, EPU, ISNB 978-954-2983-12-5.
36. Dimitrov A.V. (2013), Practical Handbook of Finite Element Method for Civil Engineers, EPU, ISNB 978-954-2983-11-8.
37. BDS EN 437 A1: Gas testing; Pressure testing; Type of instruments (Bulgarian).
38. Rodin A.N. (1987), Gas radiant heating, Nedra, Leningrad (Russian).
39. Dimitrov A.V. (1980), The influence of the nozzle construction on the free turbulence jets, Ph.D. Disertation, ScC of Energetic and energy machines, HAC.
40. Revankar Sh., Majumdar P. (2016), Fuel Cells, Design, and Analisis, CRC Press.
41. Dimitrov A.V. (2017), Introduction to Energy Technologies for Efficient Power Generation, CRC Press, ISBN: 13-978-1-4987-9644-6.
42. Dimitrov A.V. (2017), Indoor building gas lighting installation using a remote separate gas light generator, Utility model registration certificate No 2900 U1 from 06.21.2017, Patent office of Republic Bulgaria.

ADDITIONAL REFERENCES

- Regulation No Iз-1971 from 29.10.2009 г. concerning norms of constructional-technical fire safety (OG 96, from 04.12.2009) (Bulgarian).
- Regulation of the essential requirements to and the estimation of gas device correspondence (OG 100, from 14.11.2003) (Bulgarian).
- BDS EN 1555-1: 2010 Plastic pipe line systems for gas supply. Polyethylene (PE). Part 2: Pipes (Bulgarian).
- BDS EN 1555-4: 2010 Plastic pipe line systems for gas supply. Polyethylene (PE). Part 4: Valves (Bulgarian).
- BDS EN 1555-5: 2010 Plastic pipe line systems for gas supply. Polyethylene (PE). Part 5: Usage suitability (Bulgarian).
- BDS EN 1775: 2010 Gas supply. Building pipeworks. Maximal operational pressure lower than or equal to 0.5MPa. Functioning recommendations (Bulgarian).
- BDS EN 12327: 2013 Gas supply systems. Pressure test and procedures of introduction into exploitation. Functional requirements (Bulgarian).

Collection of monographs and textbooks by the author on thermal energetic and building engineering installations.

Printed in the United States
By Bookmasters